Fish-Eating in Greece from the Fifth Century B.C. to the Seventh Century A.D.

A story of impoverished fishermen or luxurious fish banquets?

Dimitra Mylona

BAR International Series 1754
2008

Published in 2016 by
BAR Publishing, Oxford

BAR International Series 1754

Fish-Eating in Greece from the Fifth Century B.C. to the Seventh Century A.D.

ISBN 978 1 4073 0193 8

BAR Publishing is the trading name of British Archaeological Reports (Oxford) Ltd. British Archaeological Reports was first incorporated in 1974 to publish the BAR Series, International and British. In 1992 Hadrian Books Ltd became part of the BAR group. This volume was originally published by Archaeopress in conjunction with British Archaeological Reports (Oxford) Ltd / Hadrian Books Ltd, the Series principal publisher, in 2008. This present volume is published by BAR Publishing, 2016.

Printed in England

BAR
PUBLISHING

BAR titles are available from:

BAR Publishing
122 Banbury Rd, Oxford, OX2 7BP, UK
EMAIL info@barpublishing.com
PHONE +44 (0)1865 310431
FAX +44 (0)1865 316916
www.barpublishing.com

CONTENTS

LIST OF TABLES

LIST OF ILLUSTRATIONS

ACKNOWLEDGMENTS

This thesis has benefited from the help and support of many people over the years, who have contributed either directly or in more subtle ways. My supervisor, Yannis Hamilakis, has been steadily supportive and a great help in my discovering of new intellectual paths and the re-tracking of old, familiar ones. I would like to thank him not only for being such a good teacher but also for being a constant calming and encouraging power during the difficult phases of this long-distance, part-time research and thesis writing. His flexible, imaginative supervision program was an extra bonus. At the first exploratory stages of this research I also benefited from the wisdom and guidance of Clive Gamble, who was my supervisor at the time. On issues of intellectual guidance I would also like to thank Naomi Sykes and Lyn Foxhall for their instructive comments as upgrading examiners. These comments have led to some very rewarding areas of research.

Certain members of the faculty have been of great practical help, in my short, repeated visits to Southampton over the years. Special thanks are due to Mary Stubbington who has been an unfailing help in disentangling my administrative affairs and keeping me up to date with my obligations. Tim Sly is also to be thanked for his help with computer related matters. Dale Serjeantson has been there for me, generously offering her academic expertise but also an exceptionally welcoming living place. Adrienne Powell, as a F.R.U. zooarchaeologist and a friend has supported me practically and psychologically on many occasions during the first phase of this research. This goes also for several of my fellow PhD students who were always welcoming and caring, making me feel at home at Southampton despite my long absences.

I also feel deep gratitude to many people in Greece, where I conducted my research and lived all these years. Many of the fish remains used in this thesis have been collected during long excavation summers with the practical support and enthusiasm of excavation directors and teams. Special thanks are due to all members of the Kalaureia project (Swedish Institute at Athens). Collaboration with Berit Wells and Arto Penttinen was crucial not only in the collection of the fish remains but also in placing them in context. They also have been very generous in commenting on various parts of this thesis lending me their expertise in Classical archaeology A large number of people from the Kalaureia excavation are also to be thanked, for giving a hand with the water-flotation and sorting program but also Dale Serjeantson, Sven Isaksson and Anaya Sarpaki for sharing generously with me their ideas and the unpublished results of their research. Special thanks are due to Emanuel Savini for providing custom made site plans for the excavated features in the Poseidon Sanctuary that are discussed in this thesis. Members of the Itanos excavation project (Institute for Mediterranean Studies and French Archaeological School at Athens) were equally helpful and supportive. I specially thank Thanasis Kalpaxis, Emanuel Greco and Maria Xanthopoulou, for their help and support. And last, but not least, I am grateful to Efi Poulaki-Pantermali, the excavator of Krania for offering me the exciting archaeo-ichthyologicla material from that site, to Evi Margaritis who undertook the tedious water-flotation and sorting task and shared with me the results of her ethnobotanic research and to Eleni Klinaki for facilitating access to this material. I would also like to thank David Reese who allowed me access to unpublished work of his and valuable off-prints, Liana Ouranou for translating for me excerpts of the Russian texts consulted for this thesis and Marie-Françoise Billot for bibliographical guidance regarding the "Sacred Lake" on Delos.

The world of living fish and ichthyological matters has been a particular challenge to me but I was lucky to have much generous help. Prof. Panos Economidis (University of Thessalonike, Ichthyology Department) has been extremely helpful and kind to me, answering all sorts of questions, many of them undoubtedly naïve, about fish, fish habitats and sources of ichtyological information. Manos Koutrakis and Argyris Kallianiotis (Institute for Fisheries Research) have also been very helpful in answering questions, but mostly in introducing me to the world of the modern fishermen and to the complex, delicately balanced relations between fishermen, economy, politics and the aquatic resources.

This research would not have been possible without the help of a special group of people. These are the Library staff of the University of Crete (Rethymno), who have been there for me, facilitating every aspect of my library research, even without me being a member of their institution. Special thanks are due to Kalli Kardaki, who has been particularly patient in tracking and making available via interlibrary loan the most obscure books and papers on ancient fishing. Similarly, practical support has been generously given by the Laboratory of Geophysical - Satellite Remote Sensing & Archaeo-environment - Institute for the Mediterranean Studies, by providing laboratory facilities for the analysis of the fish bones. Special thanks are due to Apostolos Sarris for his help with customizing the satellite maps used in this thesis

This work has benefited greatly from discussions and debates with many colleague friends against whose expertise and good will I have been testing my ideas and hypothesis on ancient fish, fish eating, or extravagant archaeological interpretations. Anagia Sarpaki, the colleague who first introduced me to the beauty of fish bones has encouraged me the longest, and I thank her deeply for that.

My gratitude also goes to the two examiners of this thesis, Andrew (Bone) Jones from the University of Bradford and Vedia Izzet from the University of Southampton. They both, each in his or her own special way, helped me to finely tune this work.

This work however owes the most to my family, who have been there for me unfailingly, offering encouragement, practical support and a strong connection to real life! Antonis and our children, Myrto and Nikos not only offered me a beautiful and challenging life outside my research, but made this thesis part of their own life, and fish, ancient and modern, became a personal favorite of them all. Grandmothers and grandfathers gladly stood in my shoes taking care of the children every time I had to be away for research. To these people I am especially grateful and dedicate this thesis.

ABBREVIATIONS

A.P.	Beckby, H., 1957-8. *Anthologia Graeca (Palatina)*. Griechisch-Deutsch, München, Heimeran.
FGrHist	Jacoby, F., 1923-1930. *Die Fragmente der Griechischen Historiker*, Berlin .
FHG	Müller, C. and Müller, Th., ed., 1841-1870. *Fragmenta Historicorum Graecorum*, Paris.
IEphesos	Wankel, H., 1980. *Die Inschriften von Ephesos*, V. Inschriften griechischer Städte aus Kleinasien, Bonn: Habelt.
IG IV2[1]	Hiller von Gaertringen, F., 1977. *Inscriptiones Epidauri*, editio minor, Chicago: Ares Publishers (reprinted of the Berlin edition 1929).
IG IV[2]	Hiller von Gaertringen, F., *Inscriptiones Epidauri*, editio minor, Berlín 1929.
IG XII[3]	Hiller von Gaertringen, F., 1904. *Inscriptiones Symes, Teutlussae, Teli, Nisyri, Astypalaeae, Anaphes, Therae et Therasiae, Pholegandri, Meli, Cimoli, Supplementum*, Berlín.
IKyzikos 1	Schwertheim, E., 1980. *Die Inschriften von Kyzikos und Umgebung. Teil I. Grabtexte*, Inschriften griechischer Stadte aus Kleinasien 18, Bonn.
K-A	Kassel, R. and Austin, C., eds. 1983. *Poetae Comici Graeci*, Berlin, New York, W.de Gruyter.
Kock	Kock, T., 1976. *Comicorum Atticorum Fragmetna.* HES Publishers: Utrecht (reprinted of the edition Leipzig, Teubner, 1880-1888).
Liddell-Scott	Liddell, H.G. and Scott, H.G., 1925. *Mega Lexicon tis Ellinikis Glossis.* Athens: Ioannis N. Sideris.
Loeb	Loeb Classical Library, available from http://www.hup.harvard.edu/loeb/author.html (accessed 16-November 2006)
OGI	Dittenberger, W., 1982. *Orientis Graeci Inscriptiones Selectae*, Hildesheim, New York: G. Olms.
SEG	*Supplementum Epigraphicum Graecum*
SIG[1-3]	Dittenberger, W., 1982. *Sylloge Inscriptionum Graecarum*, Hildersheim, Zürich, New York: Georg Olms Verlag (reprinted from the Leipzig edition 1920).
SVF	Ioannes ab Arnim, 1903-1924. *Stoicorum Veterum Fragments*, Lipzing: Teubner.

The following Classical works have been used in this thesis:

Aelian	*On the Characteristics of Animals*	Loeb
Aristophanes	*Pease*	Loeb
Aristophanes	*Wasps*	Loeb
Aristophanes	*Knights*	Loeb
Aristophanes	*Clouds*	Loeb
Aristophanes	*Lysistrata*	Loeb
Aristophanes	*Frogs*	Loeb
Aristophanes	*Pease*	Loeb
Aristophanes	*Acharnians*	Loeb
Aristotle	*History of Animals*	Loeb
Aristotle	*Rhetoric*	Loeb
Aristotle	*Politics*	Loeb
Athenaeus	*Deipnosophistae*	Loeb
Demosthenes	*De Falsa Legationes*	Loeb
Diodorus Siscilus	*Universal History*	Loeb
Galen	*On the Powers of Foods*	Powell 2003
Herodotus	*Historiae*	Loeb
Hesychios	*Lexicon*	Schmidt 1965
Lucian	*De Dea Syria*	Loeb
Oppian	*Halieutika*	Loeb
Pausanias	*Description of Greece*	Loeb
Plato	*Republic*	Loeb
Plato	*The Laws*	Loeb
Plato	*Charmides*	Loeb
Plutarch	*Table-Talk*	Loeb
Porphyry	*Vita Pythagorii*	des Places 1982
Porphyry	*On Abstinence from Animal Food*	Bouffartigue and Patillon 1977
Rufus of Ephesus	*Medical Questions*	Gärtner 1970
Strabo	*Geography*	Loeb
Suidas	*Lexicon*	Alder 1967-71
Theophrastus	*Enquiry into Plants*	Loeb
Xenophon	*Anabasis*	Loeb
Xenophon	*Hellenica*	Loeb

CHAPTER 1

INTRODUCTION: DEFINING THE PROBLEM OF FISH-EATING IN GREECE BETWEEN 5TH CENTURY B.C. TO 7TH CENTURY A.D.

This book is about fish and fish-eating in Greece from the Classical period to the Late Roman period (5th c. B.C. – 7th c. A.D.). In Classics fish-eating appears to be one of the most prominent and ideologically charged activities related to food. This work is an investigation of the way in which fish and its consumption was incorporated in the economic, social and ideological life of Greeks. It is also an exercise in the integration of different classes of data in dealing with questions about ancient societies. Although the focus of this work is archaeological, it takes advantage of historical, philological, anthropological and ichthyological data and methodologies.

The research on fishing and fish consumption in the Aegean of the historical era could be summarily characterized as fragmentary and often contradictory, both in terms of the research methodologies applied, and the outcome of the various approaches. These have developed around two polar positions which dominate the field and create tensions and deep misunderstandings around the issue of fish consumption in Greece in the centuries under study.

The first position, within the domain of Classics, places fish-eating at a position of prominence. This has been based on a very rich body of literary references to fish-eating, which have mostly been taken as a direct documentation of life in the past. The most extreme exponents in this field claim that fish, fresh or preserved, was central in the diet of people, both rich and poor, being however principally an object of trade. Furthermore, according to the same tradition, fish in these societies were highly symbolic. Focusing the fish-related research on issues of trade, symbolism and representation, very little attention has been paid to the availability and production of fish. The actual consumption, the material organisation of this act, as well as its context-specific social and ideological parameters remained elusive. Classics focus on the general instead of the particular, on the prominent instead of the mundane (for details of this approach see chapter 2.1).

Contrary to this approach, and as a direct reaction to it, a totally different hypothesis has been developed. At the core of that approach has been placed the idea of a supposedly poor sea, of an inefficient technology and the theoretical, rather artificial question of whether ancient Greeks could subsist on fish or not. The most extreme exponents of this approach used a range of seemingly robust "scientific" methodologies to prove the unimportance of fish as a regular food source. Instead, they viewed fish as an emergency resource, tapped in periods of need (for details of this approach see chapter 2.2).

Recent developments in archaeological field and laboratory methodologies, the accumulation of fish-related data over the years, and an innovative shift in the theoretical approaches of both archaeology and ancient history enable and in fact demand a reconsideration of the issue. Individual, small-scale attempts in the form of short papers or reports have demonstrated the potential richness of an effort to bridge the polarities in the research on ancient fish-eating (for further discussion see chapter 2.3).

Assemblages of fish bones, the actual remains of fish-eating, are found in considerable quantities in excavations, whenever certain field techniques are applied. Fishing tools, various fish-related structures and relevant inscriptions are also quite common. These new or previously neglected sources of information open up new spheres of knowledge on past fishing and fish-eating. Furthermore, a much richer knowledge of the hydrographic and biological character of the Greek waters, and of the manner in which fish populations behave and react to fishing, is also a potential rich source of understanding. Finally, a re-examination of the literary sources, not as mere documentary material but as products of their own cultural context, also produces a wealth of nuanced information which is highly relevant, not to a generalised treatment of past fishing and fish-eating, but to a context specific approach which focuses on particular humans/groups, their actions and ideas in relation to fish.

An exploration of the type described above demands not only the adoption of new or advanced methodologies, but also the construction of new or at least specially modified frames of reference and theoretical frameworks. Several options are open for consideration, but this book focuses more firmly on a range of anthropological ideas which are modified and tested in the process of this research. This book, rather than viewing fish just as an economic resource, an object of trade or as a symbolic food, emphasises its connection to every aspect of human life, being it economic, social or ideological. As such it is relevant not only to body maintenance and subsistence, but also to the social life and world views of the consumers (for further discussion of these ideas see chapter 3). By viewing fish as food in this sense, we acquire the means to bridge the gap described at the beginning of this chapter. Once fish is perceived thus, then a whole range of new, little explored issues, becomes relevant to our study.

The way fish-eating involves the senses and imprints itself on the body links a food resource (fish), a subject traditionally dealt with through economic approaches, to the individual, the consumer and his/her cultural environment. Furthermore, a juxtaposition of the naturally available fish resources to the culturally perceived usable fish categories (i.e. available edible matter versus food to be consumed) is also of great potential in exploring how material and cultural factors are interwoven in the way a society constructs its everyday life.

The analysis attempted in this study is based on the fundamental assumption that all these aspects of fish and its consumption vary and change from context to context. The shifting nature of the parameters of fish-eating is quite hard to capture in a single narrative. Instead of trying to do this, this work is organised around a type of a "cultural biography" of fish. Cultural biographies are usually built for distinct objects, which are visible archaeologically, and their moving through time and space can be followed relatively easily (Gosden and Marshall 1999). Food items (such as fish) are not usually treated in this manner (for a detailed discussion of the idea of cultural biography see chapter 3). However, an examination of fish consumption through a cultural biography of fish (in the generic sense), permits the unfolding of the innumerable variations and shifts in the human-fish relationships, emphasising their interconnections.

With these thoughts in mind, this study will address the following specific questions:

What was the actual contribution of fish as a food resource to the economy between 5th century B.C. to 7th century A.D., in light of recent developments in recovery techniques and the accumulation of a varied corpus of empirical data on past fishing, such as fish bone remains, related implements and structures?

What was the social context of production and consumption of this resource? The economic and social position of the fishermen, the role of fish as a commodity and the relationship between the status of fishermen and their products, the differences between fresh and preserved fish, the identity of fish consumers and the character of consumption contexts are some relevant issues.

How did the act of fish incorporation, i.e. eating, affect the cultural categorisation of fish? How were the senses, taste, smell, sight, hearing, but also memory involved in this process and how did they influence fish preferences and consumption practices? What was the social and ideological impact of this type of fish categorization?

What were the ideological implications of fish consumption? How did these vary in accordance with different circumstances, historical, political or social? Was there any symbolism around fish and how did it affect its consumption? What was the relevance of the fish-related symbolism to broader world views and ideas? How were the ideological and symbolic aspects of fish-eating connected with the economic and ecological aspects of its production, marketing and consumption?

What is the most efficient way to handle such complex questions about past societies? More specifically, how can the diverse bio-archaeological, artefactual, textual and other data be effectively integrated?

The research problems stated above will be dealt with in the following manner. After setting the problem (chapter 1) there will be a critical review of the relevant literature to date, with a brief discussion of the theoretical trends that underline the existing research (chapter 2). An alternative approach will be proposed which is based on modern advances in archaeological, historical and anthropological thought (chapter 3). Methodological issues will be discussed and analytically presented in the following chapter (chapter 4). The research questions at hand will be dealt with by constructing a cultural biography of fish, articulated around several key issues. These are the catching of fish (chapter 5), the role of fishermen (chapter 6), the distribution and marketing of fish (chapter 7), the social context of fish consumption (chapter 8), the criteria by which fish might have been chosen for the table (chapter 9) and finally the impact on the senses of the fish-eating and its importance (chapter 10). The last chapter of the book summarizes the results of the research and highlights the important or innovative elements of this work (chapter 11). The main text is followed by a number of tables and figures as well as by a number of appendices which present the raw data used in this book.

Chapter 1 is an introduction to the subject of the study, which summarises the rationale concerning the choice of the specific research topic and the proposed approaches to it.

Chapter 2 reviews all literature on ancient fishing in Greece, underlying the theoretical trends which, implicitly or more directly, supported these approaches. These works are presented according to the dominant epistemological paradigm in which they were created, and their merits and liabilities are discussed. Thus philological and ancient historical works are presented together (chapter 2.1), as are the works which are based on the modelling and quantification of ancient fishing (chapter 2.2). Finally, more recent and synthetic works which view fishing in the economy and beyond are also discussed (chapter 2.3). This review facilitates the understanding of our current knowledge about fish consumption in Classical Greece and serves as a background against which a new approach will be articulated in the following chapters.

Chapter 3 offers a discussion of the approach to the issue of fish-eating which is proposed in this study. First it presents and discusses theories and ideas about food consumption which, getting inspiration from anthropology, open up new and largely unexplored avenues to the understanding of food phenomena in the past, such as the fish consumption. After a presentation and critical evaluation of the classical anthropological theories on food consumption (chapter 3.1) and related ideas that treat food as a social phenomenon (chapter 3.2) the chapter proceeds by discussing the advantages of viewing fish as a commodity and approaching it by constructing a cultural biography (chapter 3.3). Finally the chapter concludes with the discussion of notions such as food embodiment and the role of the senses in this experience (chapter 3.4). The relevance of the ideas presented in this chapter to the study of ancient fish consumption in Greece is highlighted throughout by reference to examples or relative works.

Chapter 4 deals with methodological issues. It discusses the main theoretical notions which underlie the methodological choices made in this book, which are in fact relevant to any subject falling within the range of historical archaeology. The methodology adopted in this study is discussed in detail. It mostly deals with the way the fish bones and literary sources are treated in order to offer comprehensive information about past fish-eating and with the problem of handling and integrating disparate data. Finally, a commentary is offered on several choices made in this study such as the terminology used or the assumptions made.

Chapter 5 explores the actual productive potential of the Greek waters, both fresh and marine, in light of recent work in the field of oceanography and ichthyology and available archaeological and historical data. The treatment of the issue is done in a geographically specific manner. Greek aquatic resources are examined geographically, from the point of view of aquatic zones rather than modern geo-political divisions (chapter 5.1). Different types of fish resources available to the ancient fishermen are presented, with an emphasis on those characteristics of the fish which are considered most relevant to the articulation of ancient fishing and fish-eating (chapter 5.2). After setting the physical framework, the chapter develops by presenting the geographical distribution of the aquatic resources which had been exploited in the period under study. This discussion is based on a combination of all available data, zoo-archaeological, artefactual and literary (chapter 5.3). The chapter concludes by highlighting the variability of the fish resources exploited in the past and their wide geographical distribution and by offering a commentary on the relevance of these observations to the exploration of ancient fishing and fish-eating.

Chapter 6, remaining in the realm of production, shifts the focus from the fish resources to the fishermen, an archaeologically elusive, yet crucial factor in the exploitation of the available fish resources. After discussing those technical and cultural characteristics that constitute a fisherman and, by extension, a fishing community (chapter 6.1), the chapter embarks on a search for the ancient Greek fisherman. An analysis of the multitude of fish names which are preserved in written sources (chapter 6.2) and an exploration of myths and beliefs about fish, fishing and the aquatic environments (chapter 6.3) offers the opportunity to speculate on the existence of distinct, articulated fishing cultures and ideologies in several parts of Greece, which were in contact with but also in opposition to similarly articulated ideas of land-bound cultures about the sea. The points which are relevant to this research are summarised at the end of the chapter (chapter 6.4).

Chapter 7 discusses aspects of the distribution and marketing of fish, in other words, it explores the ways in which fish is transformed from an exploitable resource into food. Two distinct processes are discerned and discussed separately; one involves fresh fish (chapter 7.1) and the other preserved fish (chapter 7.2). In situ consumption of fresh fish or their transportation over short distances (chapter 7.1.1), fresh fish kept alive for later consumption (chapter 7.1.2) and fresh fish for the urban market are three crucial aspects in the marketing and distribution of fresh fish. Discussion of the technical parameters of fish preservation (chapter 7.2.1), the evidence for such preservation in Greece (chapter 7.2.2) and issues related to preserved fish as an object of trade (chapter 7.2.3) illuminate the special role of preserved fish. This chapter concludes with a synopsis of the main features of the fish distribution and marketing mechanisms and their differences, as well as of their social impact. The issue of the cultural distance between fishermen and fish consumers appears again as relevant and is further elaborated.

Chapter 8 discusses the social context of fish-eating, by focussing on the circumstances in which fish-eating is documented and on the identity of participants in these events. Two major social domains are explored. Fish-eating in a cultic context, with the gods or the dead (chapter 8.1), appears as a multi-dimensional act. Starting with the analysis of a fish-bone assemblage from a dining deposit in the Sanctuary of Poseidon on Kalaureia, Poros (chapter 8.1.1) the discussion proceeds to investigate the role of fish in sacrifice, in ritual dining and other cultic affairs, placing emphasis on the consumers and supplicants rather than the gods (chapter 8.1.2). The discussion of the secular context of fish-eating focuses mostly on fish consumption in the house and in the tavern, with a rather stronger emphasis on the later (chapter 8.2).

Chapter 9 focuses on the analysis of different criteria by which fish might have been chosen for the table. Starting with the analysis of the Akraephia fish price list (chapter 9.1), a range of edible fish categories are discussed as an alternative to modern Linnaean taxonomy (chapter 9.2).

These categories appear in our sources as antithetical pairs, which are formulated on the basis of different physical or cultural characteristics of fish. Edible and inedible fish (chapter 9.2.1), large fish versus small fish (chapter 9.2.2), local versus non local and imported fish (chapter 9.2.3), fresh-water versus marine fish (chapter 9.2.4), fresh versus preserved fish (chapter 9.2.5) and fish of dry flesh versus fish of moist flesh (chapter 9.2.6) are the categories discussed. Although they do not cover the whole range of possible different taxonomies, yet their articulation reveals something of the priorities in food choice and human needs. These issues are discussed at the end of this chapter.

Chapter 10 moves the discussion to the table, to the actual consumption of fish. It approaches the eating of fish through an analysis of a range of sensual experiences involved in the act, highlighting the importance of a shared cultural context of consumption for the formulation of particular taste preferences and sensual experiences (10.1). The importance of considering the role of the senses in fish-eating in order to understand a range of food-related cultural phenomena is explored through the analysis of the Classical fish plates (chapter 10.2).

Chapter 11 is the final chapter of the book and gives a synopsis of the preceding analysis by way of answering the research questions set in the first chapter. This synthesis highlights the contribution of this book to research not only on the issue of fish-eating in Classical Greece, but also to the role of food consumption in ancient Greek society. This final chapter concludes with a brief discussion of areas which could prove particularly interesting and rewarding in future research.

CHAPTER 2

FISH-EATING BETWEEN THE HAMMER OF CLASSICS AND THE ANVIL OF ECONOMY

The literature that has been produced on ancient fishing and fish-eating is quite extensive. Chronologically it spans over a century and it represents a wide array of ideological and methodological positions. Its presentation in this chapter is intended as a means to explore the ideas behind these approaches and to evaluate their contribution to our present understanding of fishing and fish consumption in ancient Greece.

A decision has been made concerning the limits of this review. The following presentation includes works that refer both to prehistory and the historical past, and also works that fall within the domain of both archaeology and ancient history. Although these fields are traditionally viewed as distinct in terms of research interests and methodology, in the case of fishing and fish-eating there has been a considerable interaction. The examination, therefore, of relevant works in terms of ideas and methods, rather than discipline, is probably more appropriate in this book.

This review is wide but not exhaustive. It includes works that I consider groundbreaking, influential or just typical. I have attempted to discuss as wide a range of literature as possible and to examine old and less accessible publications. Although some works have not been accessed, due to limitations of resources and language, the following review is quite representative.

2.1 Philologists, ancient historians and the archaeologists of the early 20th century on ancient fishing and fish-eating.

The first modern works on Classical and Roman fishing and fish-eating date back to the 19th century. They are studies compiled within the field of Classics, mostly by German philologists (e.g. Eberl 1892; Rhode 1892; Schneider 1892; Keller 1909; Dölger 1922; Kalitsounakis 1926), although, some British (e.g. Badham 1854) and American contributions (e.g. Smidth 1875; Scott 1916/17; Shewan 1927) were also published. These works were produced in the spirit of "*bildung*" – the secular spiritual regeneration through education in Classics – which first developed in Germany at the beginning of the 19th c. and was later emulated by others (for a detailed discussion see Bruford 1975; Morris 1994b, p. 28; Morris 2001, p. 11). In this tradition, the humanists sought to explain and appreciate the variety, uniqueness, complexity and originality of what humans are capable of doing (Morris 2001a, p. 12). They looked for the highest expressions of human achievement in the Classical past and ancient texts were the primary sources in this quest. In this framework a whole range of activities, formal or more casual, were examined. Fishing was just such a phenomenon and Rhode's detailed account on tuna fishing and fish preservation in antiquity (Rhode 1892) forms one example of this approach.

The 19th century was also the era when large corpora of data were produced (Pfeiffer 1976). Although no corpus of data as such had been compiled on ancient fishing, all works written in that era consisted of a detailed citation of all the available evidence (mostly literary references) on the issue at hand. Dögler's massive work *ΙΧΘΥΣ* (1922) forms a typical example. In dealing with the symbolic nature of fish in the Eastern Mediterranean, he brought together a vast number of mostly literary and to a lesser degree epigraphic citations. This tradition lasted for almost a century, with works like Bucholz's, *Jagd und Fishfang* (Bucholz *et al.* 1973), following the same paths, towards the closing of the 20th century (see also Kallergis 1953).

By the beginning of the 20th century the subject of fishing had also become popular among the English-speaking writers. It is interesting to note that some of the most often cited works on the subject of ancient fish and fishing have been written not by classicists, but by members of the British upper class, who had received an education in Classics. Radcliffe, for example, a military official, wrote his *Fishing from the Earliest Times* (1926) "at intervals between war, work and illness" (Radcliffe 1926, p. vii). According to him, his book is "the first attempt to examine Classical and other ancient writers on fishing *from the standpoint of a practitioner (pisciculturalist and angler) and of one who has been taught (though somewhat forgotten) his Greek and Latin*' (Radcliffe 1926, p. vii, italics are mine). Radcliffe himself defines the audience of his work and justifies its compilation by writing that: "for those who have their business on the waters, it is certainly of interest to essay the tracing of fishing's pedigree" because "it is abnormal for one not to seek genealogical garments, wherein to hide his nakedness" (Radcliffe 1926, p. 3).

The fact that fishing (or some type of fishing) was considered by the educated as a genteel sport which seemed to possess a long and noble ancestry (Danier 1813; Wigglesworth, 1996, p. 15) was perhaps a forceful motive for a plethora of works which were philological and highly subject-specific. It is interesting that much work has been dedicated to fishing in Homer, especially in light of the fact that almost all these works conclude that fishing and fish consumption was unimportant for the Homeric heroes or Homeric society (e.g. Scott 1916/17; Shewan 1972; Cough 1935/36). This preoccupation with fish in Homer, which is still echoed in recent works (e.g. Sakellarakis 1974; Trantalidou 1990) can be understood, considering that for the classicists of those days the Homeric era was the earliest historic period. Therefore, in the search for genealogies, the *Iliad* and the *Odyssey* were the earliest written sources which could provide evidence (for other aspects of the phenomenon see Davidson 1996).

Most early works on fish-related issues within Classics fall in two major categories, depending on the questions they are called to answer. One approach is linguistic and is mostly concerned with the names of fish (and their identification) as these are cited in ancient literary sources (Papendic 1926; Stromberg 1943; Thompson 1947; Andrews 1949; Georgacas 1977). This approach falls well within the philological tradition in Classics, but it is interesting that two of the most renowned works in this field were not produced by classicists. D' Arcy Thompson, the author of *A Glossary of Greek Fishes* (1947) was a biologist (O'Connor and Robertson 2003), while D. Georgacas, the author of *The Ichthyological Terms for the Sturgeon* (1977) was a linguist (Zgusta 2002). They both, however, had received a robust training in Classics. Thompson's work in particular has been very widely used by virtually everyone who deals with ancient fishing, as it is a compilation of all available information found in ancient literary texts on every fish species known in ancient Greek literature, including physiology, ethology, medicinal use and cooking instructions.

The other approach falls within the realm of economic history. These works focus on aspects of production, especially fish preservation and the commerce of pickled fish (Rhode 1892; Schneider 1892; Bunsmann 1911; Blumer 1920; Jardin 1961; Corcoran 1963; Buchholz *et al.* 1973; Dumont 1976/7; Longo 1989). One of the main features of this approach is the fact that these works perceive fish production as an activity on an industrial scale, and discuss issues such as the fishermen's associations and fish prices, presupposing the existence of a market of modern, western type (e.g. Rostovtzeff 1926; 1941).

Studies such as these were produced in the midst of a fierce controversy over the nature of ancient economy, which began in the 19th century and intensified in the second part of the 20th century. This came to be known as the "primitivist - modernist" debate. According to the "primitivist" view, advocated by the 19th century German economist Karl Bucher (1893), the ancient economy was "primitive", characterised by a pursuit of self-sufficiency and lack of modern economic traits, such as large-scale manufacture of goods or trade. The opposing view, the "modernist", claimed that the ancient Greek economy was similar to modern capitalistic economies but perhaps of a lesser degree of complexity. As a consequence, the ancient economy could be analysed in terms of modern economics, with issues such as international trade or industries becoming central (for a synopsis of the primitivist – modernist arguments see papers in Scheidel and von Reden 2002 and Austin and Vidal-Naquet 1977; also Finley 1979 and Burke 1992). Evidence for large-scale fishing, fish processing establishments and trade which emerged from works such as the ones referred to earlier, do clearly support the idea of a "modern" economy where fishing was an important source of revenues. In this approach, however, the aspects of fishing which involved the fish as a natural resource, its habitat, availability and so on were totally ignored.

It is interesting that other views of the nature of ancient economy, such as those advocated by Polanyi and his school, never dealt with fishing. Historians in this school emphasize that the economy as a distinct concept, whether "primitive" or "modern" is only pertinent to the modern world, while in other types of societies, phenomena, processes and functions which we identify as economic, may operate through other structures, such as religion, and not be perceived as independent (Polanyi 1957; Terray 1969; Godelier 1972; Meillassoux 1974). This "embededness" of what in modern word are considered as economic phenomena (in this case fishing) is exactly the feature which may prove a particularly useful notion in this book.

The end of the 19th century and the beginning of the 20th also saw publications on ancient fishing in Greek. These form a small, but historically interesting body of work, published by two Greek archaeologists, Christos Tsountas and Antonis Keramopoulos. Although their arguments focused on the interpretation of prehistoric archaeological finds, they nevertheless followed the Classical tradition of their time and provided a review of relevant classical literature on the subject.

Tsountas, at the end of the 19th century was a member and an official of the Archaeological Society of Athens, an institution which aimed at the salvage of antiquities and was entitled to excavations alongside the official Archaeological Service in the young Greek state (Petrakos 1987, pp. 18-23). Tsountas became renowned as the founder of Prehistoric archaeology in Greece, with extensive excavations in Neolithic Dimini and Sesclo in Thessaly, as well as in the Cyclades (Petrakos 1987).

In a book published in 1893, entitled "*Mycenae and Mycenean civilisation*", he discussed fishing from an archaeological point of view with the aid of ancient texts. His argument was based on archaeological data, or rather the lack of it. He used the evidence to support his argument on the racial affinity of Greeks and Italians of the Classical era. According to him, both the Mycaeneans and the Italiots of the north did not know of fishing. Fishing appeared suddenly in both areas, perhaps brought by the Northern races who were the progenitors of both Classical Greeks and Romans (Tsountas 1893, pp. 230-1). This book reflects a predominant trend in the intellectual life of Greeks at the time, namely the effort to prove a cultural affinity with Western Europe. Greek social scientists were preoccupied with, among others, racial issues (Kotsakis 1991, p. 67) and Tsountas's treatment of Mycenaean fishing was part of this discourse.

Another work by Tsountas, a paper entitled "*Cycaldica*" also dealt with fish and fishing, but this time for the bronze-age Cyclades (1898). This particular work is very unusual in that it gives what is perhaps the first systematic discussion of certain aspects of the economy in a site- and time-specific context. Archaeological evidence, taphonomy, ecology and ethnographic paradigms, were all used as tools to explore the role of marine resources in the life of an insular community in the bronze-age Cyclades (Mylona 2003a). This approach was completely at odds with the German, British and French research interests of the time, be it Classical or Prehistoric (for the nature of these interests see Morris 1994b). However, in such a work one can trace not only a bright inquisitive mind, but also the influence of Greek folk studies, which at that time were characterized by an interest in proving unity between modern Greeks and the ancients on the basis of similarities in several phenomena and institutions of their lives. Archaeology of the type practiced by Tsountas in this paper could have served this goal (Kyriakidou–Nestoros 1978, pp. 89-110; Herzfeld 1982; Danforth 1984).

The third work of this group of studies in the Greek language is a different case. By the paper's title "*Ichthyophagia*" (fish-eating), Keramopoulos defines his subject very precisely (1918). Reviewing the literature available in his day and starting from Tsountas's position about the absence of fish from the Mycenaean diet, he produced a short history of fishing in Greek antiquity with emphasis on prehistory. Keramopoulos in this paper integrated masterfully the literary and archaeological evidence, resembling German treatises in thoroughness, but being quite innovative in the way he combined literary with archaeological data.

Although these works have been motivated by the very strong intellectual quests of their time, in a sense they are quite idiosyncratic. Some of the ideas explored in them, such as Tsountas's analysis of the role of fishing in the Cyclades, read as very contemporary work even today. However, none is cited in the subsequent literature on fishing and fish-eating in the Aegean.

The trends described above, may have been born within the intellectual spirit of the end of the 19th century and the beginning of the 20th, but perhaps due to the inertia and conservatism characterizing Classics, they can be traced even in recent works. After the 1980's, however, fish and fishing leave the domain of Classical philology and enter the ancient historical and archaeological arena, explored through theory-oriented perspectives, where ideas about the environment and the economy prevail.

2.2 Modelling and quantification of ancient fishing.

Environmental archaeology and model-oriented approaches to the past entered the archaeology of Greece in the late 1960's with two major research projects, one by the British and one by the Americans (British: Higgs 1972; Higgs and Jarman 1975; American: McDonald and Rapp 1972; Jacobsen 2000; for a discussion of the historical and intellectual framework of these projects, Fotiadis 1995). A range of ideas developed in the following years as various research projects took place in various parts of Greece. Two of these ideas have been central to the study of ancient fishing and fish-eating.

One of the focal research points has been the interrelation between humans and their physical environment (e.g. Clark 1972, p. vii; McDonald 1972, p. 6) and especially the investigation of the "economic base" of the societies under study, with emphasis on subsistence. This term "subsistence" was mostly used as covering all activities that aim to satisfy "basic" human needs, such as feeding, shelter etc. It was mostly linked to survival (Ellen 1982). Fish in this intellectual framework could only be viewed as a food resource.

Related to the above was the adoption by archaeologists of the biological concept of optimality (Jochim 1979; Keene 1983; Salmon 1989; Jochim 1998). The principle of optimality claims that in environments with finite resources, natural selection will favour those organisms the behaviour of which facilitates easier access to resources (Foley 1985). In an archaeological context, optimality has been viewed as a natural tendency of humans to achieve the maximum profit in their interaction with the resources of their environment. In this framework, different classes of activities such as agriculture, pastoralism and fishing, are evaluated in terms of their efficiency in providing what is arbitrarily considered an adequate quantity of food (measured as energy), and ultimately survival and security to a predefined human population.

Fishing entered recent archaeological discourse on the prehistoric Aegean through the work of John Bintliff (1977). In an extensive book which explored how the several constituents of the environment have largely determined the location and subsistence basis of settlements in the prehistoric Aegean, he advocated that the location of early settlements, trade and sea-faring were linked to the systematic, seasonal exploitation of migratory fish around the Aegean. He postulated that the pursuit of tuna along coasts and between islands by migrant fishermen (*transmerance*) provided opportunities for cultural interaction (Bintliff 1977, p. 121).

Bintliff gained inspiration for this idea from the two fish-bone assemblages that were partly published at the time, one from the Mesolithic Franchthi Cave in the Argolid and the other from the Late Neolithic Saliagos on Antiparos (Renfrew *et al.* 1968; Payne 1973). His argument was based on information collected by the extensive interviewing of fishermen in the areas where Bintliff conducted his fieldwork. On this basis he constructed a fishermen's movement map (1977, part I, map 1, p. 130). Several routes shown on the map incorporated prehistoric settlements. Bintliff hypothesised that these routes were also used by mariners/fishermen in the past playing an important role in the life of those coastal communities.

Bintliff's scheme was very innovative at its time. His idea about the important role of travelling fishermen in the development of cultural contacts between different areas in the Aegean was also challenging (Warren 1979). The weak documentation and the enthusiastic, but inaccurate, assessment of the tuna migrations and availability weakened the influence that Bintliff's idea might have had in the investigation of past fishing in the Aegean (for a critique see Gamble 1979; 1982; Rose 1994, pp. 429-31 and 446-48).

The next comprehensive work on past fishing in the Aegean proved to be very influential and for this reason it will be discussed here in detail. It was compiled by a historian, Thomas Gallant, a student of ancient economic history at Cambridge. In his book *A Fisherman's Tale: an Analysis of the Potential Productivity of Fishing in the Ancient World* (1985), Gallant set out to investigate "what was the role of fish in the diet and of fishing in the economy of the ancient world?" (Gallant 1985, p. 11). To do so he attempted to integrate the literary data available to him into a wider framework, "in an attempt to set the parameters of the *potential productivity* of ancient fishing" (Gallant 1985, p. 12, italics are mine). The framework he refers to is based upon ethnographic analogy with 19th century fishing in the Adriatic (Faber 1883), statistics on modern Mediterranean fisheries (e.g. papers in the *Proceedings of the General Fisheries Council of Mediterranean*) and studies on modern peasant fishermen in Malaysia (Fraser 1960; Firth 1966). He chose those particular cases either because of their geographical relevance or, in the case of the Malay fishermen, because their technology, target-fish species and position in the market were judged to be similar to that of the "ancient world" (Gallant 1985, p. 12).

In dealing with his subject, Gallant set out to disprove what he considered to be the dominant theory at the time, "that fish, usually preserved and imported from the Black Sea, was, after cereal products, the most important element of the diet of the Greeks and that it was the food of the masses" (Gallant 1985, p. 11 and references therein). To do so, he first considered the technology available to the ancient fishermen, then the fish, which were the most important or common catches, and finally the role of fish in the diet and economy.

Ancient fishing technology was mostly inferred from literary sources (mainly Oppian and Aelian) with occasional references to iconography. It was examined in terms of labour requirements and productivity (i.e. weight of fish per fishing attempt). Gallant concluded that, given the scarcity of fish resources, which is inferred from modern statistics, the technology available to the ancient fishermen was inefficient and very labour intensive. Seine nets and set nets were, according to him, the most efficient, albeit labour intensive, fishing tools, which were used to target migratory fish, such as tunas. Making extensive use of modern statistics, he suggested that although this resource could be plentiful, it was nevertheless completely unreliable and erratic due to the unpredictability of the factors that govern their movement (i.e. water temperature, water currents, nutrient availability and so on).

A second step in Gallant's analysis was to consider the role of fish in the diet. He did so by converting fish into calories and by comparing the calorific importance of the fish in the diet to cereals. This was achieved by assuming a "reasonable, working figure" (Gallant 1985, p. 31) of daily caloric intake of 3000 kcal/person/day for a four-member family (mother, father, son and daughter of unspecified age). Through an elaborate series of calculations, and comparisons with modern statistics on fish yields in different countries, also converted into calories, he concluded that fish could make nothing more than a minute contribution to the diet of the ancients.

Along the same lines, he argued that the contribution of fish to the economy could not be significant either. Since productivity was very restricted and unreliable, it could not support fishermen on a year-round basis. Furthermore, preservation of fish could not have reached an "industrial" scale due both to the above difficulties and to further problems raised by the need for salt and additional labour. He concluded that fishing could only have played the role of a cash-crop. Just as in Malaysia, "fishing can be used like casual day labour, to give an agriculturalist a little cash or food immediately, or fill a gap in another task" (Firth 1966, p. 2 cited in Gallant 1985, p. 38).

After discussing all these issues, Gallant proceeded to formulate an alternative model. This was an adaptation of the "risk and uncertainty model" which, at the same time, was beginning to be applied to Greek prehistory (Halstead 1981a; 1981b). According to this, in Gallant's own words:

"the role of fishing in the diet and the economy would have been, on the whole, subordinate and supplementary: given the nature of the resource base and the technology employed to exploit it, it could not have been otherwise. Its main function would have been to supply a source of subsistence during periods of food scarcity due to reduced crop yields. In this way, it would have furnished a short-term solution to what would have been an endemic problem in the Mediterranean world" (Gallant 1985, pp. 43-44).

Gallant's analysis became very influential in the study of ancient fishing around the Aegean, both in prehistory and historical times. The main conclusions of his model, and some of the secondary ones, have been taken for granted and very little criticism, if any, has been put forward of the methodologies employed and the assumptions made. Only recently, more than twenty years after its publication, has some criticism been put forward, most of which is restricted to only a few sentences (Purcell 1995; Gallo 1997; Greaves 1999; Wilkins 2000, p. 300, note 154 but see Jacobsen 2005).

In order to appreciate the long lasting impact of this particular book, one has to consider it along with a large number of similarly inspired works which were produced in that era. Garnsey's *Famine and Food Supply in Greco-Roman World* (1988), Gallant's *Risk and Survival in Ancient Greece* (1991), Sallares's *The Ecology of the Ancient Greek World* (1991) and Osborne's *Classical Landscape with Figures* (1987) are some such examples. All these focused on the investigation of the ways in which ancient Greeks (and Romans) secured subsistence. Following the path opened by Moses Finley (1973; 1985), they chose to highlight "economic" aspects of the past in a systematic, structured way. Their research was model-oriented, and quantification held a central position in their reasoning. By emphasising the productive aspects of the countryside, and by singling out the environment as a determining factor in human decision-making, they offered a breakthrough in classical studies (archaeology and history) and they formulated a paradigm that is still powerful. Analogous developments in prehistoric archaeology (e.g. Forbes 1976; Halstead 1987) created, perhaps for the first time, a common frame of reference for Prehistoric and Historical archaeology.

The Fisherman's Tale was among the earliest attempts to apply the new approach to the study of the Classical past. It reflects both the enthusiasm generated by a new field of possibilities and the awkwardness of a largely untested methodology. Here follows a short critique of some crucial points of *The Fisherman's Tale.* Although this will be a review of a single piece of work, it is of wider relevance because it touches upon ideas which permeate subsequent views on past fishing, either stated explicitly or not.

The idea which runs through the entire work is that of optimisation. The goal of optimisation here appears to be security [thus the extensive discussion of the issue of unreliability of fish resources (Gallant 1985, pp. 27-31] and the currency used is spent in labour or received as calories energy (Gallant 1985, p.31). It has been assumed that the ancients were aiming for a maximum return from fishing; the fishing technology, the choice of fish targets and the role of fishing and fish in the economy have been judged against what was perceived by Gallant as optimal conditions.

In this framework, the Aegean and Eastern Mediterranean in general were judged as poor, without any consideration of the remarkable variety and perenniality of most of their fish resources. A large body of technological applications, which according to the sources cited by Gallant himself, were very common in antiquity, e.g. hooks and lines, small nets and fish traps, have been dismissed as inefficient, with no consideration of other parameters of their use. The social demarcation of their use, their suitability in particular environments, the ease of their manufacture and use and their complementarity are some aspects of these technological applications that might have counterbalanced their inefficiency in the amount of fish caught. Technology is viewed strictly as a set of tools, with no consideration of the social environment in which these tools become operational or the vast practical knowledge on the part of the fishermen on a wide array of relevant topics, such as the ethology of fish and the morphology and specific features of fishing grounds, weather patterns and so on (see chapter 6)

Furthermore, the contribution of fish to the diet has been calculated as if it was aimed to be a staple food, against all the evidence from the ancient sources which suggests that cereals and pulses were the staple food around the Mediterranean (Foxhall and Forbes 1982) and of course, on the basis of the standards used, fish proved very inefficient. The analysis of fish in the diet made by Gallant constitutes a classic case of "calorific obsession" (Vayda and McCay 1975). Food, in this case fish, is reduced to calories. Diet is regarded as ingestion of calories, proteins etc. for survival (for a discussion of the issue see Denell 1979; Hamilakis 1999b; Sarpaki 1999) stripped of its social and ideological vestiges. The fish in this model do not smell, are not ingredients in simple or elaborate recipes. They are not desired for their taste or the social messages they convey, and they do not participate in a world of proverbs, similes and myths. These aspects of fish are ignored, despite the wealth of literary reference which strongly point in that direction (such information can be found in Thompson 1947; the same issues are extensively analysed in chapters 7-10).

Fishing in the economy has again been measured in terms of optimal standards in technology and richness of resources and also in relation to an optimal standard of a large scale "industry". On this basis, fishing as a professional activity with considerable revenues and the function of specialised markets with recognizable products is proved impossible, despite all evidence to the contrary (e.g. Curtis 1991; Rostovtzeff 1926; 1941). Fishing, according to Gallant, seems to participate in a simplified economy, which is all about survival. Fish is just food, and preserved fish serve as a food item, which in good years could be exchanged in areas which had a bad crop year. Only the tastiest morsels would be preserved as items in a luxury trade of restricted scope. Even when fish are viewed in a market, in a discussion of the fish prices (Gallant 1985, pp.39-40), these are compared with wheat prices, and ultimately converted to food and calories, again due to the homogenising effect of the subsistence paradigm which has been discussed above (Hamilakis 1999b). Although discussions about the nature of the ancient economy, whether embedded or monetarised, were already heating up, especially in Cambridge, when the *Fisherman's Tale* was published (Scheidel and von Reden 2000, pp.270-71), and despite the fact that there was already ample evidence for a recognisable contribution of fish in a market system (see Rostovtzeff 1926 and 1941 for relevant sources and references), these issues have been either dismissed or belittled.

No regard was paid to particular environmental, social, ideological or other conditions in specific contexts, either spatial or chronological. Although this disregard for specificity is partly a feature inherent in abstract models (see Morris 1999a and references therein), it is at the same time, along with the miss-appreciation of variability and flexibility of human responses, one of Gallant's model's major liabilities. The recent research on communities of small-scale inshore fishermen is, I think, instructive. This research on present day, state- and market-regulated fishing communities in Greece reveals that the fishermen's preference for specific tools and catch varies considerably among different areas in the country as does the social and economic profile of the fishermen (Anonymous 2003). If such variability is evident today, one would expect it to be even more distinct in less homogenised communities of the pre-industrial past.

It is clear that this treatment of ancient fishing is permeated by a rationality of a modern western type, which is then applied to the motives of people in the past. In such approaches it is often forgotten that there is no rationality as such in absolute terms, and that every social and historical context defines its own rationality (Godelier 1972, p. 317). To impose, therefore, our own concept of rationality when explaining the behaviour of past social actors is misleading, hindering our understanding of the past and what is more "it is an unacceptable form of intellectual chauvinism and ethnocentrism" (Hamilakis 1995, p. 26).

Methodologically, the model proposed by Gallant has been formulated mostly on the basis of analogies either with modern Malaysian, Black Sea, Aegean or other European fisheries or with the 19th c. Adriatic fisheries. Gallant explains why these particular analogies have been considered suitable (Gallant 1985, p. 12), but we can find a detailed discussion of his ideas on the proper use of analogy in a later work of his (Gallant 1991, pp. 1-5). Although ethnographic or historical comparisons may be very useful in constructing historically and contextually informed frameworks of understanding, the use of comparative data to build analogies may be problematic. The assumption of analogies between two different cultures emphasizes the similarities between the compared cases disregarding the differences and thus ignoring the importance of the particular social and historical conditions in the shaping of a given society (Wylie 1985; Garnsey and Saller 1987, p. 45; Halstead 1987 for the misuse of ethnographic analogy in the study of ancient pastoralism in Greece).

The consideration of the Malay fishermen for example, in the study of ancient Greek fishing, could be instructive if viewed as offering one possible scenario for the incorporation of fishing in the life of an insular community. If it is viewed as a directly analogous case, then the social and historical uniqueness of two distinct cultures is automatically cancelled. Furthermore, by assuming that the two societies use fishing in the same way because they are both on a similar level of social complexity or exploit similar fish species (Gallant 1985, p. 12), either an environmental or an evolutionary determinism is indirectly imposed.

Gallant's treatment of ancient fishing relies on the use of statistical data originating from two types of work. On the one hand, he uses the results of ichthyologic surveys and studies on specific issues such as the factors controlling tuna migrations from the Black Sea to the Sea of Marmara in Turkey (Demir and Acara 1955). On the other hand, he draws data from more general works of an anthropological (Fraser 1960; Firth 1966) or economic nature (Faber 1883). Almost all these works are based either on data collected by the National Statistic Services of the respective countries, or on data collected with specific project questions in mind. Consequently, Gallant's data are for the most part secondary. Although they are potentially useful in exploring specific questions about past fishing, in order to make full use of the comparisons, one has to disentangle the data from the research agendas of each source-work. The problem appears more severe when the ethnographic analogies take the place of proof of the argument without any consideration of the above issues. Finally, the use of fisheries statistics has to be complemented with a knowledge and understanding of aquatic ecosystems, the nature of a fishery and its biological parameters. Jacobsen's critique on Gallant's use of statistics (2005) emphasises how the statistical values of annual catches are drastically influenced by the intensity and type of fishing in any given sea, concluding that modern statistics can not be used in themselves as a base for the estimation of the productivity of ancient fisheries.

The *Fisherman's tale*, with its pessimistic conclusions, which were based on a seemingly solid base of scientific method and common sense, has had an overpowering effect on subsequent research on ancient fishing and fish consumption. For over fifteen years, subsequently, no systematic work on the nature of fishing and its importance in the ancient Aegean, either historical or prehistoric, had been attempted. Despite the gradual accumulation of fish-bone assemblages, and a very rigorous discussion of several aspects of life in rural Greece and its relevance to the interpretation of the past, research on ancient fishing was mostly absent. Ethnographic and ethno-archaeological studies in the Aegean have mostly focused on agro-pastoral communities and on land bound activities. Very little research focused on the sea, and when it did, it dealt with phenomena that are mostly linked to trade (e.g. Bernard 1976). The few ethnographic and anthropological works that were produced did not make an impact, perhaps because they were not produced within the dominant paradigm in Aegean paleoeconomy or ancient economic history. Furthermore, they were published in French (Guest–Papamanoli 1984; 1985) and in Greek (e.g. Paraskeuopoulou, date unknown; Lambadaridis 1973; Garioti 1977; Zachariou – Mamaliga 1986) becoming thus relatively uninfluential in the English language literature.

The idea that fish was an emergency resource, poor and erratic, and that it was exploited as a means of diversification of the subsistence base of agricultural communities has been recurrent in the literature. Most of the works dealing with the subject of the ancient economy of pastoral or agricultural communities at that era were careful to emphasize the variability of human decision-making as a response to varied environmental constraints, and were sensitive to the use of ethnographic comparisons. These principles, however, were never applied to fishing. Fishing was simply not considered an issue, both in ethno-archaeological and historical research. Since the publication of Gallant's book, the major theoretical issues put forward in it have been discussed and criticized in both archaeological and anthropological discourse. The *Fisherman's Tale* however, remains a recurring reference in every work dealing with fish and fishing in Classical antiquity, whether historical or literary, showing perhaps the need, on the part of the Classicists, to understand the material aspects of relatively unfamiliar activities such as fishing in the Aegean.

2.3 Fish in the economy and beyond. The recent work.

The two approaches described above, the philological and the economic, share a common lack of material evidence in support of the positions taken. The work of philologists and ancient historians (chapter 2.1) was based almost exclusively on texts, while those of the economic archaeologists and historians (chapter 2.2) were mostly general, theoretical, based on analogy and on modern presumptions. A new wave of works on ancient fishing, mostly after the 90's, deals exactly with this problem. They share a more or less openly expressed interest in vigorous documentation, methodological robustness and data criticism. Also, despite the predominant focus on issues pertinent to the economic aspects of fishing and fish consumption, they nevertheless, attempt to explore social and ideological parameters as well. Two major trends can be discerned.

The first approach sprung from archaeology. It set out to explore the issue of ancient fishing based on archaeological evidence in a space- and time-specific context. A major innovation has been the use of fish bones as a source of information, mostly on past fishing and to a lesser degree on past fish consumption. Fish bones had, of course, been reported in earlier days (e.g. Bates 1936; Becker 1986; Bockonyi 1986; Reese 1984; Renfrew *et al.* 1968; Williams 1979, refs in Mylona 2003a), but in almost all cases they consisted of lists of species, with an occasional minimal commentary on the fishing methods they represented. Two exceptions were the tuna bones reported from Saliagos, on Antiparos (Renfrew *et al.* 1968), and from Franchthi cave in the Argolid (Payne 1973), both of which had been over-interpreted on the basis of preliminary or inadequate publishing (Bintliff 1977; for a discussion see Rose 1994, p. 434-8). Fishing implements and representations in art have also been noted and discussed or catalogued from quite early in the history of the discipline. Such discussions, however, rarely surpassed the level of formal description and aesthetic analysis (e.g. Robinson 1941; McPhee and Trendall 1987).

It is interesting that fish bones have only recently been used in broader discussions on fishing and fish consumption, and even in discussions of broader paleo-economic schemes, mostly dealing with subsistence. This development is apparently related to the delayed application in Greece of efficient collection methods during excavation that permit the accumulation of reliable fish-bone assemblages (see discussion on the issue by Rose 1994 and Mylona 2003a). This explains perhaps the chronological gap between the emergence of an interest in fish bone analysis in other parts of the world (Casteel 1976; Desse-Berset 1982; Wheeler and Jones 1989) and the first comprehensive applications in Greek archaeology.

Mark Rose has been a pioneer in this field of archaeology and his work can be taken as representative of this trend. In a series of papers, but more comprehensively in his PhD book (1994), he set out to explore the role of fishing in Bronze Age Aegean. He amassed all available fish bone data, whether water-floated, sieved or hand collected, and analysed them against two frames of reference. One is ecological and it was constructed on the basis of fish's ethology and biology, modern hydro-biological conditions in the Aegean and the Mediterranean in general and the types of fisheries developed in them. The second is technological and was constructed with reference to ethnographic and archaeological comparisons with pre-modern Aegean, ancient Egypt and ancient Greece, both prehistoric and historical. The data used in this context were artefacts, iconography and texts.

Rose identified certain common trends in Bronze Age fishing in the Aegean, most prominent of which was that fishing was predominantly coastal, targeting small and medium size fish on a year-round basis. Migratory fish, albeit potentially important at certain favourable locations, were far less systematically explored. He concluded that for Bronze Age Aegean economies, fish could only have been a secondary resource, important as a source of valuable nutrients in a cereal dominated diet. Fishing, according to Rose, could fit well within the work schedule of littoral farmers, thus becoming supplementary to their subsistence practices.

Methodologically, this work borrows heavily, but also critically from Gallant, but theoretically it is based on the "risk and uncertainty" model, which had been a dominant paradigm in the field of environmental archaeology and Palaeoeconomy in Greece after the mid-80s. In summary, this model that had been initially formulated in anthropological studies (e.g. Scott 1976) is based on the idea that variability, especially in food supplies, is a determining factor in decision making concerning subsistence (Cancian 1980; Halstead and O'Shea 1989). In areas of high ecological diversity and with a highly fragmented terrain, scattering of resources and unpredictability of weather pattern variation, variability in the means of subsistence is inherent and scarcity of resources is recurrent and difficult to predict and calculate precisely. The model, as applied in archaeology, aims to predict the mechanisms employed in the past by the humans involved to ensure a minimal availability of food under the worst possible conditions, in order to survive. These mechanisms may vary considerably, in both character and intensity (Halstead and O'Shea 1989, p.3). Halstead and O'Shea have grouped them in four basic categories: mobility, diversification, physical storage and exchange (for a discussion and critique of this model see Hamilakis 1995, ch. 9). In Rose's work, fishing is clearly perceived as a mode of diversification of the subsistence base, in an attempt of the Bronze Age inhabitants of coastal settlements around the Aegean to achieve a measure of security. Nevertheless, no explicit, systematic discussion of theoretical principles and assumptions has been offered, despite occasional commentary on specific topics.

An alternative approach to the archaeology of fishing in the Aegean, which shares several features with the one presented above, focuses on fishing-related artefacts, namely tools and iconography. A representative of this approach is Judith Powell, whose work *Fishing in Prehistoric Aegean* (1996) forms the most comprehensive synthesis of such data. Powell explored the Bronze Age fishing technology and methods in the Aegean, setting all the available artefactual evidence against the ecological and ichthyological context in which fishing operated. She supported her arguments with ethnographic comparisons from both contemporary and ancient fishing. Again, as in the case of Rose, although not explicitly stated, the "risk and uncertainty" model can be discerned behind the rationale of this work, and Gallant's model is considered basically valid although strongly criticised.

Rose's and Powell's works have set a notional and methodological framework that accommodates most of the subsequent archaeological research on the issue. Specialised work on fish bones is presently mushrooming, covering several periods of Greek Antiquity, but mostly prehistory. The abundant Mesolithic fish remains are analysed in detail (Rose 1995c; Mylona 2003c; Powell 2003) as are Neolithic and Bronze Age assemblages throughout Greece (Rose 1994; 1995a, 1995b, 1998; Mylona 2000; 2004; Theodoropoulou, forthcoming). Work on material from historical periods is far less common (Bookidis *et al.* 1999; Rose 2000; Mylona 2003b; 2006; in prep.). This marked periodisation of research reflects both the distinct methodologies applied to prehistoric and historical excavations, and the different interests of each branch of archaeology in Greece (Mylona 2003a). Artefacts and iconography remain marginal in the research on ancient fishing, with contextualisation of finds still mostly absent. Works such as Sparkes' short paper on narrative scenes related to fish and fishing on Classical pots (1995), however, hints at the usefulness of a more comprehensive approach to iconography, but makes the need for contextualisation more evident.

All these works have made a significant contribution to research on the ancient Aegean, especially its maritime aspects. They have enriched our knowledge of the types of fish that contributed to human diet and on the way they were caught. They have proposed ways in which fishing might have been incorporated into the economy of ancient communities and they have informed us of activities indirectly related to fishing, such as trade. The most important contribution of these works, however, is that they attempted to bring together a wide range of evidence and use it critically. Despite occasional failures to do so rigorously, they nevertheless manage to use these data to raise details of ancient fishing, making the issue of fishing and fish-eating relevant to the standard archaeological discourse. Another important point is that they attempted, along with other branches of research on bio-archaeological remains, to explore one type of human activity (fishing) not in isolation but within its physical context. Although this notion echoes the preoccupation with the physical environment of Environmental/Economic archaeology of earlier decades (chapter 2.1), what is novel and more potent in the most recent approaches is the specificity with which they deal with the physical parameters of fishing.

What is mostly lacking from this type of archaeological work, however, and is in fact their major liability, is that fishing and fish consumption are viewed as being almost totally disconnected from the social context in which they operated. Following the tradition of the predominant paradigm in Environmental/Economic archaeology in prehistoric Aegean, they implicitly assume that fish-bones and related artefacts "talk for themselves". Furthermore, most of them assume that fishing is a technological application, and fish-eating is about nutrition. The human agents in the fishing — fish-eating circle remain obscure and ideas about the world of fish and its relation to the world of humans are hardly explored.

At this point, developments in the research around fishing and fish-eating in antiquity, in other regions around the Mediterranean and Europe in general, should be mentioned. The analysis of fish-bone assemblages has become a fairly common procedure in excavations around the Mediterranean coasts (for a fair sample of such publications see the ICAZ - Fish Working Group web page, 2007), contributing to a better understanding of several so-far 'unknown' fishing aspects of this sea. Various alternative ways to explore the importance of fish in the diet form another domain where rapid developments take place. Stable isotope analysis on human bones or other tissue offers exciting possibilities (e.g. Barrett *et al.* 2001; O'Connell *et al.* 2003) although such applications in Greek case studies are still few (Bourbou and Richards 2007, p. 66 for a review of all Greek case studies). Along similar lines, the chemical analysis of food residues on pottery (e.g. Evershed *et al.* 2001; Isaksson forthcoming) could offer another set of possibilities, which also need to be more widely tested.

A second approach to past fishing, which fills in some of the gaps described above, was born in the realm of ancient social-economic history. A number of works attempt to explore fishing and fish consumption as an economic, social and ideological phenomenon. They use the traditionally perceived historical sources (literary texts and epigraphy) and occasionally archaeology and anthropology. Researchers who work within this framework begin their explorations with the economy of fishing and fish-eating, and often broaden their scope to include issues such as the ideology of fishermen and fish consumers, the symbolism of fish-eating etc. This type of work modifies and expands on previous discussions on the issue (see chapters. 3.1 and 3.2), exploring new data and offering fresh insights and interpretation of the known ones.

A typical example of the economy-oriented side of it is Robert Curtis's work *Garum and Salsamenta* (1991). In an impressive survey of the whole Mediterranean, Curtis brings together a huge body of evidence on the processing and trading of fish and fish products. He combines, very successfully, a broad range of archaeological finds (e.g. architecture, pottery, fish remains) with a variety of text types (e.g. stone inscriptions, amphora labels, literature) to produce a picture of an ancient (mostly Roman) industry based on fish and its by-products. The situation described by Curtis, i.e. the existence of massive fish processing establishments in various areas along the coasts of the ancient Mediterranean, has been perceived by scholars as rather exaggerated, directly opposite to the reductionist position taken by Gallant (Purcell 1995). But even if the intensity of fish processing activities around the Mediterranean had in fact been lower than that described by Curtis, works like the *Garum and Salsamenta*, and others along the same lines (Braund 1995; Bekker-Nielsen 2002; Mylona 2003b) do highlight the existence of a distinct, identifiable sector in society which was making a living from the sea and its fruits.

All the archaeological and historical works presented above have made a major contribution, towards documenting the existence in the ancient Aegean of a distinct set of activities related to fishing and fish consumption. By exposing the variety, complexity and wide distribution of these activities throughout the Aegean and further into its adjoining seas, they give rise to a number of interesting questions. For example, if there was a distinct fishing and fish-consumption sphere, what was the mentality around it, how was it created and how was it perceived by other sectors of the ancient population? How did it affect the practice of fishing, fish-trading and fish-eating?

Researchers have recently attempted to answer these types of questions. Most of these attempts originate from ancient historians, who begin their explorations based on the traditional historical sources, i.e. texts. They do so by placing the texts in their social, political and ideological context, informed by anthropological ideas and occasionally by archaeological findings.

Nicolas Purcell's paper "Eating fish. The paradoxes of seafood" (1995) is such an example and perhaps one of the most successful attempts. Purcell explores the antithetical ideas which surrounded fish, fishing and fish-eating in classical Greece. These ideas are very convincingly taken to express wider social and cosmological contrapositions. In the common perception, fish and the activities which surround them (capture, marketing, consumption) were vehicles that expressed the antithesis between land and sea, poor and rich, secular and sacred, death and survival. The formulation of these antithetical ideas, often simultaneously expressed or felt by ancient Greeks, is clearly demonstrated through a perceptive analysis of a variety of sources.

Moreover, works in this framework focus on fish consumption as a vehicle that communicates social and political ideas, thus totally removing fish from the sphere of subsistence into that of representation (e.g. Davidson 1993; 1995; 1996; 1997). Davidson's work provides one of the most enchanting examples of such an approach. In his book *Courtesans and Fishcakes, the Consuming Passions of Classical Athens* (1997) and a series of papers (1993; 1995; 1996) he set out to show how the consumption of fish, along with other activities of pleasure (eating, drinking, sex), was in Classical Athens laden with political and ideological meanings.

Davidson bases his arguments on a compilation of extracts from Classical texts of public character, mostly comedies, orators' speeches and pamphlets on various subjects. Most of these texts are extracts of lost works preserved in Athenaeus's *Deipnosophistae*, a work produced within the intellectual movement of the "second Sophistic" in the Roman era, which aimed at the revival of the values of the Classical era (Bowie 1970; Braund and Wilkins 2000). By examining what ancient Athenians were saying about fish-eating (and other pleasures) in different contexts, he proceedes to explore what they considered to be a proper way of behaving as citizens of democratic Athens.

Fish-eating and its related discourse, which also involved the fish-market, the fishmongers, fish anecdotes and various other themes, is viewed as one of the pleasures enjoyed by the Athenians, with drinking, music and sex being the others. He explored what Athenians thought and said about fish-eating. Starting from the particularities, from the comments in the texts, he proceeded to connect these with broader issues of politics and ideology. More specifically he demonstrated that the fish-market was the locus of distinction in a discourse on class and wealth, where the desire for large and expensive fish went beyond the need for sustenance and into the realm of political debate and confrontation.

A search for meaning in fish-eating in antiquity is not new in Classics. The role of fish as a religious symbol or sacred animal had repeatedly been noted (e.g. Dölger 1922). Davidson's major contribution to our understanding of fish-eating in Ancient Greece lies in the contextualisation of this meaning. He stresses the fact that the meaning attributed to fish consumption depends on the social context in which it takes place. A second, no less important contribution of Davidson's work, is his demonstration that ancient texts are not just a depository of information about the past, but rather tools of persuasion in their time, and as such actively involved in negotiations of power.

Despite the fascinating insight into the Classical fish-eating that we gain from Davidson's work, it does have a serious shortcoming in the understanding of fish consumption as a phenomenon. Davidson focuses on discourse. He analyses stories, words and nuances of expression around fish-eating. He explores representation and builds on the interplay of complementary or antithetical ideas about fish. What is completely absent from his treatise is the materiality of fish and fish-eating. This materiality is as much part of the situation as any verbal representation of fish. What fish were really available in the fish-market, what was actually eaten, how and by whom are some aspects of this materiality which served, in my opinion, to create a frame of reference in which the texts acquired their specific meaning. In this way this fresh approach to fish-related issues in antiquity that is exemplified by Davidson, exhibits a liability which is exactly opposite, but in a way similar, to that found among the works of the archaeologists described above. Ancient historians' focus on texts, even in a critical way, leads most works to concentrate on the way fishing and fish-eating had been conceptualised, on the way they served as symbols in the social and political arena, disregarding the importance (or occasionally even the existence) of the practical aspects of it.

2.4 Conclusions

Fishing and fish-eating in the ancient Aegean are issues which have been discussed in archaeology and ancient history in a disparate manner. On the one hand, emphasis has been placed, especially by philologists and ancient historians, on the written sources. This research reached diverse conclusions which range from an exaltation of the importance of fishing and fish-eating to its total downgrading. Much of the research within this field has been based on theoretical schemes for the study of past societies which were formulated to deal with ancient economy. Research based on the material remains of fishing and fish consumption within the realm of archaeology (e.g. fish bones, fishing-related artefacts, trace element analysis), although methodologically innovative, mostly operated within the theoretical frameworks mentioned above, thus providing new data, but largely failing to exploit their full potential in better understanding past fishing and fish-eating.

The critical review of the various approaches to past fishing and fish-eating in the Aegean that has been presented in this chapter offers an obvious conclusion and points to a possibly more insightful path of research. The obvious conclusion is that fishing and fish-eating can (or must) be explored from different angles. The natural environment and the available technology obviously offer possibilities or pose limitations which ancient fishermen could/would or could not/would not exploit. The fishermen and fish consumers themselves acted upon these preconditions, occasionally altering them, or adapting to them. However, their decisions and actions were influenced as much by the environment and technology as by their ideas about this environment, about themselves and the world in general. Finally, their ideas about the aquatic world and its relevance to their everyday life could not be an abstract construction, irrelevant to the realities of fishing and fish-eating. These are very space and time-specific ideas, and in order for us to appreciate them, we have to be able to judge them against the ordinary and the extraordinary conditions that generated them. Therefore, we need to go back and explore the practicalities of fishing and fish-consumption.

Investigating ancient fishing and fish-eating in ancient Greece in this cyclical manner, the issue of the predominance of "economic" or "cultural" motivation becomes irrelevant. On the contrary, emphasis is placed on the interrelations and the unifying threads between the way people used to act and to think in relation to the aquatic environment and the fish. In my opinion, to explore this particular aspect of the past is a very promising path.

The review presented in this chapter, and the conclusions drawn from it, demand a suggestion. How can one go about exploring these interrelations? How can the practical and the ideological come together without any hierarchical ordering? It seems that there is a domain where "ecological", "economic" and "cultural" merge and where the types of evidence available to us are most relevant. This is the domain of food consumption. The following chapter will present and discuss in detail a theoretical framework which could accommodate such an approach to fish consumption.

CHAPTER 3

THEORETICAL FRAMEWORK ADOPTED IN THIS STUDY

The ideas emerging from the works presented in the previous chapter (chapter 2) and more particularly whether fish-eating is all about meaning and representations or about calories and nutrition, places us squarely in the heart of an ongoing anthropological debate on whether food in general is good to eat (Harris and Ross 1986) or rather, good to think with (Levi-Strauss 1963, p.89). The relative importance of the nutritional, economic and environmental dimensions of food compared to the symbolic, communicational and cultural aspects of it have been the focus of a burgeoning corpus of anthropological and sociological research for the past few decades. The ideas thus developed are now increasingly permeating archaeological thinking, pointing to novel approaches to past societies.

Examining fish consumption in such an anthropologically inspired theoretical framework may prove quite rewarding. This chapter will present and evaluate a variety of issues emerging from contemporary anthropological and sociological research on food consumption. These range from the classic theories in the anthropology of food to a discussion of food as a commodity and finally to the theories of embodiment in the anthropological and archaeological research. Most of this discussion refers to food in general. However, some adaptations of these ideas to the peculiarities of fish as food are also suggested. The approaches put forward here have been chosen because they could be adapted to an archaeological framework and prove useful in the interpretation of archaeological and literary evidence. In this chapter, I attempt not only to offer a theoretical framework, which is broad and flexible enough to accommodate the varied and disparate evidence for past fish consumption, but additionally to provide a theoretical framework for achieving the merging of anthropology and archaeology.

3.1 Food consumption in anthropological thinking

Hunger is thought to be the most imperious and constant need in the life of humans. We must eat regularly, for as long as we live. Therefore, food - its acquisition and consumption - is an ever-present element in our life and for this reason anthropologists have thought that it is a good vehicle by which to explore a multitude of other aspects of living.

Even a cursory review of the anthropological and sociological literature on food and eating shows that the relevant research is positioned around one of the most dominant controversies of western thought; that between nature and culture, which have often been regarded as two distinct (often conflicting) aspects of human nature (Hinde 1991; Ellen 1996; Ingold 2000b; Hamilakis 2001a). In this framework, food consumption has been approached and analysed according to a polarised scheme, with some alternative approaches emerging in recent years. In anthropological literature, these works could be grouped in three broad categories. The following presentation of these categories does not aim to be exhaustive but rather to set the basic framework for the subsequent discussion.

3.1.1 *Structuralist and cultural –materialist view on food consumption.*

The structuralist–cultural approach stands at the one extreme of the range of approaches to food and food eating (Douglas 1966; Levi-Stauss 1969; Sahlins 1976; Barthes 1997; Douglas 1997). Its first exponent was the French anthropologist Claude Levis-Strauss, who used food preparation, along with language, as case studies by which he demonstrated his ideas about the existence of common "deep structures" in the human mind that shape the way humans express themselves culturally. He explored one universal cultural phenomenon, cooking, using a linguistics methodology (Levi-Strauss 1969). Levi-Srauss aimed at the recognition of the deep structures of human mind in the way cooking is organized and practised within any given society.

By way of illustrating these relationships, he devised the "culinary triangle', or in his own words "food in a triangular semantic field" (Levi-Strauss 1997). The "culinary triangle" is a graphic way to illustrate the relationship between the natural and cultural, the elaborate and unelaborated character of food. By assuming that the raw food represents its natural condition, he then took the cooked food to represent a cultural transformation, while the rotted stands for the naturally transformed food. Boiling and roasting are incorporated in this scheme as forms that express more refined relations between the raw, the cooked and the rotted, always in a form of binary oppositions between any two of these categories. Taking the argument further, Levi-Strauss related these forms of transformation to a number of other contradictory notions, such as public-domestic and male-female.

Although Levi-Strauss's analysis of cooking has been severely criticised, mostly because of its limited support by empirical data (e.g. Harris 1980, pp. 188-194), its a-historical treatment of the phenomenon and the untested methodology of classification (Leach 1970, p. 32; Mennell 1985, p.9; Murcott 1988, pp. 6-10), it nevertheless contributed a breakthrough in anthropological thought about food. It placed an emphasis on the semiotic character of food, its social significance as a code, as an information system and as a symbol.

Other researchers who worked in the structural tradition chose to focus on exactly these features of food, underplaying the binary oppositions, thus producing a distinct body of ideas. Roland Barthes (1997) searched for a code or "grammar in the way food is prepared and consumed". He sees food as a sign as well as a need, claiming that "to eat is a behaviour that develops beyond its own ends, replacing, summing up and signalling other behaviours and it is precisely for these reasons that it is a sign" (Barthes 1997, p. 25). Mary Douglas, in her work on the structuring of the meal (Douglas 1997) shows that in British households meals are constituted in certain repeated ways. The position of each food in the meals encodes meanings of social boundaries. For her, food and eating are symbolic of a particular social order, thus the patterns of meals stand for much more than themselves (Caplan 1994, p. 8).

The structural-cultural research on food habits has clearly demonstrated that food is not a static element in human life, neither is it just fuel for the human bodies (Douglas and Isherwood 1996, p. 8), but it is a field where social or symbolic meanings are expressed and understood by all members of a society. Such an approach, however, has not been without drawbacks. The most serious of them is that it assumes that societies and cultures are homogenous and a-historical. They do not recognise, for example, that different groups within societies may express themselves in different and often conflicting ways, exhibited in distinct food habits within the same society. Neither do they recognise that they may perceive the messages transmitted by food habits in different ways (Goody 1982, p. 33). Such approaches perceive cultures and societies as static arenas where nothing changes (Goody 1982; Mennell 1985, p.11-12; Mintz 1993) and they neglect or treat superficially the interactions between production, distribution and consumption, between the biological, social and semiotic aspects of food (for an analytic discussion see Goody 1982; Lupton 1996).

At the other extreme of the research related to food habits stand works that follow the cultural-materialist tradition (Harris 1986; Harris and Ross 1986; Farb and Armelagos 1980). In this framework, researchers emphasise the practical reasons behind human food choices (Harris, 1986, p. 14). They accept that people eat what they do because it is both nutritionally and ecologically sound. Marvin Harris, perhaps the most famous representative of this approach, demonstrated the credibility of his argument by showing the adaptive character of some of the most famous, "strange" food habits in the world, such as the avoidance of pig eating by Muslims and Jews (Harris 1986) or the Aztecs' cannibalism (Harris and Ross 1986, pp. 124-5). Researchers in this paradigm accept the existence of a symbolic dimension to food (Harris 1986, p. 61), but they insist that the practical precedes the symbolic and that meaning is the consequence of activity. Major food complexes have been shaped by biology, ecology, economy and politics.

The cultural-materialist approach emphasised the importance of the material dimension of food, thus countering, to a certain degree, the emphasis placed by structuralists on food's communicative value. The potential gains of this contribution have, however, been negated by its persistent neglect or underplaying of these other factors. Furthermore, the cultural-functional approach to food eating tends to focus on distinguished, peculiar and strange case studies, ignoring the everyday, "normal" practices related to food.

3.1.2 *The eclectic approaches to food consumption.*

The strengths and flaws of the works described above have been noticed and discussed by various researchers who set out to formulate more satisfactory approaches, eclectically combining ideas and methodologies of both the structural and materialist approaches. Although they vary greatly in focus and subject matter, nevertheless, these works share some common features. In short, they accept (or take for granted) that food habits, preferences and tastes transmit social and symbolic meanings, but also accept that material conditions formulate them to a certain degree. According to this approach an explanation of food habits may develop by placing these "meanings" into the material and social context in which they operate, and by viewing both in a historical, comparative perspective.

The American anthropologist Sidney Mintz (1985) discussed the history of sugar consumption in Western Europe from the 18th c. A.D. onwards. Influenced by Marxist thought, he focused on issues such as the interdependence of demand and supply and most importantly he placed emphasis on the historical dimension of food habits.

The work of Jack Goody, one of the best-known exponents of this "middle ground" type of approach is, in a way, typical (1982). In his treatment of the reasons behind the development of high cuisine in some cultures, and its absence in others, he does not give primacy to either matter or culture. Instead, he recognizes that structuralist and materialist approaches are not necessarily contradictory but may be supplementary. He places emphasis on the role of class and power in the shaping of food habits and stresses the need for history to be incorporated into the anthropological analysis. He proposes a discussion of food habits in terms of the diversified structures of household and class.

The absence of homogeneity within cultures and the importance of distinguishing between social strata is another theme developed in this field, with the introduction of notions such as the consumers' emotions as these are manifest in the changing food preferences and emerging cuisines (Murcott 1988, pp. 29-34; Mennell 1997). A characteristic feature of this type of work is the quest for an explanation of the formation of particular food preferences and habits. The main question asked is "why" certain food habits have developed in particular ways? What are the motivating factors in shaping the particular developments?

3.1.3 Post-structuralist ideas.

More recent developments in research on food habits overcome causal reasoning, and focus on the exploration of processes. They tip the balance in favour of cultural explanations of food habits again, mostly ignoring, but not explicitly denying, the material aspect of the process, focusing instead on the representational. Deborah Lupton's statement perhaps best encapsulates the basic orientation of these works. In stating the central theoretical position of the book "*Food, the Body and the Self*" she writes "I am interested in exploring the changeable and contextual nature of meaning...to understand the ways in which preferences for food develop and are reproduced as socio-cultural phenomena" (1996, p. 12). Lupton (along with other researchers working with a similar agenda) takes for granted that food preferences and habits are generated, reproduced and diffused through specific social contexts, and that there is a strong link between these habits and the cultural environment that generates them. Furthermore, according to this line of thought, food preferences and tastes are linked to such issues as gender and power relations. The unifying thread between all these issues is the body. Food habits and preferences are central practises of the self, directed to self-care via nourishment of the body with foods that are culturally deemed appropriate, comprising a source of pleasure and acting symbolically, to present a persona to oneself and others (Lupton 1996, p. 15). In this framework, food-related emotions and memories are also relevant.

These specific issues have, in the recent years, given rise to a proliferation of anthropological, sociological and human geography studies that deal with the way the embodiment of food (in all its dimensions) acts as a medium of experiencing the world. These have explored and enhanced the way in which subjectivity may not only reflect the manifold ways in which individuals understand themselves, in relation to others, through food but also the way in which they experience their life and perhaps modify it to suit themselves (Lupton 1996, p. 13). Such approaches have introduced to the study of food discourses the element of the "practice of everyday life", shifting focus from the special, the unique, the unusual food-related phenomena of the previous generation of thinkers to "the triviality of many of the things people do... (that) ... blocks them from consciousness, since they constitute acts of utter common sense" (Cowan 1991, p. 324).

This summary presentation of the anthropological ideas about food and eating is permeated by certain unifying ideas. Perhaps the most important of them is that there is not one way to look at food and food habits, but instead there are different levels and perspectives of analysis. The structuralists' insistence on the role of food as a communicative sign and the post-structuralist emphasis on subjectivity in the formulation of certain attitudes towards food could all be equally valid, depending on the focus or level of analysis. Another unifying theme, which emerges from the case studies or from the various critiques of certain approaches, is the realisation of the vast diversity of cultural expressions which are related to food. Although this variability seems to be quite an obvious observation, it is however, one commonly neglected in the archaeological treatment of food-related issues, which mostly focus on general trends and a regional scale, tending to obliterate variability and individuality.

On account of the above, the present book takes an eclectic position in relation to these theoretical approaches to food consumption, accepting that not only one idea is appropriate, but several, at different levels of analysis. In this sense, the material and the communicative nature of food, or the historicity and the contemporaneity of food habits do not have to negate each other, but are viewed as complementary, as different aspects of the same phenomenon. During the analysis or discussion of the consumption of a particular food, such as fish in this case, certain of the above aspects may be highlighted more than others at different points of the discussion.

3.2 Food consumption as a social phenomenon.

"Food consumption" may appear, on first reading, as a fairly straight-forward term to describe the eating of food. Starting from there however, the consideration of various parameters relating to food eating shows that this view is quite simplistic. It is perhaps instructive to begin this discussion by first considering the issue of consumption in broader terms. By starting, along with Mary Douglas and Baron Isherwood (1996, p. 3), from the basic question "why people want goods" one is lead to further ask, what is the meaning of consumption and whether consumption is a phenomenon of a strictly economic nature (for a presentation and critique of this view see Douglas and Isherwood 1996, pp. 11-35) or is it broader, touching upon issues of social and cultural structuring (Bourdieu 1990; Hamilakis 1995; Douglas and Isherwood 1996).

From the discussion in the previous section it becomes apparent that through consumption (of food but also other goods and services) people satisfy certain biological needs but also enter into social relationships where communication of meaning takes place. Consumption not only reflects these relationships but also establishes, maintains and changes them (Orlove and Rutz 1989, p. 5). Therefore, by studying the way consumption is structured and transformed, we have a way to explore broader issues of social relationships. In this process a certain feature of consumption is crucial. Consumption is a unifying concept in the sense that similar rules can apply to the consumption of a diverse range of goods and services (Hamilakis 1995, p. 174). Consumed organic vegetables and "haute couture" for example, share the fact that they both act as communication signs and also as signifiers in social relationships. Consequently we can analyse them both using similar methods, and study them in relation to each other.

Transferring the above reasoning to archaeology, we could view the disparate archaeological evidence (food leftovers, pots, architectural remains, texts) as consumed objects, which not only satisfied some immediate biological needs but also did so in a specific social context, each transmitting messages through their use, complementing or contesting each other. By accepting the existence of such a relationship between consumed objects (the remains of which are found archaeologically), we are able to have a clearer picture of the situation under study. To take a specific example, Hamilakis, in his analysis of the role of wine and olive oil in the development of power in Bronze Age Crete, analysed the consumption patterns of these products, and tested his hypothesis and results against evidence for the consumption of pottery in the same context. His work demonstrated how both these food and non-food consumables played a parallel role in the context of conspicuous consumption and power relations in Bronze Age Crete (Hamilakis 1995).

If we turn to food consumption, there is a central issue to deal with. What can be categorised as food? Anthropological and ethnographic work has demonstrated that not all edible substances are considered food neither are all ingested substances nourishing. Certain plants and animals, for example, acquire the identity of the edible food under specific conditions. Ethnographic work in several areas of Greece, for example, has shown that certain foods such as common and bitter vetch (*Vicia sativa* and *Vicia ervilia*), are normally considered animal fodder but in conditions of hunger they turn into human food (Forbes 1989, p. 36-41; Halstead and Jones 1989; Garnsey 1999, pp. 36-41). Among the aLuund in Zaire, foodstuffs become edible only when they are acquired through reciprocity and exchange and when they relate to reproduction and life transmission. Outside this context foodstuffs are inedible, regardless of the occurrence of hunger that may threaten the aLuunds' survival (Boeck de 1994). A third example involves taboo foods, foods that are not considered edible, for a specific sector of the population, or for individuals at certain life stages or in particular contexts. Such, for example, is the case of a number of fish in ancient Egypt, that were considered divine or polluting and their consumption was either restricted or forbidden (Chouliara-Raiou 2003, p. 99-113). It is interesting to note that these prohibitions were not universal, but only applied to certain fish in certain areas of Egypt, or certain sectors of the population, such as the priests.

What these examples show is that the categorisation of a substance as edible implies that it is accepted into the community and participates in a broader system of values and rules which are not only time and space specific, but potentially flexible (Falk 1994). It also implies that it is accepted as suitable for incorporation; to cross the boundaries between the body and the world and become self, become part of who we are (Lupton 1996, pp. 15-19; Hamilakis 2002b, pp. 123-127).

The same is true of another categorization of foodstuffs, that of luxury and subsistence food. Although such distinctions are observed, it must be clear that this categorisation is fluid, depending once again on local and temporal cultural conditions. Mintz's analysis of the changing role of sugar through its history, clearly illustrates this idea. Sugar, from being a scarce, precious substance of restricted, privileged use during the early centuries of its appearance in Europe, became an overtly common substance that is regularly consumed by almost everyone, regardless of class or status (Mintz 1985).

Related to the above distinction is a misunderstanding that only luxurious, prestige foods carry meaning (e.g. lobster and caviar in modern western society) while it is absent from common, everyday foods, such as a glass of milk. Although the first category of foods have attracted more attention in the literature, perhaps due to its heightened visibility (for a discussion of luxury foods see van den Veen 2003), in fact all kinds of food transmit messages, which are closely related to the context in which they operate and which ascribes to them value.

Many beliefs about food are culturally reproduced from generation to generation, through the membership of the individual from early childhood in one's household and other broader social-cultural groups. Sharing the act of eating brings people into the same community: they are members of the same food culture (Lupton 1996, p. 25). So, for example, participation in a sacrificial meal in Classical Athens, where red meat of domesticated animals was consumed in a precisely structured manner, certified and maintained the coherence of the social and political body of the democratic citizens (Detienne 1989a; Garnsey 1999).

Historically food practices have served to emphasise ethnic, cultural, ideological and political distinctions (Lupton 1996, pp. 25-27). Authors from all periods of antiquity for example exploit the dichotomy between the civilised, sedentary farmers who live off the cultivated land and the uncivilised, pastoral nomads who are "eaters of meat and drinkers of milk" (Shaw 1982-3), thus constructing an image of the "other", the barbarian on the basis of food habits. Similar use of food is applied to other types of distinction. Mystics' abstinence from food of animal origin, such as that of the Orphics' and the Pythagorians' is just one such example (Detienne 1989b; Garnsey 1999, pp. 108-111).

By being a factor of distinction, food is also functioning as an arena of competition and negotiation of power. Food consumption may be a way of exhibiting status and wealth or the potential of power and authority (Dietler 1996; Wiessner and Schiefenhovel 1996; Dietler and Hayden 2001). Such is the case of conspicuous consumption in institutions such as ancient symposia and sacrificial meals. But even in less conspicuous contexts, within the household and the family, power relations between age and gender groups are evident in the structuring of meals (Bradley 1998; Garnsey 1999, pp. 108-11).

If we turn to the subject of fish consumption in Classical Greece with these thoughts in mind, a whole range of possibilities opens up for investigation. What is needed is a theoretical framework (or more than one) to help us navigate through them, explore their interconnections and perhaps investigate the reasons behind certain choices and actions.

3.3 The social life of things – the social life of fish.

As research stands at the moment, we have at our disposal a wide range of potential data about fishing and fish-eating in Greece between the 5th century B.C. and the 7th century A.D. On the one hand we have material remains: fish bones as leftovers of meals and storage, artefacts related to fishing, fish preparation and marketing and in some cases chemical signatures of fish on pots and human bones. On the other hand we have images generated in the representational arts and written sources, some pragmatic, recording events or intentions, and some imaginary accounts of situations that have been deemed worth recording and reproducing. There is however, a theoretical as well as a methodological difficulty for the research on past fish-eating to move between different classes of data and different contexts of reference. Anthropological thinking once again may offer some way out of the deadlock.

Starting from the observation that despite the dominant tendency of the archaeology of food remains (i.e. zoo-archaeology, ethnobotany etc.) to generalise and deal with food phenomena of the past in broad terms (chronological phases instead of events, categories of edible substances instead of specific foods or dishes, geographical areas instead of individual households etc.), for any consumer, a fish is, on an elementary level, a physical entity. An object caught, removed from its natural environment, sold in the market or exchanged and eventually cooked and eaten in specific, singular conditions. Furthermore, this very same fish might continue to exist as a memory, recalled either in the private sphere (recounts of past food events) or in public (through writing or fables about extraordinary catches).

In this sense fish can be seen as a category of objects, of things. Objects, mostly in the form of commodities have attracted some attention. Economists and anthropologists, who adopt Marxist and Weberian ideas on economics, regard commodities as special kinds of manufactured goods (or services). These are considered either as the cost of maintaining the individual in order for him/her to be able to produce or as the end stage, the objective of a linear process of production-distribution-consumption (Douglas and Isherwood 1996, pp. 7-13). An alternative view that recognises the role of goods in relation to demand and consumption, like those of Sombart and Veblen, has remained much less popular (Sherratt and Sherratt 1991; Hamilakis 1995, p. 170; Appadurai 1986). A view of consumption as an activity on its own, however, that is not the final stage but just one episode in the circular process of production - distribution - consumption (Sraffa 1972; Hamilakis 1995, p. 171) may prove helpful in understanding the way goods move and are consumed in various contexts, contemporary as well as ancient.

Goods as objects participating in a circular process of production, distribution and consumption, can be explored through a biographical approach that seeks to understand the way objects become invested with meaning through the social interactions they are caught up in. Objects change through their existence, and they often have the capability of accumulating histories, acquiring their significance from the persons and events to which they are connected (Gosden and Marshall 1999, p. 170). Some objects accumulate their own inherent meaning through their biography while others may contribute to the meaning of the context in which they participate, e.g. a ceremony or a body of knowledge (Gosden and Marshall 1999, p. 176).

Kopytoff, suggested that things can not be fully understood at just one point in their existence and that processes and cycles of production, exchange and consumption, in short, their life history, have to be looked at as a whole (Kopytoff 1986). This can be done by looking at issues such as the specific and general conditions that prevailed at each stage of the objects life, or its changing use and meaning through time and in different contexts (Kopytoff 1986, pp. 66-7).

At the heart of the notion of "biography" are questions about the links between people and things, about the ways meaning and value are accumulated and transformed. There are many ways of understanding these links and many ways of conceptualising the objects which lie at the heart of them (Gosden and Marchal 1999, p. 172). One way of such a conceptualisation, especially suitable to a food item such as fish in the context of Classical Greece, is through the notion of commodity.

By the simplest definition, commodities are objects of economic value (Appadurai 1986, p. 3). These are materially produced as things, but are culturally marked as being a certain kind of thing which is invested with value (Kopytoff 1986, p. 64). More detailed definitions have been proposed, most of which emphasise the exchangeability of the object-commodity and its intentional production for this purpose (discussion in Appadurai 1986, pp. 6-9).

Appadurai however, suggests that to view commodities in a less static way may be more rewarding. According to him, commodities are:

> "things in a certain situation that can characterise many different kinds of thing, at different points in their social lives. This means looking at the commodity potential of all things rather than searching fruitlessly for the magic distinction between commodities and other sorts of things. It also means breaking significantly with the production-dominated Marxian view of the commodity and focusing on its total trajectory from production, through exchange /distribution to consumption" (1986, p. 13).

An implication of the above definition is that an object can become a commodity at a certain (or various) stage in its life. In this case questions about when, where and how a thing becomes a commodity are crucial (Kopytoff 1986, p. 64). Fish for example may become commodities, when bought in the market or exchanged as gifts, but they may be just "products" when eaten by the fishermen or the buyers.

Furthermore, any thing has a "commodity candidacy", which refers to the standards and criteria (symbolic, classificatory and moral) that define the exchangeability of things in any particular social and historical context (Kopytoff 1986, pp. 13-14). In this sense, tuna fish could be characterised as commodities in the context of transactions between mortals and Olympic gods, thus being suitable for sacrifice, due to their bloodiness. Other, less bloody fish, like the groupers or the sea breams did not fit this category, did not possess this "commodity candidacy" and they were not formally sacrificed to the gods (Durand 1989, p. 127 and note 37; Davidson 1997, p. 12).

Finally, of relevance in this discussion is the commodity *context*, a notion that "refers to the variety of social arenas within or between cultural units, that help link the commodity candidacy of a thing to the commodity phase of its career" (Kopytoff 1986, p. 15). A fish fair on the beach for example may bring together a fisherman and a wealthy cosmopolitan gourmet, who might share very little understanding of each other's worlds, moral values or cosmologies and who might only agree on the terms of the fish trade. The urban fish-market, a different commodity context, might, on the contrary, not have permitted such a contact thus separating the fish from its producer (for an analysis of this phenomenon see chapter 7).

Fish in Classical Greece, at least as far as the literary sources go, appear and function mostly in the market as commodities (chapter 7). Therefore, I consider these three notions, along with the biography of things as very useful for the present study. This is because they enable the free logical movement between different planes of meaning and function of fish. Also because they link production, distribution and consumption in an organic whole, and permit the viewing of fish consumption in several different, yet connected contexts (e.g. fish production, fish marketing and distribution, fish-eating in a social context, fish symbolism etc.).

This seems to be a particularly promising way of approaching archaeological finds. Although most of the applications so far (archaeological, anthropological and art-historical) refer to prestigious artefacts, objects of art and in general objects which are long lived and incorporate considerable artistry in their making (see papers in Gosden and Marshall 1999), food, with its ephemeral, short lived nature offers a challenge. By being ever-present in almost all contexts, and by combining a strong material and a communicational nature, the tracing of food's biographies may prove very instructive. Fish in particular, much like other kinds of meat, presents some special problems.

Fish are unlike artefacts in two respects: unless processed and preserved, they have very short life spans (one or two days at the most in the East Mediterranean climate) and each fish has very few distinctive characteristics (for an analysis of such characteristics see chapter 9) to set them apart from other fish of the same species. Even so, a casual review of the available evidence proves that the accumulation of past histories for a food item such as fish is possible.

One such history, for example, might refer to the fish's actual material life. Remains of fish associated with Punic amphorae found in the centre of Classical Corinth offer the material for such a life history, this is discussed in detail in chapter 7.2.3. Slices of preserved sea bream and tuna were imported from the Atlantic coast of Spain to be sold in a central retail shop in Corinth, participating in a series of commercial and social transactions (Appendix 4). Such is also the case of a single large vertebra of a grouper (Serranidae) found at Pyrgouthi, in a Late Roman farmhouse in the inland valley of Berbati in the Argolid (Appendix 4, Mylona 2005). Such a find, speaks not only of the distant seas, but also possibly of massive catches, of coastal fish processing establishments, of travelling and storing and also of farmers/herders, maintaining a connection to this realm through the consumption of a salty/smoky dish of grouper.

Most often however, our sources show that fish acquire a life history, often a very impressive one, as generic entities, as "fish", "large fish", or "uniquely tasty fish" etc. Such life histories are mostly referring to the fish's reputation. These are stories of representations. Some fish for example acquired a reputation of excellence, as did the dog-fish from the island of Rhodes (Table 5.3), others had become royal gifts involved in intricate stories of lost and found treasures (Appendix 1: Hdt III, 41-3). Others were depicted on pots, perpetually reminding their viewers of past dishes or fishing deeds in the context of a symposium, or else a context of conspicuous consumption and competitive display (McPhee and Trendall 1987; Sparkes 1995; for a detailed discussion of this case see chapter 10.2).

3.4 Food embodiment and the senses.

The notions and ideas about food consumption discussed so far highlight the communicative character of food. They also highlight the need to acknowledge the diversity and variability of ways in which different groups of people, in different temporal and geographical contexts construct and practice food-related choices and preferences. A question to be asked, however, is how to approach the way in which the actual actors of these choices, practises and changes perceive the different aspects of everyday or special food consumption practises, and how the transmission and perception of messages through food takes place. In other words, we need to turn to the phenomenology of food consumption, attempting to understand and describe the phenomenon (of eating and drinking) as it is expressed by the subject (Tilley 1995, p. 12; Lupton 1996; Tarlow 1999; Hamilakis *et al.* 2002b). By shedding light on such issues, we may end up better equipped to approach the archaeological material remains of these actions.

Based on the idea that people construct and represent their world through experiences and shared meanings, anthropologists and social scientists working within the post-processual paradigm, focus on the way the human body, through its senses, mediates between the surrounding world (material and cultural) and the mind (discussion of several aspects of the issue in Lupton 1996). The discussion concerns the embodiment of experience and the role of the body in this process. The relationship between body and mind is fiercely debated across the epistemic field, with prominence or emphasis given either to one or the other (Ingold 2000b, p. 243-278). Tim Ingold's ideas, however, offer an escape from such a dilemma. According to him, the body enables the sensory involvement of the mind with the world. At the same time the body undergoes continual growth and development within its manifold environmental relationships. "Body and mind are not two separate things but two ways of describing the same thing- or better process, namely the activity of the organism-person in their environment" (Ingold 2000a, p. 240).

For the purposes of the present study, i.e. the analysis of the consumption of a particular foodstuff, fish, this idea appears to be a promising starting point. Food is the embodied substance "par excellence". Eating provides the body with a very clear set of "environmental relationships", to use Ingold's terminology. It enables the body's growth and development (Falk 1994; Hamilakis *et al.* 2002b) and at the same time it requires the involvement of the mind through the senses. Eating involves taste, smell, touch, vision, hearing and also memory (a metasense according to Seremetakis 1994). It also involves emotions, pleasures and feelings (e.g. Messer 1984, p. 221; Falk 1994; Lupton 1996, pp. 30-36; Hamilakis 1999a; Tarlow 2000).

We can't begin to understand food consumption without first accepting, and then exploring the importance of these relationships. As Kus puts it, referring to the relevance of the 'sensory human experience' to archaeology, "the physical and the emotional are part of our social theoretical discussions just as much as are the cold, calculated motives and logic. I can not see how it can be otherwise without reducing humans to automatons or cerebral essences" (Kus 1992, p. 172).

But how can these ideas be applied in archaeology? How feasible is it to explore and understand the experiences and meanings of past people? Are these experiences archaeologically accessible? (Barrett 1993a; Hodder 1998, pp. 11-23; Tarlow 2000; Hamilakis *et* al. 2002b, p. 8). And more specifically, how do we expect such an approach to help us in our exploration of fish consumption in Classical Greece (see research questions in chapter 1)?

Starting from the philosophical positions of thinkers like Heidegger, Merlau-Ponty and Foucault who emphasise the role of the body in the individual's perception and experience of the world, a dynamic discussion in archaeology about the phenomenologies of the body focuses on the applicability of the senses-oriented view to archaeological materials (see references in Ingold 2000b; Hamilakis 2002; Tsamis forthcoming). Researchers who work with this agenda not only chose to focus on different aspects of the past, but also use different theoretical and methodological tools (see collection of papers in Montserrat 1998; Hamilakis *et al.* 2002a). Although most works favour vision over other senses (Thomas 1993; Hamilakis 2001), some attempts to alternative approaches are also made (Goodman *et al.* 1995; Hamilakis 1999a; 1999b; 2002; Tsamis forthcoming).

Fish consumption in Greece, during the centuries under study, appears particularly suitable domain for the exploration of the senses in an archaeological context. First because, in food consumption, the whole range of sensory experience is at play, thus permitting, at least in principle, its exploration. Secondly, the issue can be approached simultaneously through a variety of sources, such as the actual remains of fish, related structures and artefacts, literary accounts of fish consumption and representations in art. Many of these, especially the archaeological evidence, can be tied to specific consumption conditions and viewed in relation to other classes of evidence, which provide a physical and notional context within which fish consumption took place. The physical remains of a distinct fish-eating event, preserved in the closed context of a destruction layer in a specific excavated architectural feature, for example, could be a promising field for the type of exploration suggested here. By using the evidence on fish consumption in such a combined, context specific manner, in exploring the senses, the researcher may overcome the danger of imposing subjective contemporary perceptions about the issue and achieve a more reliable understanding of past fish-eating.

From the extensive body of ideas and approaches to the archaeology of embodiment and the senses, the notion of the consuming body and the power processes that operate through it is of particular relevance (Dietler 1996; Hamilakis 1998). Fish consumption took place in several contexts in the Classical world, both private and public, domestic, ritual and others, as it is attested by the recovery of numerous fish remains from excavations (see references in Mylona 2003a, Table 5.1). Yet a very rich body of literary works elaborate mainly on public, conspicuous fish-eating (e.g. Davidson 1997). Such a conflicting picture of fish consumption probably indicates that the embodiment of fish (especially large fish) was for the Classical Greeks, or at least for some of them, an act of exercising some kind of power, quite distinct from that represented by the consumption of other foodstuffs, such as red meat (Detienne 1989; Davidson 1997, pp. 12-16). The conflicting picture briefly referred to above, is just one of the fields in which "bio-politics" around fish consumption were constructed (for a discussion of the term see Hamilakis 2002).

Another notion which may also prove very helpful in the present analysis is that of the connection between memory and food. The crucial role of food in evoking memories, of an even totally unrelated nature, has recently entered the anthropological research (Sutton 2002). Seremetakis defines commensality (eating together) not simply as the social organisation of food and drink consumption, but as the exchange of sensory feelings and memories and of substances and objects incarnating remembrance and feelings. Memory is a metasense, which is activated by embodied acts, such as food and drink consumption (Seremetakis 1994). Sutton's anthropological work on present-day island of Calymnos in the Aegean fully supports these ideas. The close relation between food and memory has recently entered the archaeological discussion (Hamilakis 1998), but the issue is still largely unexplored. The use of fish, or fish-dishes, already consumed or about to be so, as mnemonic devises in the Classical world emerges as a promising venue of research.

3.5 Conclusions.

Much of the discussion in this chapter focused on ideas about food/fish and its consumption that have developed or been explored in the fields of anthropology and sociology. Their unifying theme is that they all emphasise the role of fish as food that acquires meaning from the social context in which it is consumed. Other dimensions of fish, such as the fish as an economic resource, fish as sources of nutrients in a biological sense, fish as a symbolic or aesthetic object are all considered relevant but are all viewed as aspects of fish which become meaningful for any given society only if connected to fish consumption.

The application of the above ideas in archaeology is not a straight-forward task. The theoretical framework for the study of Classical fish-eating that has been proposed in this chapter is in a way explorative. How fish-eating conveyed diverse meanings in different contexts, how it performed different functions at various stages of its short life as a commodity and how the incorporation of fish touched upon issues of identity, memory and ultimately world order are some pertinent issues. An attempt will be made throughout this book to analyse these ideas and test them against the archaeological and literary evidence for fish-eating in Greece from 500 B.C. to Late Antiquity. Therefore, the shaping of a methodology which will be able to link data and theory is a vital part of the current project. Methodological issues will be discussed in the following chapter.

CHAPTER 4

SHAPING THE METHODOLOGY

In chapter 2 it has been demonstrated that our understanding of Classical fish-eating is seriously hindered by the existence of deep "divides", to borrow Renfrew's term (1980), in the treatments of the issue by various scholars. These divides refer to the raw data on which studies of the issue have been based (texts versus archaeological finds), the theoretical framework within which they develop (idealism versus functionalism) and the discipline within which the study of Classical fish-eating has taken place (philology/ancient history versus archaeology). In chapter 4 a unifying theoretical scheme has been proposed that could potentially bridge these gaps. In this chapter I will describe the methodology by which such bridging will be attempted.

I have chosen to base my research mainly on two types of data. One, which has been repeatedly used in the past for the study of ancient fishing and fish-eating, is textual, both literary and epigraphical. The other is fairly new in the context of Classical archaeology; that is fish bones, which are found in archaeological excavations. Other physical remains of fishing and fish consumption activities, such as the fishing tools and ceramic vessels related to storage, cooking or eating, will only be used sporadically, in a non-systematic manner. Iconography, despite its relevance to the present study will similarly be underused, for reasons of space and time within the limitations of a book.

The choice of the above mentioned classes of data brings this particular research into the heart of a thorny debate in historical archaeology. A brief discussion of this debate here is relevant because it clarifies the methodological choices that will be described below.

4.1 Objects versus texts: the scope and methods of historical archaeology.

The last few decades have witnessed a wide-ranging debate on the relationship of historical archaeology to ancient history. The argument has focused mostly on the issue of definition of subject matter and type of evidence on which each discipline relies, on the type of insights each can offer about the past, on the solidity of the evidence each relies on and ultimately, on the relevant importance of each discipline. Although I will not go into an extensive discussion of this debate, which has been developed by numerous archaeologists and historians (Arnold, 1986; Leone and Porter 1988; Small 1995a; Small 1995b; Vermeule 1996; Andren 1998; Johnson 1999; Small 1999; Moreland 2001; Snodgrass 2002, Foxhall 2004; Sauer 2004), I will refer here to certain points which are relevant in the formulation of the methodology for the present study.

It has been a long-held axiom that history (ancient history in our case) deals with texts, while archaeology deals with material remains, with some overlapping in certain marginal areas. Archaeology has often taken its lead from history, both investigating places and events already known from the texts, or issues that are thought to be inadequately covered by them (Small 1999; Moreland 2001, pp. 1-32; Snodgrass 2002, p. 83). Many ancient historians and archaeologists share M. Finley's contention that:

> "It is self-evident that the potential contribution of archaeology to history is, in a rough way, inversely proportional to the quantity and quality of the available written sources" (Finley 1985, p. 93).

Beliefs like this have led to a quest for a research "space" for historical archaeology. It has been suggested, in various, often theoretically different contexts, that because of the incomplete nature of the archaeological record, only inferences about basic production and consumption activities can be easily drawn, and that the 'superstructure' of past societies, the social rules and regulations, as these are expressed in fields such as politics and laws, is beyond archaeology's reach, in the heart of historical jurisdiction (Hawkes 1954, pp. 161-2; Binford 1972, p. 94; Lloyd 1986, pp. 42-47; Moreland 2001, pp. 13-16).

Conversely the argument is modified, shifting the emphasis from the incompleteness of the archaeological record, to the selective, 'biased' character of the written accounts. According to this rationale, written accounts are the products of an educated "elite", intentionally excluding those who were socially, economically or geographically remote from the power centres of the elite culture. Archaeology, by concentrating on material remains can give a voice to those 'silenced' by exclusion from these "elite", biased, written accounts (Tarlow 1999, p. 263; Moreland 2001, p. 19). Approaches to history and archaeology such as the above maintain a division between the two disciplines and demarcate the territory of each. History seems, in any case, to set the agenda, by or against which archaeology works.

Some reaction to a direct or more veiled predominance of history over archaeology has come from archaeologists and historians who deemed written accounts as just one source of evidence about the past among others, with ethno-archaeological observation or experimental archaeology valued as more crucial in the understanding of the past (Moreland 2001, pp. 21-28). Documents have been deemed idiosyncratic or partial and material remains as more straightforward sources of information.

In this case, as in the previous ones, some hierarchical ordering of disciplines is implicated but most importantly both objects and texts have been viewed merely as providing information about the past.

It is at this point that the problem lies. By accepting that documents and other material remains are just bearers of information about the past, we only view them in the present, disconnecting them completely from the context of their production and consumption in the past (Moreland 2001, pp. 21-32 and references therein). But both objects and texts were once active in the production, negotiation and transformation of social relations (Moreland 2001, pp. 28-31). They were "implicated in identity and the expression of power, whether the domain was the everyday life on the farm ....or the grand schemes of house form, garden design and the layout of towns and cities" (Hall 2000, p. 26). By denying these qualities to both ancient texts and objects, we render them incapable of illuminating the means by which society and the self were produced and transformed.

We should perhaps view them both in the way C. Briggs views texts "as eloquent, historically contextualisable and contextualising artefacts (2000, p. 398). By accepting such qualities of both archaeological remains and texts, historical archaeology can be defined in a new way, which blurs the dividing lines with history and emphasizes instead their common goal and complementarity of sources (Sauer 2004, pp. 40-42). According to a definition given by Moreland:

> "Historical archaeology is a practice which recognizes that artefacts and texts are more than just sources of evidence about the past; that they had efficacy in the past; and which seeks to determine the ways in which they were used in the construction of social relationships and identities in historically specific circumstances" (2001, p.111).

The above discussion revolves mainly around the relevant importance of textual and archaeological evidence and the optimal way of viewing them in order for them to make sense. At this point however, a word of caution is required. Although texts and archaeological material could be seen as broadly complementary in exploring certain periods of the past, especially with a social perspective, they, however, can not, in most cases answer the same questions. Foxhall, discussing the issue, identifies this as a problem of contextualisation (2004). Texts and archaeological materials reflect different activity domains and timescales in the past, each having acquired different 'biographies' over time. According to Foxhall, even though in some privileged cases, texts and archaeological data could be used to construct a comprehensive narrative, in the case of Classical Greece it probably can't (Foxhall 2004, p. 83). This certainly is true in the case of the investigation of Classical fish-eating in Greece. The methodology followed in this book (as described below), is oriented towards turning this discrepancy between our different kinds of sources into an advantage rather than hindrance. The questions are asked and the data, textual or archaeological, are used differentially, depending on the case.

4.2 Objects and texts: shaping a methodology.

Here follows a discussion of the ways in which fish bones and texts, along with other evidence are to be used in order to illuminate the ways in which fish-eating had been incorporated in the economic, social and ideological life of different communities in Greece in the period between 500 B.C. to the seventh century A.D. Also some clarifications are given on choices made in this book regarding a range of issues such as the basic terminology used, the treatment of time etc.

4.2.1 Analysis of fish bones.

The fish-bone assemblages that will be used in this book are of two kinds. On the one hand are fish remains, which have been systematically collected by water-flotation of soil samples from specific archaeological features (Itanos, Kommos, Kalaureia, Krania, Sanctuary of Demeter and Kore-Corinth, Pyrgouthi, see Table 5.1 a,d for details on these assemblages). These assemblages vary in size, from a few to several thousand bones, depending mostly on the proximity of the excavated site to the sea or other aquatic environments and the type of the excavated deposits. They include bones as small as 1-2 mm in length and an extensive variety of anatomical parts, reflecting the meticulous sieving and sorting processes applied (Wheeler and Jones 1989, pp. 50-51; Gordon 1993; Shaffer and Sanchez 1994). The largest of these assemblages are particularly rich in variety of taxa (Mylona 2003a). On the other hand are fish remains that have been hand-collected during the normal process of excavation (Corinth–various sites, Kassope, New Halos, Delphes, Pyrgouthi, Halieis, Messene, Athens-various sites, Delos, Thera, Mytilene, Knossos and Eleutherna-various sites; Table 5.1b-e). These assemblages consist mostly of selected bones of larger fish and are, in most cases, small in size and of a restricted variety of both anatomical parts and species. Some of the fish-bone assemblages used in this book have been analysed by the author in a more or less consistent manner, while others have been analysed by various researchers to a varied degree of elaboration, with different research priorities in mind. Another factor of variability among the fish-bone assemblages used here is the degree of their contextualisation. In some cases, such as the Kalaureia assemblage this degree is high, while in others, it is minimal.

Although the fish-bone assemblages under study present several problems of comparability, in a situation such as

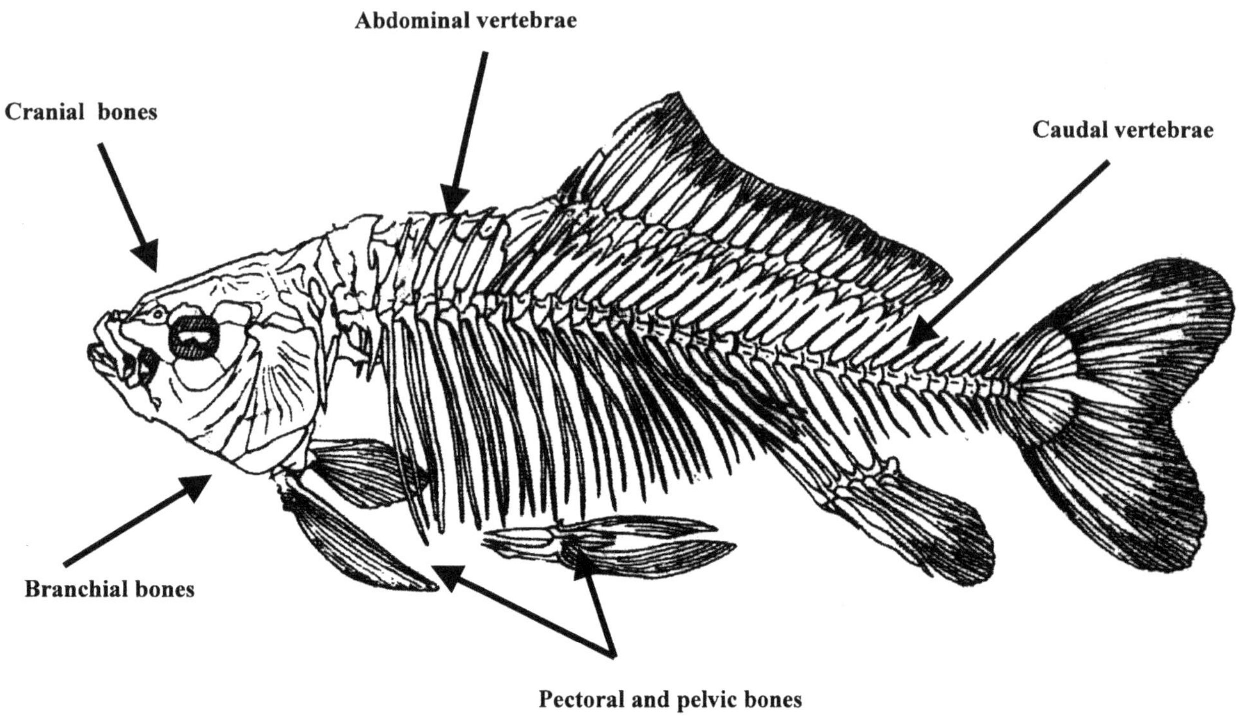

Fig. 4.1 Fish basic skeletal groups: Cranial, branchial, pectoral and pelvic bones, abdominal vertebrae and caudal vertebrae (drawing from Bristow 1992, p.10)

that of Classical archaeology in Greece, where water flotation programs are still rare, we can not be very selective. It seems that a methodological strategy is needed in order to deal with the aforementioned problems. This strategy is to adopt several, alternative or complementary analytic approaches (see below). In other words each fish-bone assemblage is used differentially depending on the research questions at hand. Although, for example, almost all fish remains under consideration can be used to explore the nature of the exploited aquatic resources, only the largest assemblages, and those better collected, recorded and analysed could be useful in the discussion of issues related to methods of food preparation, or spatial distribution of different fish preparation or consumption activities. Decisions are made throughout the book on which fish bone data are more suitable in each case.

The detailed recording and analysis of the fish-bone assemblages under consideration are not done within this book, but in independent reports which have either been already published or are in the process of being published (see references in Table 5.1). A summary of the analysis of the most substantial assemblages along with contextual information and a short description of the relevant sites are given in Appendix 4. Information on particular aspects of each assemblage is given throughout the book in relevant tables. The case of the Krania fish bones should be especially commented upon. The analysis of this exceptionally large fish-bone assemblage has not been completed. Its richness in taxa requires the use of an extensive reference collection for these fish. Such a collection was not available in time. Therefore, some aspects of this assemblage can be discussed in this book but others not. The flexible, multilevel approach to the fish bone data adopted here however permits the inclusion in the data base even of this idiosyncratic assemblage.

The analysis of fish bones in general, is to a large degree dependant on methods devised by each of its practitioners, which mostly find inspiration in parallel works on mammal bones. No robust, systematic set of analytic "steps" has been suggested so far, apart from the fundamental work of Wheeler and Jones (1989). The international archaeo-ichthyological literature ranges from works which comprise in compiling a simple taxonomic list to others which are extremely elaborate, in terms of the quantification and analytical procedures used. The Global Rachidian Profile (Desse and Desse-Berset 1989) or the meat yield estimates based on bone weight (Barrett 1993b) are examples of the later cases. The analytic procedures followed by the present author in the analysis of the Itanos, Kalaureia, Krania, Pyrgouthi and Eleutherna assemblages are presented in Appendix 5. The methodologies used for the rest of the assemblages are described, more or less comprehensively, in the corresponding publications.

Taking into account the disparate nature of the available fish bone data, the use of any elaborate quantification and analytic methods appears quite unsuitable for this book, except in specific cases. The analysis of the fish bones that is described below, keeps quantification and statistical analysis at its simplest, in an attempt to make use of as much of the fish remains as possible. Different quantification and presentation methods are used at each step of analysis depending on the research question asked.

On a first level, the criterion of "presence" (of specific fish taxa) is preferred to that of "frequency" (which is the most common way to present fish remains in zoo-archaeological works), in order to make archaeo-ichthyological data of diverse origins and qualities comparable. This happens for two reasons. First, the visibility of archaeological fish remains is very strongly affected by taphonomic processes such as soil erosion, scavenging, field collection method etc. (e.g. Jones 1986; Nicholson 1995; Vale and Gargett 2002). As the fish-bone assemblages at our disposal (Table 5.1a-e) are not comparable on these grounds, any transformation of the data into statistically complex forms would be meaningless. Secondly, the fish-bone assemblages at our disposal are few, and their size and character vary, depending on the scale of the excavation, the nature of the excavated deposits and the proximity to a fish source. Again, for these reasons the pooling and homogenization of their data would be problematic. The "presence" (but not the "absence"), however, is a measure which is simple and flexible enough to help the current investigation by revealing some basic trends. "Presence" has also the added advantage of permitting the comparison of the fish bone data to data derived from a totally different source, the written texts.

Although "presence" is useful in certain cases (e.g. to define the range of consumed fish in different contexts), it is nevertheless incapable of providing any measure of the relative importance of certain fish taxa as opposed to others. The Minimum Number of Individuals (MNI) has often been used in fish bone studies for such purposes (e.g. Rose 1994). I find this quantification method, however, of restricted usefulness, in the present circumstances. MNI calculations require assemblages with relatively well preserved elements, which can be confidently identified, both anatomically and taxonomically (Wheeler and Jones 1987, pp. 149-153). In most contexts around the Aegean, the cranial and branchial elements, the most suitable bones for MNI counts, are the rarest, while few of the vertebrae, usually the most common elements, can accurately be identified to their exact position on the skeleton. Furthermore, the elements on which MNI is more firmly based are the ones which are expected to be removed in cases of fish preservation or preparation for cooking, when the fish head is removed. If this method is applied, any such feature of the assemblage, instead of being highlighted, would go unnoticed. Therefore, MNI counts, will be used very sparingly.

Perhaps the most flexible and easily adaptable method of quantification in our case, besides the recording of "presence", is the calculation of the Number of Identifiable Specimens (NISP). This takes into account all available identifiable remains. It does present a problem of over- or under-representation of certain species with exceptionally few or many preserved elements (e.g. elasmatobrachians and eels). Also it is severely affected by possible processing activities, which might have removed part of the skeleton. The solution to this problem is probably the calculation of the NISP for several anatomical groups of bones for each taxon. Head bones, the gill zone bones, the abdominal vertebrae and the caudal vertebrae are such categories (Fig. 4.1). In this manner, several NISP figures for each taxon or group of taxa will be combined to provide a more representative figure for the relative frequency of each taxon. Furthermore, such a method will highlight possible fish processing practices (e.g. Mylona 2003c).

Throughout this book and in the fish bone publications used, fish are initially identified on the basis of modern Linnean taxonomy. In the main text of the book common names are mostly used (as these are given in the current ichthyological taxonomic corpora – e.g. FishBase - Froese and Pauly 2006), with the Scientific name given when necessary. Table 4.1 provides a list of the Latin and common English name for the fish discussed here. When the various marine (and euryaline) fish taxa are presented in plain taxonomic tables, they are arranged according to the standard taxonomic order, following Papakonstantionou's *Check-List of Marine Species of Greece* (1988). The fresh-water fish are not incorporated in this taxonomic scheme because no such combined species-list has been located for the ichthyofauna of southern European fish. Alternatively, fish taxa are presented in tables according to habitat or size.

One of the main concerns of this book is the contextualisation of the fish remains. If we are to discuss fish consumption and its implications in any meaningful way, we need to be able to do so for specific cases, in specific conditions and restricted periods in time. To do this it is of primary importance to clarify the taphonomic history of each assemblage, in other words to clarify the alterations of the fish remains pre-and post-depositionaly. In this manner, not only can we speculate on the degree to which a fish-bone assemblage is representative of the initially consumed fish but we can also detect features of the assemblage which are related to food preparation and consumption. Issues related to the taphonomic history of each assemblage are in some cases, but unfortunately not all, considered in the respective reports and whenever available they are summarised in Appendix 4 and discussed throughout the book.

Part of this attempt to contextualise the fish remains is to examine them on the micro-scale of the particular archaeological feature to which they are connected archaeologically. A refuse pile, the contents of a vessel or the floor of a particular room could be such units of analysis. The temporality of each assemblage, i.e. its timing within the life rhythms of the humans who

Table 4.1 Scientific and common English fish names.

Elasmobranch	Cartilaginous fish
Squatina squatina	Angel shark
Dasyatis sp.	Stingrays
Clupeidae	Shads, herrings and pilchards
Engraulidae	Anchovies
Anguilla anguilla	Eel
Muraenidae	Morays
Congridae	Conger eel
Belone belone	Garfish
Gadidae	Haddocks and cods
Gaidropsarus mediterraneus	Shore rockling
Zeus faber	John Dory
Serranidae	Sea basses: groupers and combers
Epinephelus sp.	Groupers
Seranus cabrilla/scriba	Combers
Dicentrarchus sp.	Sea bass
Pomatomus saltator	Blue fish
Carangidae	Jacks and pompanos
Seriola dumerili	Amberjack
Trachurus sp.	Horse mackerel
Sciaena umbra	Brown meagre
Umbrina cirrosa	Shi drum
Mullidae	Red mullets
Sparidae indet. sp.	Sea breams
Boop boops	Bogue
Dentex dentex	Common dentex
Diplodus annularis	Annular sea bream
Lithognathus mormyris	Striped sea bream
Oblada melanura	Saddled sea bream
Pagellus erythrinus	Common pandora
Pagellus erythrinus/ Pagrus pagrus	Pandora/Common sea bream
Pagrus pagrus	Common sea bream
Sarpa salpa	Salema
Sparus aurata	Gilt-head sea bream
Spondyliosoma cantharus	Black sea bream
Centracanthidae	Pickarel
Chromis chromis	Damsel fish
Labridae indet. sp.	Wrasses
Sparisoma cretense	Parrot fish
Trachinidae	Dragon fish
Uranoscopus scaber	Stagazer
Scombridae	Mackerels, tunas and bonitos
Euthynnus alletteratus	Little tunny
Sarda sarda	Atlantic bonito
Scomber japonicus	*Chub mackerel*
Scomber scombrus	*Atlantic mackerel*
Thunnus sp.	Tunnas
Xiphias gladius	Sword fish
Gobius sp.	Gobies
Blennius sp.	Blennies
Sphyraena sphyraena	Barracuda

produced the remains under study is obviously relevant. An assemblage formed through long term accumulation is connected to a different temporality than material originating from one time deposition. This type of information in relation to the assemblages under study however is not always available. The connection of the fish bones to other material remains such as architecture, pottery and tools is also relevant, because it offers yet another means of contextualisation. Differential rates of finds processing and analysis at each excavation, and different priorities in the publication of archaeological sites, result in unequal detail for this type of information.

4.2.2 *Analysis of textual evidence.*

The written documents that will be used in this book are of two kinds: inscriptions and literary texts of various genres. I have chosen not to confine my sources to any particular type or genre (as in Wilkins 2000) but to expand my research on as broad a range of sources as possible. This choice is dictated by my research questions. If we are interested in understanding how the Greeks experienced fish-eating, and how they constructed a whole range of ideas and practices around it, then a multitude of view points, as these are represented by a multitude of text types and literary genres gives us a better chance to achieve it. The contemporary literature on the subject provides useful insights on the issue. Davidson's work on fish-eating as it is represented mostly in Athenian comic drama and orations is exciting in the insights it provides (Davidson 1995; 1997), but Purcell's approach to the same general topic (Purcell 1995), through the exploitation of a range of literary genres and text types is, in my opinion, more illuminating.

Among the written sources used, the selection of literary texts is the most problematic. Choosing those sources that might be more suitable or useful in illuminating the issue of fish consumption in Greece (for a definition of the geographical unit of Greece see chapter 4.3), during the centuries covered by this book, requires the consideration of the cultural context in which these texts were produced and consumed. One obvious choice would be to use texts produced by Greek authors, who wrote for a readership brought up within the Greek milieu. The situation, however, can not be that simple. Many of the sources used in this book date to the Roman period. During the first three centuries of the common era, the intellectual trend of the "second sophistic" with its deliberate archaism and preoccupation with the Classical, mostly Attic, literary production and language, made the Classical texts relevant to scholars of various cultural backgrounds across broad geographical areas (for the phenomenon of the "second sophistic" and its various aspects see Bowie 1970; Said and Trédé 1999; Alcock 2002, pp. 41-42).

In this intellectual context texts were written in Greek, in the traditionally Greek-speaking regions (e.g. Plut. *Mor.*), but also in far-away and culturally diverse places, such as Rome (e.g. Ael.), Egypt (e.g. Ath.) or the East (e.g. Opp. *Hal.*). Although the readership of such works came from different cultural backgrounds, these texts share an interest for and also a deep familiarity with Greek scholarship and literary tradition. It is due to this phenomenon that numerous ancient Classical and Hellenistic works, many of which are cited in this book, are only known from excerpts cited by "Roman" writers (eg Theopompus in Ath.). For these reasons, works such as the ones described above are considered relevant for this book. For the purposes of the present study, however, even works written in Latin may prove a useful source of information, not only because they contain a lot of material originating from Greek sources, but also because of their wide readership (Morton Braund 2001).

Having set these broad limits for the selection of written sources for this book, a more closely defined set of criteria is needed in order to make use of the hundreds of references to fish, fishermen and fish-eating within the relevant literature and epigraphic record. Taking into account the research questions and theoretical framework of this book, both of which require a context specific treatment of the issue of fish consumption (see chapters 1 and 3), I have made some choices regarding the kind of information that I find useful in this research. I have selected those texts that refer to:

- contexts of fish consumption (e.g. symposium, domestic, religious)
- qualitative comments on fish, fish consumers, places reputed for a
- particular fish product etc.
- anecdotal stories about fish
- fish in mythology, dreams, magic etc.
- medicinal uses of fish
- organization of fishing
- qualitative comments on fishing as an occupation and on fishermen
- fish in the market, prices of fish, fishmongers etc.

Not all fish-related data are considered relevant. Technical issues on fishing methods have been excluded, unless they fall into the above categories. The description of the way tuna are caught, for example, by communal effort of a group of fishermen (Appendix 1: Ael. 15.3-6, Opp. *Hal.* 620-648), besides the technical information, provides insights into the organization of fishermen and as such it is included in my data base. The various, often bizarre, ways by which parrot fish were caught on the other hand (see Thompson 1947, pp. 238-41 for a collection of such references), are not. The large body of references to fish processing and the relevant trade is also excluded from my listing, because the very comprehensive work by Curtis, *Garum and Salsamenta* (1991) can be used as a source, whenever needed. Fish recipes as such are not systematically collected, even though they are obviously relevant to the issue of fish consumption. If, however, a fish recipe is part of a commentary on the social context where consumption of the particular dish takes place, it is considered relevant to this study.

One way to achieve a contextualised analysis of the textual evidence is to choose fish-related information that is geographically specific. Placing the inscriptions in their geographical context is a fairly straight-forward task. The contents of each inscription or their dedication/find spot usually clarify the issue of provenance or geographical relevance. The fish price-list inscription found in Akraephia, for example, must clearly be viewed in the context of an inland town that lies some distance from the sea, with access to the elsewhere documented (see various references in Appendix 1) rich fish resources of Lake Kopais. Placing literary texts in a spatial context is less straightforward. In many of the instances catalogued here (Appendix 1), a geographical dimension is given through reference to specific place names. Therefore, we can assume that these excerpts provide fish-related information about particular areas (and often for a specific time, see above). A large number of selected excerpts, however, do not refer to a specific location. In those cases, our guide in determining the geographical context in which they had a meaning, in the past, is to determine the provenance of the author and the audience/readers of the work the excerpts originate from. The reference, for example, to the foul character of the fishmongers (see various references in Appendix 1) abounding in comedies of various dates, has to be understood in relation to the audience of these plays, which was initially Athenian. Such qualitative judgements about the fishmongers might or might not have been valid elsewhere.

Some comment is required at this point regarding the temporal contextualisation of the texts used in this book. Inscriptions are again the most straight forward sources in this respect, because they were produced to serve spatially and temporarily specific needs. Their dating is often quite precise. Literary texts are more problematic in this respect. Some, such as dicanic speeches or comedies can be dated with some accuracy. Others however cannot. Oppian's *Halieutica* for example, a long poem on the art of fishing, rich in detail about fish and fishing methods, is a work which was in use not only as an entertaining text or manual but also as a school teaching tool, for many centuries after its production and as late as the Byzantine period (Keydell 1939, pp. 703; Hunger 1978, p. 115). Such texts and their contents clearly had some relevance to the public perception of fish-related matters for long periods of time. Other works have an even more complicated history. Many of the excerpts used in this book originate from a variety of 6th to 3rd century works which, however, have been preserved in one single text dating to the 2nd century A.D. That is Athenaeus's *Deipnosophistae* (Wise Men at Dinner) a work which was produced in a specific intellectual environment which demanded a recollection of Classical works (Bowersock 1974; Anderson 1993; Swain 1996; Porter 2001; for the work of Athenaeus see various papers in Braund and Wilkins 2000). Excerpts from this source clearly have a double chronological placement and, apparently, a differentiated meaning in the context of the time of their original production and the time of their reproduction. Problems such as those discussed above are treated by viewing the information provided by the texts on different levels. References to excessive fish-eating and its connection to politics in Old and Middle comedies for example are clearly relevant to the mentality of Athenians in the 5th and 4th century B.C., while the impression of excess created by the accumulation of a large number of disparate yet superficially similar such cases in *Deipnosophistae* is related to the mentality of the Roman readers of this work. An attempt is made throughout this book to provide a short biography of such cases whenever necessary.

A final issue should perhaps be clarified at this point. It has been repeatedly pointed out that one of the dangers of using texts in the study of past societies is that they are often viewed as bearers of direct information about the past, neglecting their, perhaps primary, role in the construction of a discourse (see chapter 4.1). The position taken in this book is that we should not ignore either of these qualities of ancient texts. Texts do transmit information, but this needs to be contextualised and interpreted in order to be understood. In this book a multilevel approach is applied in this case also. The commonly repeated motif of the cheating fishmonger, for example, who wants to sell his half-rotten fish for as high a price as possible, can be taken, on a first level, as a direct information on the existence of middle men in the fish trade, on the inclusion of fish in the urban food market, on the problem of keeping fish fresh and probably, on the need for or existence of control mechanisms related to the market functions. The same texts, however, on a secondary level, can be viewed as part of a discourse, laden with moral judgements, which comments on the ethical and social position of people like the fish mongers in Athenian society of the 5th, 4th and 3rd century B.C. In this book an attempt is made to view textual sources in such a way that both their documentary and discursive qualities are taken advantage of.

The literary text excerpts that have been collected as raw data for this book are presented in table form in Appendix 1. This provides information on the author, title and date of the work, a reference to the source of the excerpt and the literary genre to which the work belongs. Additionally it includes a short description of the content of each particular excerpt with emphasis on the aspect of it that is deemed relevant to the present study. The presentation of the inscriptions is organised in a similar manner in Appendix 2. This method of presentation is complementary to the direct referencing to the original works within the book. Reference to the texts throughout the book follows the standard abbreviations cited in the Oxford Classical Dictionary (Hornblower and Spawforth 2003). For works not included in it, most of which are preserved in fragmentary form, citing follows the standards set by the corpora where the fragments have been published. The Kassel and Austin's *Poetae Comici Graeci* (1983) for example is the source for the excerpts of Middle and New comedy plays while Jacoby's *Fragmenta Historicorum Graecorum* (1841-1870) is the source of various fragments of historical works. In each case, an abbreviated reference to these corpora is given in the Appendices and a full reference to the work is cited in the Abbreviations list. In the case of those excerpts that have been preserved in the work of Athenaeus, an additional reference to his work is given in Appendix 1, to facilitate those readers, who might not have access to the more specialised publications referred to above.

Appendices 1 and 2, present a wide variety of literary and epigraphical evidence to past fishing and fish-eating in Greece. These lists however do not intent to be exhaustive, covering every possible relevant case. They mostly serve to present a fair range of relevant sources (according to criteria described above), along with some additional ones, which serve as comparable cases from different cultural contexts, in a comprehensive form, which might facilitate the reader by reminding her/him of the content and character of each source used.

4.3 Definitions of terms and limits.

This book focuses on fish consumption in the geographical area of modern Greece. It is of course understood, that this entity is a modern construction which geographically includes an important part of the ancient Greek world. Large expanses of the ancient Greek world however lay outside this geographical area, with their extent fluctuating over the centuries. This particular geographical choice has been made mostly for reasons of convenience. Although in principle it would make more sense to include in this study Turkish Thrace, Asia Minor, and probably the coasts of the Black Sea or the colonies of Southern Italy, this has been avoided because of the difficulties in accessing archaeological and especially archaeo-ichthyological material from these regions. Selected evidence from these areas are, nevertheless, used in the discussion of specific issues.

Although the definition of the geographical limits of this book is a fairly straight forward process, when archaeological remains (fish bones, artefacts and structures) are discussed, the issue becomes more complicated when one deals with written sources. In that case the geographical limits of the research are more fluid (see discussion above).

The chronological scope of this book ranges from the 5th century B.C. to about 7th century A.D.). The choice of these particular chronological limits is arbitrary to some degree. Earlier periods, within the historical era, are excluded from this book because they are too poor both in archaeo-ichtyological and written data on fishing and fish-eating, while later periods fall within the realm of Byzantine archaeology and are excluded for obvious reasons. The centuries under study however, an admittedly long time period, lack homogeneity in historical, political, social or ideological conditions. Furthermore, the available data on fishing and fish-eating are not scattered uniformly throughout this period. On the contrary, certain time spans, such as the 4th century

B.C., are rich in data while others, such as the 5th 6th and 7th centuries A.D., are much poorer.

The chronological and geographical breadth of this research creates a problem of terminology. Two ways are used to describe the spatio-temporal framework of this book as outlined above. Sometimes reference is made to "Classical Greece", a rather inaccurate, but convenient shorthand, which does not coincide with the Classical period that roughly covers the 5th and 4th centuries B.C. Alternatively, the accurate but rather awkward "5th century B.C. to 7th century A.D. Greece" is used. Although none of these terms is satisfactory, they are nevertheless used here for lack of any more appropriate alternatives.

Taking into account the problems of chronologically and geographically uneven distribution of data as well as this thesis's central aspiration for a context-specific approach to the issue of fish consumption, it becomes apparent that the compilation of a history of fish-eating within the geographical area under study, based on the standard chronological divisions commonly in use in archaeological and ancient historical studies is not possible or desirable. It has already been discussed that a useful way to analyse fish-related data in a manner that serves the aims of this book is to construct a type of a cultural biography of fish (chapter 3.3). This will be done by discussing the multiple manifestations of the role of fish in relation to human affairs at several stages in its "life", as a resource to be exploited, as a marketable object and through its selection, preparation and consumption.

Because in this scheme emphasis is placed upon the variety of ways in which fish and fish-consumption were incorporated into human affairs, the temporal ordering of these expressions is of secondary importance in the present study. In the discussion of the religious context of fish-eating, for example, in chapter 8.1, the various cases attested in our sources (archaeological and literary) are presented without any chronological ordering, but are arranged thematically instead. It is understood, however, that time is an important factor. The chronological positioning of each presented case is important because it ties it to specific historical conditions which might have had a bearing on particular choices related to fish consumption. Temporality, might be a very usefull notion, because it ties fish consumption to the rhythms of human lives and makes it relevant to a context-specific approach to the issue (Foxhall 2000; for a more extensive discussion on time and archaeological interpretations see Murray 1994; Karlsson 2001 and references therein). Given the uneven geographical and temporal distribution of the data discussed above, certain measures have been taken to deal with the problem of time. First, each case discussed in this book is determined chronologically not only in the data base (Table 5.1a-e and Appendices 1 and 2) but also within the main text (e.g. 4th century B.C., Hellenistic, etc). Secondly, issues of temporality, i.e. tying particular fish-related events to the temporal macro- or micro-scale of an archaeological feature or consumption context, are selectively discussed for some cases which provide such opportunities or for which such a discussion is of relevance. Finally, chronological developments of particular fish-related phenomena are also discussed sporadically.

There are certain issues which, albeit directly pertinent to fish consumption (and food consumption in general), have been downplayed in this book. The historicity of fish consumption is such a case. It is obvious that in a geographical area and time-span as wide as those covered by this book, the social, political and historical conditions varied and changed. It is also clear that such conditions have a profound impact on food ways, which could be visible on a local scale or more widely. This dimension of the issue of fish-eating however has been little touched upon in this work, due to the time and space limitations imposed by book writing. Furthermore, certain other topics relevant to the consumption of fish (and food in general), although important, have also been downplayed in the present analysis. These are some of the issues which have been extensively discussed by ancient historians (e.g. fish-eating in a discourse of luxury, sexuality and politics; e.g. Davidson 1997; Wilkins 2000). In this book priority has been given to issues which can be better understood through a combined analysis of archaeological and textual data. Such an approach, as research stands at the moment, is more successful for some topics than others.

The use of the literary references to particular fish species requires some comment. The fish names as they survive in the sources are notoriously difficult to identify (see chapter 6.2). In this book I use data from Thompson's *A Glossary of Greek Fishes* (1947), with some reservations. I only use those fish names which appear to be reasonably safely identified, and often, I connect them to modern taxonomy only on the family or genus level. I have chosen to discuss in this book only those fish for which there is clear evidence either of their catching or of their consumption. Several fish species are discussed in the literary sources in terms of their physiology or bizarre behavior. These are excluded from the following discussion.

Finally, a comment should be made on the ethnographic and archaeological data other than fish bones that are used in this book. These consist of fishing tools, fish-related structures, pots and representations of fish as well as ethnographic cases which are pertinent to the discussions at hand. These data are apparently important in the study of ancient fishing and fish-eating, but for reasons of space and time limitations they are not analysed systematically here. It is acknowledged, however, that their analysis could offer important insights into the issue of fishing and fish-eating in Classical Greece.

4.4 Conclusions.

The methodology devised for this book combines some standard zoo-archaeological processes with textual analysis. It aims to serve two main purposes. One is the integration of diverse lines of evidence on fish consumption in the most comprehensive way. Such integration does not simply aim to increase the information on the issues at hand. The juxtaposition of the material dimension to the discourse around fish-eating is hoped to reveal something of the complexity and tensions of activities and relations involved in such aspects of everyday living in Classical Greece as fish-eating. The second purpose of this methodological strategy is to achieve some degree of contextualisation of fish consumption. Contextualisation is a notion which is central to this book. Methodologically it is approached through a consideration of the information provided by each source (whether archaeological or literary) and also of the function of this particular source in its time.

CHAPTER 5

CATCHING FISH – PRODUCTION

Any understanding of the role of fish consumption in Greece from the Classical to the Late Roman period requires the clarification of certain issues related to the availability of fish as a food resource and to the degree of its exploitation in different geographical areas. In this chapter this issue is approached in a new way which takes advantage of recent developments in archaeological recovery techniques, and of the accumulation of a varied corpus of empirical data on past fishing, such as fish bone remains, related artefacts and structures. These data are combined against a background constructed on a basis of the physical parameters of fishing (geography, ecology etc.) in a geographically specific way, and according to the results of recent oceanographic and ichthyologic research in the Hellenic area.

5.1 The productivity riddle: How rich in fish were the Greek waters?

The most appropriate beginning of an investigation of the actual contribution of fish in the economy of Greece between 5th century B.C. to 7th century A.D. is perhaps to discuss what has become an axiom in research on ancient fishing in the Aegean. According to this, the Aegean Sea is oligotrophic, i.e. very poor in nutrients, unable to sustain large fish populations, therefore, unable to support a sizable and complex fishing economy. In this framework, fishing has been viewed as an opportunistic, part- time activity, or an activity which occasionally involved individuals but not whole sectors of the population of coastal settlements (see discussion in chapter 2.2). Fishing has been perceived by modern scholars as an almost exclusively marine activity.

In view of recent developments in the understanding and recording of marine, riverine and lacustrine morphology and ecosystems in the Hellenic area, a re-evaluation of the above statement is called for, along with a re-assessment of its relevance to the study of ancient economic activities.

5.1.1 The physical characteristic of the Aegean and Ionian.

Due to its long term geodynamic evolution and active tectonism, the Aegean Sea displays a complicated physiography in terms of sea bed morphology and island configuration (Stanly and Perissoratis 1977). The Aegean is characterized as a complex system regarding certain features which are directly pertinent to its fisheries profile, i.e. its hydrology, water mass circulation, biological, chemical and sedimentological processes.

In these terms, the Aegean can be divided into two broad areas (Fig. 5.1). The northern Aegean, a semi-enclosed sea, is characterized by an extensive continental shelf, formed by sediments provided by six rivers (Axios, Aliakmon, Pinios, Strymon, Nestos and Evros) (Astaras and Sotiriadis 1988; Lykousis and Chronis 1989). The surface water temperature and salinity of the north Aegean Sea is determined by the outflow of these rivers and by large amounts of cool, low-salinity waters entering the Aegean from the Black Sea via the Dardanelles (Fig. 5.1; Unluata *et al.* 1990; Yuce 1995; Kallianiotis *et al.* 2004 and references therein). This is not only the most nutrient-rich area of the Aegean Sea but it can be definitely characterized as eutrophic (Bousoulenga *et al.* 1998).

The southern Aegean is a more open sea, communicating with the Levantine basin (eastern Mediterranean) and the Ionian Sea (western Mediterranean) through straights and sills of varied morphology. The water salinity is generally high with variations (Lykousis *et al.* 2002) and the continental shelf is limited, with steep depth gradients (Souvermezoglou *et al.* 1999). The southern Aegean basin is generally deeper, with certain areas, such as the Cyclades, forming shallower plateaux (Lykousis *et al.* 2002). It is considered one of the most oligotrophic seas within the Mediterranean and in comparison to the world's oceans (Souvermezoglou 1989; Souvermezoglou *et al.* 1999; Lycousis *et al.* 2002).

The Ionian Sea, on the western coast of Greece (Fig. 5.1) is characterized by a narrow continental shelf and great depth with salinity and temperature levels between those of the western and eastern Mediterranean. It is three times as eutrophic as the mean values for the Aegean Sea (Souvermezoglou *et al.* 1999). The fauna of the Greek Ionian Sea includes several Atlantic species along with Mediterranean ones, but lacks many of the Eastern Mediterranean fish of tropical or subtropical character (Papakonstantinou 1988, pp. 17).

Particular conditions regarding the water composition, nutrient availability, salinity etc. are highly variable, depending on sea bed morphology and prevailing weather conditions both locally and in the areas of adjoining seas. Such data are more abundant for the Aegean than the Greek part of the Ionian Sea. A measure of the spatial and temporal variation in the productivity of the Aegean Sea (Fig. 5.2) is given by satellite imaging of the average surface fluorescent pigments for spring and autumn, for the years 1979-1985 (images taken from Lykousis *et al.* 2002, p. 316). An examination of this variation, which is in essence an index of euthophisation, is instructive. In this picture, the more eutrophic areas within the Aegean, even in spring when productivity is lower, are

Fig. 5.1 Satellite image of Greece (TalentCruiser, www.cruiser.gr/en_main.html accessed 25 November 2006).

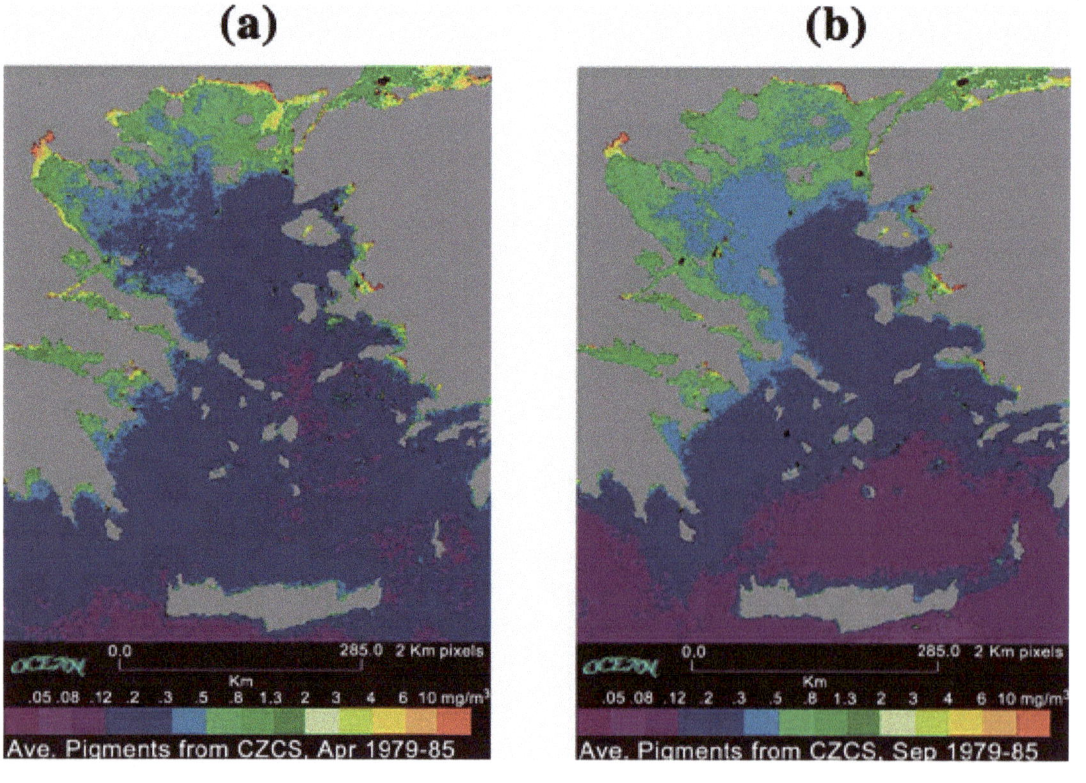

Fig. 5.2 Spatial and temporal variation of productivity in the Aegean Sea as it is indicated by images of the surface fluorescent pigments (average) during April (a) and September (b). The colour scale moves from the lower concentrations of nutrients (purple) to the higher (red) (from Lykousis *et al.* 2002, fig. 2).

concentrated in the northern Aegean, covering even broad areas of open sea. The Thermaikos gulf is perhaps the richest part of the northern Aegean. Relatively eutrophic areas are also the eastern coasts of mainland Greece, and the seas around Euboea, with Pagasetikos gulf in the area of Volos being particularly rich. The Saronic Gulf, the Corinthian Gulf, and the recesses of the major Peloponnesian gulfs (Argolic, Laconian, Mesenian) are also fairly eutrophic as is a broad sea band along the Asia Minor coast and the nearby islands. Almost all other coasts, mainland and insular, exhibit some degree of nutrient richness only very close to the coast.

Research on the composition of the ichthyofauna of the Aegean, both northern and southern, as revealed by controlled trawling catches (e.g. Tsimenides *et al.* 1991; Tserpes *et al.* 1999; Kallianiotis *et al.* 2000; Kallianiotis *et al.* 2004) have shown that here applies what characterizes the Mediterranean fisheries in general, i.e. a multi-species composition, where many commercial species appear seasonally in the catches (Kallianiotis *et al.* 2004 and references therein). Also, it appears that certain combinations of species occur with higher frequency in zones of certain depths (Kallianiotis *et al.* 2000), or in certain geographical areas (Kallianiotis *et al.* 2004). Furthermore, especially in the southern Aegean, shallow waters generally exhibit the highest densities in biomass and in variety of fish species (Kallianiotis *et al.* 2000). 447 fish species have been recorded so far, belonging to 129 families (Papakonstantinou 1988, p.29).

One of the additional observations of research programs such as the above has been that the fish species composition observed today is, to some (undefined) degree, influenced by the human activities in the broader area. Fishing with certain methods and intensity, obviously influences the characteristics of fish populations in certain areas with the shallower waters of the continental shelf being the most affected, because this is where most fishing activity in the Aegean is focused (Kallianiotis *et al.* 2004). Trawl fishing, for example, not only depletes fish numbers where it is applied, but it disturbs the age and sex balance within the remaining populations thus further affecting their renewal.

5.1.2 *The physical characteristics of the Greek fresh- and brackish-water masses.*

Greece, being a geographically fragmented, mountainous country, with a semi-arid climate, is characterized by a multitude of small basins, with small lakes and a large number of medium and small sized rivers (Skoulikidis *et al.* 1998). Most of the rivers run through narrow mountain valleys, descending abruptly to the coast, becoming lowland rivers very close to their estuaries. The largest rivers in Greece, which run considerable distances in flat land, thus developing lowland riverine habitats, are mostly found in Northern Greece (Macedonia and Thrace). The rivers of Greece have been divided into different categories according to features of their catchments, their hydrochemistry and climatic variation. It has been observed that the different characteristics of rivers support a distinct combination of fish species (Economou *et al.* 2002; Noble and Cowx 2002, pp. 13-14).

Lakes are also numerous in Greece, although their size varies considerably. The largest of them are found in the western and northern part of the country. Most are shallower than 50 m and many are so shallow that they dry up during extensive dry periods. Lakes are environments generally rich in nutrients, with the shallower of them being more eutrophic. Nowadays, lake environments are extremely degraded, mostly due to pollution, extraction of their water for irrigation, and over-fishing (Skoulikidis *et al.* 1998). Furthermore, large, historically important lakes such as the Kopais lake in Boeotia or the Filippoi marshes of Drama plain were completely drained in the early 20th century (Ogilvie 1943).

Both rivers and lakes in Greece exhibit a marked diversity in their ichthyofauna, which is one of the richest in Europe. One hundred and twenty-six fish species have been recorded and this number is increasing (Economidis 1991; 1995a; 1995b; unpublished data). Because of the rivers' and lakes' complex formation history, several fish species found in them are endemic to Greece and some southern parts of neighbouring countries (Smith and Dartwall 2005).

Finally, there is a third category of transitional water bodies in Greece, standing at the border between land and sea: the coastal lagoons and river deltas along with a variety of coastal wetlands. Such ecosystems, fragile and shifting as they are, are among the richest in Greece, both in terms of biodiversity and of the amount of biomass produced in them (Katsadorakis and Paragamian 2006 and references therein). Moreover, they are rich feeding grounds for fish (Koutrakis *et al.* 2005). Wetlands can be large, such as for example the Messolongi lagoons, or very small, such as the tiny marshes at the estuaries of seasonal ravines (Fig. 5.3). Areas like the ones described above are rich in available fish populations. Among the fish encountered there, seasonal migrating species (performing migrations of long or shorter distance), such as the eel (*Anguilla anguilla*) the sturgeon (Acipenseridae), the shad (*Alosa fallax*) and especially the grey mullets (Mugilidae) are very important (Koutrakis *et al.*, 2005). Although the largest of these ecosystems are found mostly in northern and western Greece, smaller ones are scattered all over the country (Heliotis 1988, 23, Table 3). It is characteristic that nowadays, the Aegean islands, many of them typically arid environments, exhibit about 200 such wetlands, distributed between 35-40 islands (Fig. 5.4) (Katsadorakis and Paragamian 2006). Considering that since the 1920's over 60% of the Greek wetlands have dried up, one can imagine that in the past such areas were

Fig. 5.3 A small wetland at the estuaries of a Geropotamos, a seasonal river east of Rethymnon (photograph by D. Mylona).

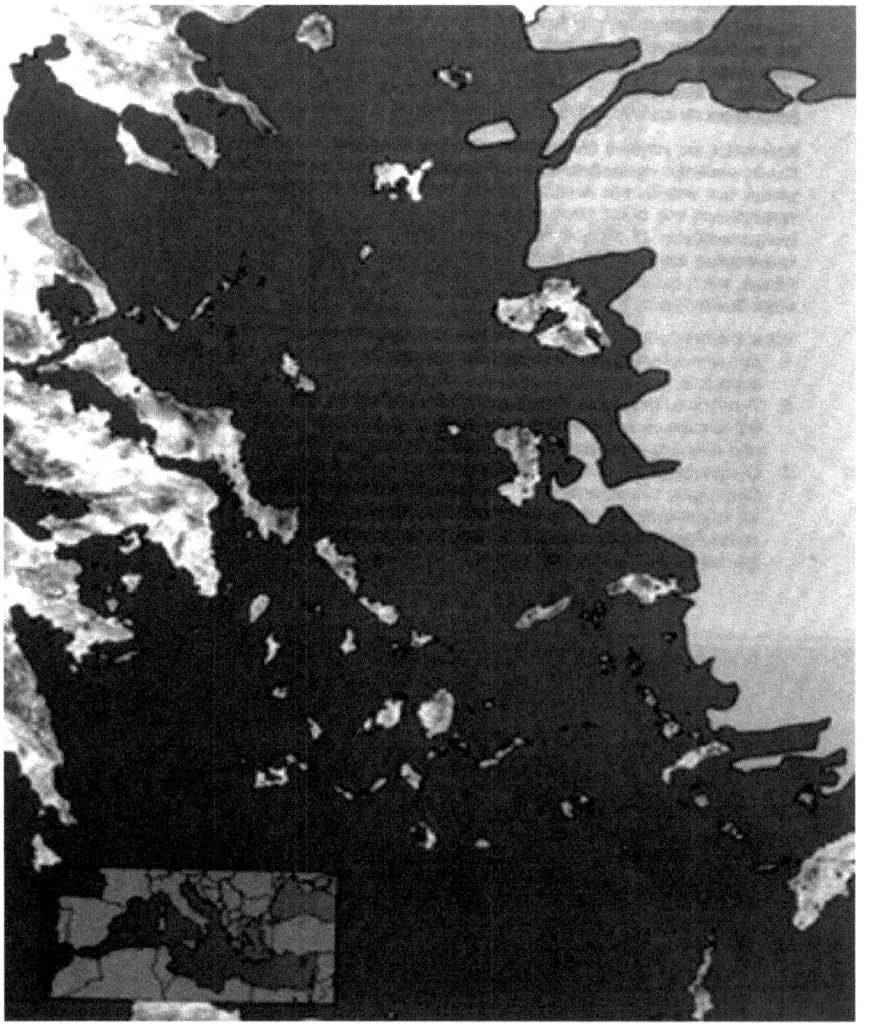

Fig. 5.4 Distribution of small wetland sites (dots) on the Aegean Islands (from Katsorodakis and Paragamian 2006, p.4).

much more common features of the Greek landscape (Heliotis 1988; Katsadorakis and Paragamian 2006).

5.1.3 *Relevance to the study of ancient fishing.*

The preceding synopsis, brief and generalized as it is, highlights certain features of the Greek aquatic environments which are directly pertinent to any evaluation of ancient fishing. These points are:

- The various physical characteristics of the Greek marine and freshwater environments are highly variable, even within the same region. They form, to use Horden and Purcell's term, a "continuum of discontinuities" (2000, p.53).
- Large expanses of Greek seas, such as the northern Aegean, are eutrophic, even by world-seas standards.
- Even in generally oligotrophic parts of the Hellenic seas, such as the southern Aegean, there are clearly eutrophic micro-regions.
- Although there is a clear variation in the productivity of different marine areas throughout Greece, some degree of richness is observed in the inshore waters of most coastal areas and around coastal wetlands.
- In relatively oligotrophic marine areas, which are generally coasts of steep depth gradients, the ichthyofauna is characterized by the presence of a large number of species, with slightly different habits and habitat requirements, the majority of which can be found in shallow waters. This offers a varied resource, a considerable part of which is accessible throughout the year. Besides, in these areas, even the ichthyofauna of deeper waters, is accessible from near the coast, given suitable technology.
- In several parts of Greece, rivers, lakes, lagoons and other wetlands of lesser extent, form important features of the landscape and are abundant sources of fish.
- Both marine and freshwater fish resources appear today to be fairly degraded in places, due to over-fishing and excessive anthropogenic alteration of their habitats. We might expect such resources to have been less affected by human activities in antiquity.

In the study of ancient fisheries, the above points are particularly relevant. They show that in an area like Greece, generalized evaluations can be very misleading. The fishing potential of each site/region should be viewed individually, taking into account the local physiography, but also the fishing technology available to the fishermen. It appears that from the resource point of view what is more important is not the total available fish biomass, but the part of it that was accessible and exploitable by the fishermen in specific locations.

5.2 What fish resources were exploited in Greece?

The type of fish resources exploited can be securely inferred mainly from two sources. One is the archaeological fish bones themselves (Table 5.1a-g and 5.2) and the other is the references to certain fish in the written sources (Tables 5.2, 5.3 and Appendix 1). A third potential source, iconography, will not be considered here. Although some instances of fish representations, such as the fish-plates (see chapter 10.2) are very rich in fish and sea-food imagery, the complicated relationship between representation and meaning makes any straight forward inference from them problematic. Furthermore, in the present discussion an attempt is made to tie the information on the exploitation of fish resources to specific geographical areas. Such an approach to artistic fish representation is rarely possible.

Table 5.1a-g shows that fishermen in Greece in the period under study were exploiting a variety of fish resources, or fishing grounds and a long list of different fish. From the 129 fish families reported in Greek seas (Papakonstantinou 1988), which include commercial fish but also rare or deep water creatures and even recent colonizers, 41 families have been identified archaeologically or encountered in literature. Taking into account that several archaeological fish bones and fish names remain unidentified for several reasons, this number is likely to be higher.

The majority of the fish caught in Greece during the period under study are marine inshore species, typical of the Aegean and Ionian Seas. They are both cartilaginous (with a skeleton made of cartilage instead of bones) and bony fishes (Table 5.1a-g). Some of them, such as certain sharks and dog fish, are considered open-sea creatures. However, leading a fairly complicated life, they migrate close to the shore, in relatively shallow waters for part of the year or at certain stages in their life. We may assume that they could all be caught near the coast. This group includes fish that are either solitary or live in small groups, again depending on their age. Large as well as tiny fish were caught in inshore waters (Table 5.4). These fish are present, in different combinations, throughout the Aegean and Ionian Seas, especially in relatively closed gulfs. A review of both ancient fishing manuals such as Oppian's *Halieutica* and more recent accounts of traditional fishing in the Aegean (e.g. Aposotolides 1883; Athanassopoulos 1925; 1926; Zappas 1968; Zachariou-Mamaliga 1986) indicate that in the catching of this type of fish a wide range of multipurpose and specialized tools is employed, but also a large variety of crafty tricks (decoys, special baits, poisoning etc.). A vivid image of just such a variety can be seen in the earlier (6th century B.C.) but highly illustrative red figure cup interior, attributed to the Ambrosios Painter (Fig. 5.5). There a boy, perched on a rock by the sea is line-fishing while at the same time a creel has been placed in the water (Beazley 1963, p. 173, no9). At least three

Table 5.1a Fish taxonomic representation at various sites.

	Kommos S. Crete Temple C mostly floor deposits	**Itanos E. Crete domestic complex**	**Krania Piereia harbor sector tavern?**	**Kalaureia Poros sanctuary of Poseidon dining deposit**
	Hellenistic	Late Antiquity	Hellenistic	Hellenistic (ca 165 B.C.)
Elasmobranch	1		+	
Squatina squatina			+	
Dasyatis sp.		1	+	
Clupeidae			+	
Engraulidae		1		
Anguilla anguilla			+	1
Muraenidae	1		+	
Congridae	1	10	+	
Muraenidae/Congridae				1
Gadidae		2		
Gaidropsarus mediterraneu	1		+	
Serranidae		44	+	8
Epinephelus sp.	2	5	+	11
Seranus cabrilla/scriba		6	+	
Dicentrarchus sp.	1			
Carangidae			+	2
Seriola dumerili			+	
Sciaena umbra	1	1		1
Umbrina cirrosa	3			
Mullidae		1	+	
Sparidae indet. sp.	5	121	+	34
Boop boops		22	+	2
Dentex dentex		5	+	4
Diplodus annularis			+	
Diplodus sp.	1	4	+	
Lithognathus mormyris		1	+	
Oblada melanura	1			
Pagellus erythrinus	1	10		1
Pagellus sp.		3	+	5
Pagellus erythrinus/ Pagrus pagrus	1		+	
Pagrus pagrus		8		1
Sparus aurata		1		1
Spondyliosoma cantharus		4		
Centracanthidae	1	24		2
Sparidae/Centracanthidae		24	+	6
Chromis chromis		3		
Labridae indet. sp.		2		1
Sparisoma cretense	1	73		
Trachinidae				1
Uranoscopus scaber		1	+	
Scombridae		1	+	3
Euthynnus alletteratus				1
Scomber japonicus			+	
Thunnus sp.			+	6
Xiphias gladius			+	1
Sphyraena sphyraena		1		
Mugilidae			+	2
Scorpaena sp.		1	+	3
Pleronicteformes			+	
Soleidae	1		+	
Lophius piscatorius			+	
Cyprinidae			+	
Silurus sp.			+	
Clarias gariepinus	1			

Table 5.1a (Continued)

River fish indet.			+	
Indeterminate		113	+	15
Sea-shells	+	+	+	+
Sepia	+	+		
Sea-urchin	+	+	+	+
Crabs	+	+		+
Sea turtle		+		
TOTAL fish identified	24	494		113
Fish non identifiable	155	525		1002
Collection method	WS	WF	WF	WF
References	Rose 2000; Reese 2000b	Mylona pers. analysis; Mylona 2003b	Mylona in prep.	Mylona in prep.

Table 5.1b Fish taxonomic representation at various sites.

	Kassope domestic contexts	**New Halos Thessaly domestic contexts**	**Delphi cella of Apollo sanctuary**
	Hellenistic	Hellenistic	6th-4th c. B.C.
Elasmobranch	1		
Epinephelus sp.	10		
Dicentrarchus sp.	17		
Sparus aurata	24		
Euthynnus alletteratus	1		
Mugilidae	1		
Fish indeterminate			1
Sea-shells	+	+	
TOTAL fish identified	54		1
Non identifiable	7		0
Collection method	HC	HC?	HC
References	Boessneck 1994	Prummel 2003a, pp. 201-2	Poplin 1991, pp. 699-700

Table 5.1c Fish taxonomic representation at various sites.

	Corinth Amphora building	**Corinth sanctuary of Demeter and Kore dining rooms**	**Corinth sanctuary of Demeter and Kore cistern fill**	**Corinth area east of theater butcher's and cooking facility**
	Classical (mid 5th c. B.C.)	Classical	Roman	Roman
Serranidae				1
Epinephelus sp.			1	
Argyrosomus regius			5-6	
Sparidae indet. sp.	+	24		
Sparus aurata				8
Thynnus sp.	+			1
Sea-shells		+?	27	
Sea-urchin		+		
TOTAL fish identified	Undefined	24	6-7	10
Non identifiable	Undefined	25	0	6
Collection method	HC	WF	HC	DS
References	Williams 1979	Bookidis *et al.* 1999	Rose 2000, p. 530	Reese *et al.* 1987, pp. 255-247

Table 5.1d Fish taxonomic representation at various sites.

	Argolid Pyrgouthi farmstead	**Argolid Halieis sanctuary of Apollo refuse pit**	**Kalaureia Poros Sanctuary of Poseidon cistern**	**Kalaureia Poros sanctuary of Poseidon fill**
	Late Antique	4th c. B.C.	Roman	Classical
Engraulidae			1	
Serranidae	1		1	
Sparidae indet. sp.			3	1
Boops boops			2	
Pagellus erythrinus			1	
Sparus aurata		1		
Sparidae/Centracanthidae				1
Sparisoma cretense			1	
Trachinidae				1
TOTAL fish identified	1	1	9	3
Non identifiable	2	0	24	3
Collection method	WF	HC	WF	WF
References	Mylona 2005	Rose 2000, p. 530.	Mylona in prep.	Mylona in prep.

Table 5.1e Fish taxonomic representation at various sites.

	Messene altar	**Messene bath deposit**	**Messene, monumental fountain house in the agora area**	**Messene in the area of temple of Artemis Orthia**
	Classical (post-369 B.C.)	$2^{nd}/1^{st}$ c. B.C.	2nd c. B.C. to A.D. 200	3^{rd-2nd} c. B.C.
Galeorhinus galeus	1			
Epinephelus sp.		4		
Seriola dumerili			1	
Thynnus sp.				1
TOTAL fish identified	1	4	1	1
Non identifiable	0	0	0	0
Collection method	HC	HC	HC	HC
References	Nobis 1997, p. 108; Rose 2000, p. 531	Nobis 1994, p. 303; Rose 2000, p.531	Nobis 1994, p. 303; Rose 2000, p. 531	Nobis 1994, p. 303; Rose 2000, p. 531

Table 5.1f Fish taxonomic representation at various sites.

	Attica Athens agora unspecified context	**Attica Athens temple of Aphrodite Ourania**	**Attica Athens tavern in the Agora**	**Delos The House of the lake cistern**	**Thera Pillarou Cave Zeus sanctuary dining refuse**
	Classical	about 500 B.C.	Classical	Hellenistic	Hellenistic
Elasmobranch	2				
Sparidae indet. sp.		1			
Sparus aurata					6
Thunnus sp.					1
Indeterminate			+	+	
Sea-shells			+		
Sepia					
Sea-urchin					

Table 5.1f (Continued)

Crabs					
Sea turtle					
TOTAL fish identified	2	1			7
Non identifiable	0	2			5
Collection method	HC	DS?	HC?	HC	HC?
References	Reese 1984	Reese 1998	Shear 1975	Zapheiro-poulou 1991	Becker 1997

Table 5.1g Fish taxonomic representation at various sites.

	Mytilene Villa – domestic	**Knossos – Crete Demeter sanctuary**	**Eleutherna – Crete villa- domestic**	**Eleutherna – Crete domestic contexts**
	Roman	Minoan-Roman mixed	Roman	Late Antiquity
Elasmobranch		+		
Carcharodon carcharias				1
Dasyatis sp.				1
Congridae	+			
Serranidae	+			
Epinephelus sp.			+	3
Mullidae	+			
Sparidae indet. sp.	+			
Dentex dentex			+	
Scombridae	+			
Thynnus sp.	+		+	
Xiphias gladius	+			
Scorpaena sp.	+			
Mugilidae				1
Cyprinidae	+			
Indeterminate		1		
Sea-shells	+			
Sea-urchin	+			
Crabs and lobsters	+			
TOTAL fish identified		1		6
Non identifiable		0		0
Collection method	HC	HC	HC	HC
References	Ruscillo 1993	Jarman 1973	Nobis 1993, p. 115	Mylona, pers. analysis

Table 5.2 Taxonomic presence of fish on the basis of fish remains and references in literary sources.

Habitat	**Family**	**Taxon**	**Fish bones**	**Written sources**
Inshore - offshore	(Elasmobranch)	Indeterminate	+	+
	Alopiidae	*Alopias vulpinus*		+
	Scyliorhinidae	*Scyliorhynus stellaris*		+
	Triakidae	*Galeorhinus galeus*	+	
	Sphyrnidae	*Sphyrna zygaena*		+
	Squatinidae	*Squatina squatina*	+	+
	Torpaenidae	Indeterminate		+
	Rajidae	*Rajia* sp.		+
	Dasyatidae	*Dasyatis* sp.	+	+
	Muraenidae	Indeterminate	+	+
	Congridae	Indeterminate	+	+
		Muraenidae/Congridae	+	

Table 5.2 (Continued)

	Gadidae	Indeterminate	+	+
		Gaidropsarus mediterraneus	+	
	Serranidae	Indeterminate	+	
		Epinephelus sp.	+	+
		Serranus cabrilla/ scriba	+	
		Serranidae small		+
	Sciaenidae	*Sciaena umbra*	+	
		Umbrina cirrosa	+	
	Mullidae	Indeterminate	+	
	Sparidae	Indeterminate	+	+
		Boop boops	+	
		Dentex dentex	+	
		Diplodus annularis	+	
		Diplodus sargus		+
		Diplodus sp.	+	
		Lithognathus mormyris	+	+
		Oblada melanura	+	+
		Pagellus sp.	+	
		Pagellus erythrinus	+	+
		Pagellus erythrinus/ Pagrus pagrus	+	
		Pagrus pagrus	+	
		Salpa sarpa		+
		Sparus aurata	+	
		Spondyliosoma cantharus		+
	Pomacentridae	*Chromis chromis*	+	
	Labridae	Indeterminate	+	+
	Scaridae	*Sparisoma cretense*	+	
	Trachinidae	Indeterminate	+	+
	Uranoscopidae	*Uranoscopus scaber*	+	
	Gobiidae	*Gobius* sp.		+
	Blenniidae	*Blennius* sp.		+
	Sphyraenidae	*Sphyraena sphyraena*	+	
	Scorpaenidae	*Scorpaena* sp.	+	
	Triglidae	Indeterminate		+
	(Pleronicteformes)	Indeterminate	+	+
	Soleidae	Indeterminate		
	Lophiidae	*Lophius piscatorius*	+	+
Marine schooling	Centracanthidae	Indeterminate	+	+
	Sparidae	*Boop boops*	+	+
	Sparidae/ Centracanthidae		+	
	Clupeidae	Indeterminate	+	+
	Atherinidae	Indeterminate		+
Marine migrating	Carangidae	Indeterminate	+	
		Seriola dumerili	+	
	Coryphaenidae	*Coryphaena hippurus*		+
	Scombridae	Indeterminate	+	
		Euthynnus alletteratus	+	
		Sarda sarda		+
		Scomber scombrus / japonicus	+	+
		Thynnus sp.	+	+
	Xiphiidae	*Xiphias gladius*	+	+
Euryaline	Engraulidae	Indeterminate	+	+
	Serranidae	*Dicentrarchus* sp.	+	+
	Sparidae	*Sparus aurata*	+	
	Sciaenidae	*Argyrosomus regius* (young)	+	

Table 5.2 (Continued)

		Umbrina cirrosa (young)	+	
	Gobiidae	*Gobius* sp.	+	+
	Mugilidae	Indeterminate	+	+
	Pleuronectidae	*Platiychthys* sp.	+	
	Soleidae	*Solea* sp.	+	
Anadromus - Catadromus	Acipenseridae	Indeterminate	+	+
	Clupeidae	*Alosa* sp.	+	
	Anguillidae	*Anguilla anguilla*	+	+
Fresh-water	Cyprinidae	Indeterminate	+	+
	Siluridae	*Silurus* sp.	+	+
	Percidae	Indeterminate		+
Exotic		*Clarias gariepinus*	+	
	(Various Nile fish)			+

*The table is based on data from Tables 5.1a-g and 5.3 and from Appendix 1.

Fig. 5.5 Young fisherman on the coast. Scene in the interior of cup by Ambrosios Painter, 6th c. B.C., now in the Boston Museum of Fine Arts. A young fisherman catches inshore fish and octopus using hook and line and a fish trap (Beazley 1963, p. 173, no 9; photograph from Sparkes 1995).

types of sea creatures seem to be the target of this double fishing effort, an octopus and two different fish.

Another group of marine fish caught in antiquity are those that form large, sometimes enormous, shoals. These fish can be found seasonally, relatively close to the coast. Their presence exhibits a temporal patterning, but their seasonal migration is small scale, from shallow to deeper water, within the same geographical area. The tapping of this resource can result in an abundance of fish per each fishing episode. Often some fish processing is required. Oppian in his *Halieutica* in the 2nd c. A.D. vividly portrays the exploitation of one such type of fish, the *afyi* (here it probably refers to sardines or anchovies; for the identification of *afyi* in general see Thompson 1947). After he described how enormous and dense the "*afyi*" schools could be, he added:

> "Now when the fishermen behold them (the fish) huddled together, they gladly enclose them with their hollow seine-nets and without trouble bring ashore abundant booty and fill with the fry all their vessels and their boats and on the deep beaches they pile up heaps, an infinite abundance of spoil" (Opp. *Hal.* iv. 468-503, transl. by A.W. Mair).

Such shoaling fish exhibit a distinct geographical distribution. Clupeidae (mostly sardines, sprats and sand smelts) and Engraulidae (anchovies) are more abundant in the northern Aegean, while Centracanthidae (mostly picarel) and *Boops boops* (boque) are better adapted to southern Aegean conditions. Of course all these species are found in smaller numbers in less favourable areas too (Vidalis and Tsimenidis 1997).

A third group of fish caught in antiquity are the marine migrating species. These are pelagic fish which appear seasonally, forming large schools. Their exploitation results in large concentrations of fish, and often requires some processing (Fig. 5.6). The presence of each of these species in any particular geographical area is restricted to certain, narrowly defined time-periods during the year. The pelagic migratory fish range in size from relatively small (e.g. the mackerels - *Scomber* sp.) to very large (tunas - *Thynnus* sp.) (Table 5.4). Although these are pelagic fish, some species tend to approach the coast to within meters, thus becoming accessible, even using coast-based fishing technologies (e.g. *thynnia* – permanent net arrangements and observation tower, Fig. 5.6). The basic scheme is that large shoals of these fish

Table 5.3 Geography of Classical fishing, fish eating and fish in cult based on literary sources.

	Location	Fish	Function	Authority
Northern Aegean	Thassos	Lobster Red mullet Octopus Scorpion fish	Best taste Best taste Best size Best taste under conditions	Aristotle, *HA* 549b, line 16 Archaestratos in Ath. vii 325e Archaestratos in Ath. vii 318f Archaestratos in Ath. vii 321a
	Lemnos	Crabs	Cult	Hesychios, *Lexicon* (Schmidt 1965, v.2, p. 382)
	Pella	Sciaenidae (*chromis*)	Best taste	Archaestratos in Ath. vii 328a
	Strymon river	Eel Wels	Best size Presence	Archaestratos in Ath. vii 298a Antiphanes, *Thamyras* (104 K-A) in Ath. vii 300c Ael.xii.14
	Torone	Shark	Best taste	Archaestratos in Ath. vii 310c
	Avdera	Grey mullet	Best taste	Archaestratos in Ath. vii 307b
	Lake Volve	Sea bass Danube bleak	Best taste Cult	Archaestratos in Ath. vii 311a Hegesander, *Commentaries* (FHGr IV 420)
	Samos	Tuna Large fish (generic)	Best size Myth	Archaestratos in Ath. vii 321a Hdt, iii, 41-43
	Mytilene	Salema	Best taste	Archaestratos in Ath. vii 321f
	Olynthus	*Glaukos* (unknown)	Best taste	Archaestratos in Ath. vii 295c
	Samothrace	Escort fish	Cult	Pankrates, *Occupations at Sea* in Ath. vii 283a
Northern Ionian sea and islands	Amvrakia	*Kapros* Sciaenidae (*chromis*) Sea bass	Best taste/ high price Best taste Best taste	Archaestratos in Ath. vii 305e Archaestratos in Ath. vii 328a Archaestratos in Ath. vii 311a
	Corcyra	Tuna Octopus	Economy – cult Best size	Paus. 10.9.3-4 Archaestratos in Ath. vii 318f
Corinthian gulf	Sicyon	Conger eel	Best taste	Eudox. Ib 288c Philemon, *Soldier* (82 K-A) in Ath. vii 288c
	Kalydon	Sea-bass Lake fish	Best taste Economy	Archaestratos in Ath. vii 311a Strabo 10.2.21
	Corinth	Fish-market	Economy	Williams 1993, pp. 39-40
	Bulis	Purple shell	Economy	Pausanias 10.37.3
Sterea, mainland	Copais - Boetia	Eel Eels	Best taste Cult	Paus. 9. 24.2; Ath. 297e Antiphanes, *The Sheep-owner* (191 K-A) in Ath.vii 295c Ar. *Ach.* 884cf, 894 Ar.*Pax* 1013-14 Paus. 4.24.1 Agatharchides, *European History* (FHG III, p.192) in Ath. vii 297d.

Table 5.3 (Continued)

	Tanagra	Triton	Myth	Ael.13.21
	Boetia- Akraephia	Marine and fresh-water Fish	Economy	Salviat et Vatin 1971; Feyel 1936
Euboic gulf – Central Aegean	Karystos	Tuna Sprat Dolphin fish	Best taste Best taste Best taste	Archaestratos in Ath. vii 302a Ath. 285c; Antiphanes, *The Sheep-owner* (191 K-A) in Ath. vii 295c Archaestratos in Ath. vii 304d
	Chalkis	Flat fish (*vouglossa*)	Best under conditions	Archaestratos in Ath. vii 288b
	Eretreia	Sea-breams	Best size	Antiphanes, *The Sheep-owner* (191 K-A) in Ath. vii 295c Archaestratos in Ath. vii 327d
	Aedypsos	Fish	Abundance	Plut., *Mor*. iv. 4
	Anthedon	Fishing	Economy	Heraklides Kritikos, *Description of Greece* (Schäfer 1968) in Ath. i 23d
Eastern coast of Attica	Marathon	*Afyi*	Best taste	Aristotle, *HA* 569b, lines 10-14
Attica indeterminate	Attica	*Glaukos*	Best taste	Philemon, *Soldier* (82 K-A) in Ath. vii 288c
Saronic gulf	Megara	*Glaukos*	Best taste	Archaestratos, in Ath. 295c Antiphanes, *The Sheep-owner* (191 K-A) in Ath. vii 295c
	Aixon in Attica	Red mullet Sting ray Saddled bream	Best size Best size Best size	Ath. vii 325e
	Hale Aixonidae	Tunna	Cult and economy	Krates, *Sacrifices at Athens* (FHG IV p. 362) in Ath. vii 297e
	Phalero	*Afyi*	Best taste	Ath. iv 135a, 244c, vii 285b, 285d, 293e
	Athens	*Afyi*	Best taste	Chrysippos of Soli, *On Things to be Chosen on their Own Sake*, in Ath. vii 285d
	Salamina	Whitebait	Best taste	Aristotle, *HA* 569b, lines 10-14
	Eleusis	 Flat fish (*Psitta*) Fish generic	Best taste Cult	Lynceus of Samos, *Letters* in Ath. 330a Porph. *On abst.*4.16
	Rheittoi river	River fish	Cult	Pausanias 1.38.1
	Troizen	Tuna fishing	Economy	IG [4] 1, 123; Herzog 1931, C xlvii, pp. 26-28

Table 5.3 (Continued)

Peloponnese mainland	Arcadia	Crying fish	Cult	Pausanias 8.11.2
	Argos	Fishing	Economy	IG [4] 1, 123; Herzog 1931, C xlvii, pp. 26-28
	Achaia - Pharae	River fish	Cult	Pausanias 7.12.4
Argolic gulf	Argolid	Boar-fish	Best taste	Philemon, *Soldier* (82 K-A) in Ath. vii 288c
	Halieis	Tuna	Cult and economy	Antigonus of Carystos, *On Diction* in Ath. vii, 297e
Laconian gulf	Aegiae	"Lake" fish	Cult	Pausanias 3.21.5
Cyclades	Delos	*Levias* (unknown) Sea bream	Best taste Best taste	Archaestratos in Ath. vii 301c Archaestratos in Ath. vii 327d
	Delos	Fish	Cult	Semos, *History of Delos*, FGrHist. iv 493
	Tenos	*Levias* (unknown)	Best taste	Archaestratos in Ath. 301d
	Naxos	Fish generic	Economy	Aristotle, *The Constitution of Naxos* in Ath. viii 348b
	Myconos	Fish	Cult	SIG[4] 1024
	Serifos	Lobster *(Tetix enalios)*	Cult	Ael. 13.26
Dodecanese	Rhodes	Sturgeon Fox shark	Best- taste Best taste	Archaestratos in Ath. vii 294e Lynceus of Samos in Ath.vii 285f
	Cos	Tuna fishing	Economy	SIG[4] 1000 Sherwin-White, 1978, pp. 229-235
	Cos	Fish	Cult	SIG[3] 1106. lines 177.42, 62 Paton and Hicks 1990, no 36, p. 75 Lupu 2005, pp. 86-87
Crete	Stalis	Fishing	Economy	Sylloge [4] 427 Chaniotes 1996, pp. 388-393

Table 5.4 Size range of the fish exploited in Classical Greece.

Habitat	Family	Taxon	Maximum length
Inshore - offshore	Elasmobranch		
	Alopiidae	*Alopias vulpinus*	1.2 - 6 m.
	Scyliorhinidae	*Scyliorhynus stellaris*	up to 1.5 m.
	Triakidae	*Galeorhinus galeus*	up to 2 m.
	Sphyrnidae	*Sphyrna zygaena*	0.6 – 4 m.
	Squatinidae	*Squatina squatina*	up to 2.5 m.
	Torpaenidae	Torpaenidae indet.	up to 0.60 – 1.8 m. depending on species
	Rajidae	*Rajia* sp.	up to 0.75 – 2 m. depending on species.
	Dasyatidae	*Dasyatis* sp.	wing breadth up to 1.5-3 m. depending on species.
	Muraenidae	Muraenidae indet.	up to 1 -1.3 m. depending on species.
	Congridae	Congridae indet.	up to 2.5 m.
	Belonidae	*Belone belone*	up to 0.45 - 0.9 m. depending on species.
	Gadidae	Gadidae indet.	up to 0.15-1.10 m. depending on species.
		Gaidropsarus mediterraneus	up to 0.45 m.
	Zeidae	*Zeus faber*	up to 0.9 m.
	Serranidae	Serranidae indet	-
		Epinephelus sp.	up to 0.5 – 2 m. depending on species.
		Seranus cabrilla/scriba	up to 25-35 cm depending on species.
		Serranidae small	-
	Sciaenidae	*Sciaena umbra*	up to 0.75 m.
		Umbrina cirrosa	up to 1 m.
	Mullidae	Mullidae indet.	up to 0.25 – 0.4 m depending on species.
	Sparidae	Sparidae indet.	-
		Boop boops	up to 0.35 m.
		Dentex dentex	up to 1 m.
		Dentex sp.	up to 0.4 – 1 m. depending on species.
		Diplodus annularis	up to 24 cm
		Diplodus sargus	up to 0.45 m.
		Diplodus sp.	up to 0.4 – 0.55 m. depending on species.
		Pagrus pagrus	0.9 m.
		Lithognathus mormyris	up to 0.55 m.
		Oblada melanura	up to 0.35 m.
		Pagellus sp.	up to 0.35 - 65 m. depending on species
		Pagellus erythrinus	up to 0.6 m.
		Pagellus erythrinus/ Pagrus pagrus	0.6-0.9 m. depending on species.
		Salpa sarpa	up to 50 m.
		Sparus aurata	up to 0.7 m.
		Spondyliosoma cantharus	up to 0.6 m.
	Pomacentridae	*Chromis chromis*	up to 0.15 m.
	Labridae	Labridae indet. sp.	up to 0.1-0.6 m. depending on species.
	Scaridae	*Sparisoma cretense*	up to 0.5 m.
	Trachinidae	Trachinidae indet.	up to 0.15 -0.4 m. depending on species.
	Uranoscopidae	*Uranoscopus scaber*	up to 0.4 m.
	Gobiidae	*Gobius* sp.	up to 0.12 – 0.18 m. depending on species.
	Blanniidae	*Blennius* sp.	less than 0.1 m.
	Sphyraenidae	*Sphyraena sphyraena*	up to 1.6 m.

Table 5.4 (Continued)

	Scorpaenidae	*Scorpaena* sp.	up to 0.2 -0.5 m. depending on species.
	Triglidae	Triglidae indet.	up to 0.2 – 0.75 m. depending on species
	Pleuronicteformes		-
	Soleidae	Soleidae indet.	up to 0.15 -0.7 m. depending on species.
	Lophiidae	*Lophius piscatorius*	up to 2 m.
Marine schooling	Centracanthidae	Centracanthidae	up to 0.15 -0.2 m. depending on species.
	Sparidae	*Boop boops*	up to 0.35 m.
		Sparidae/Centracanthidae	up to 0.15 -0.35 m. depending on species.
	Clupeidae	Clupeidae indet. marine	up to 0.15 -0.3 m. depending on species.
Marine migrating	Carangidae	Carangidae indet.	up to 0.5-1.9 m. depending on species.
		Trachurus sp.	up to 0.6-0.7 m. depending on species.
		Seriola dumerili	up to 1.9 m.
	Coryphaenidae	*Coryphaena hippurus*	up to 2.1 m.
	Scombridae	Scombridae indet.	up to 0.35 – 4.5 m. depending on species.
		Euthynnus alletteratus	up to 1.20 m.
		Sarda sarda	up to 0.91 m.
		Scomber japonicus/ scombrus	up to 0.3-0.5 m. depending on species.
		Thynnus sp.	up to 1.3 – 4.5 m. depending on species.
	Xiphiidae	*Xiphias gladius*	up to 4.5 m.
Euryaline	Engraulidae	*Engraulis encrasicolus*	up to 0.2 m.
	Serranidae	*Dicentrarchus* sp.	up to 0.7 – 1.2 m. depending on species.
	Sparidae	*Diplodus sargus* young	-
		Sparus aurata	up to 0.7 m.
		Spondyliosoma cantharus	up to 0.6 m.
	Sciaenidae	*Umbrina cirrosa* young	-
		Argyrosomus regius young	up to 2 m.
	Gobidae	*Gobius* sp.	up to 0.1-0.18 m depending on species.
	Mugilidae	Mugilidae indet.	up to 0.15-1.2 m. depending on species.
	Pleuronectidae	*Platiychthys* sp.	up to 0,6 m.
	Soleidae	*Solea* sp.	up to 0.3-0.6 m. depending on species.
Anadromus - Catadromus	Acipenseridae	Acipenseridae indet.	up to 2 – 4 m. depending on species.
	Clupeidae	*Alosa* sp.	up to 0.5-0.7 m. depending on species.
	Anguillidae	*Anguilla anguilla*	up to 1.33 m.
Fresh-water	Cyprinidae	Carp	up to 0.45-1.20 m. depending on species
	Cyprinidae	Indeterminate	0.25 -1.20 m. depending on species.
	Siluridae	Wels	up to 3 m. depending on species.
	Percidae	Perch	-
Exotic		*Clarias gariepinus*	up to 1.7m.
		Various Nile fish	-

*The maximum lengths given in the table have been taken from record of the FishBase (Froese and Pauly 2006) and refer to maximum lengths reported to date. Commonly caught individuals are of a smaller size.

Fig. 5.6 *Thynneio*: an arrangement of permanent nets and watching towers set near the coast for the catching of migratory fish. The drawing illustrates the working of a *thynneio* at the beginning of the 20th c. A.D. (from Koukoules 1948, p. 34).

swim close to the shore following an anti-clockwise movement during the trophic migration to the Black Sea and the reproductive one back to the Mediterranean. In this yearly circle, certain areas in the Aegean and Ionian seas are specially favoured by these fish, due to coast morphology and food richness of the area (see ethnographic observations in Athanassopoulos 1923; 1924; Bintliff 1977, part i, pp.117-122; Guest-Papamanoli 1985; 1986).

The migratory fish and especially the largest of them, the several varieties of tunas and bonitoes, are among the most discussed fish in archaeological literature. Although they appear to be a rich fish resource, their importance has been repeatedly underestimated due to an emphasis placed by archaeologists and ancient historians on the unpredictability of their appearance at certain standard spots on the coast (chapter 2.2). Some comment is called for at this point. The epigraphical record, mostly of the Hellenistic period, refers to the renting or taxation of "*thynoskopeia*" i.e. watch-towers for migrating fish (Fig. 5.6) which are rented by the interested fishermen (Appendix 2: SIG[4] 1000). The fact that there is a formalized manner of managing these issues clearly implies a repeated pattern of exploitation, and apparently of migratory fish occurrence. Besides, the immobile land-based technology which is involved in both the catching and processing of these fish, targets a variety of species, which appear at slightly different times and are not all migratory. Some of them are predators following tuna shoals (sharks) or various "by catch" fish and many "errant" coastal species such as grey mullet, red mullet etc. (Vafeiadou 1974, p.153; Economidis pers. com.). Detailed ethnographic observations of the way such a fishing system may operate are available for the admittedly much richer coasts of the Black Sea (Paraskeuopoulou, unknown date). The observed pattern of fish occurrence is at least as reliable as the annual land crops. In an Aegean context, no comprehensive study of such a system exists, but this is probably more related to the state of sea-related research rather than to the lack of this type of fishing (see related evidence in chapter 5.3).

The fifth group of fish that were exploited in Greece are the euryaline fish, those living in both brackish and sea water but usually breeding in the sea. These fish are commonly found in river deltas, estuaries, and lagoons. Some, such as the grey mullet (Mugilidae) or the guilt-head sea breams (*Sparus aurata*), migrate as juveniles to the brackish waters and in such environments they are found in large numbers. There, these migrant species attain the best quality of flesh. Other fish, such as some members of the Sparidae family enter such environments only casually (strangler species) (Koutrakis *et al.* 2005). Euryaline ecosystems, usually sustain large populations of fish which are traditionally exploited in a systematic manner with the use of stationary fish entrapment devices (Guest-Papamanoli 1985; 1986).

The next group of fish includes the anadromous and the catadromous fish, which share their lives between marine and fresh-water environments. Among these, the eels (Anguilidae) are the most commonly exploited catadromous species in antiquity, although the sturgeon (Acipenseridae) and the shads (*Alosa* sp.), both anadromous fish, were also caught. Although the current evidence originates from only certain areas of Greece, future research (mostly fish bone collection and analysis) might reveal a more widespread use of this resource.

Another group of fish among those exploited by fishermen in Greece during the centuries under study are the exclusively fresh-water ones. Both lake and river fish appear to have been caught. Archaeologically only some river fish are attested so far, but several literary references, speak of lake fish as well. Here, one would expect a more intense exploitation of this resource in areas where rivers are substantial and lakes are deep enough to sustain a fish population on a year-round basis. Some scant evidence, however, suggests a familiarity with river fish resources even in cases of small rivers, hardly more than seasonal torrents. Pausanias' comment, in the middle of the 2nd century A.D., on the vocal fish (fish that produce sounds) from Cleitor River in Arcadia, or on the sacred fish in a stream belonging to Hermes at Pharae, in the same region (Appendix 1: Paus. vii.22.4 and viii.21.2) indicate an awareness of the ichtyofauna in these rivers, which could be connected to fishing activities. The situation with the lakes is much more obscure, with evidence for the exploitation of their fish being few and geographically dispersed (see instances in chapter 5.3.)

Finally fish imported from abroad, either exotic or common varieties, were also part of the consumed fish range. Only highly visible cases of imported fish are archaeologically identified in this group. Such are the catfish bone (*Clarias gariepinus*) found at Hellenistic Kommos on Crete which had been imported from either

Egypt or Syro-Phoenicia (Rose 2000, pp. 512-514), or the sea bream and tuna remains found adhering to Punic amphora fragments in Classical Corinth, which came from the Atlantic coast of Morocco (Williams 1978; Zimmerman-Munn 2003). The literary sources also provide evidence of imported fish, but refer mostly to preserved forms. In those cases, the fish themselves are not exotic, but belong to varieties which can be easily found in Greek waters. Only the preserved product acquires an exotic character (see discussion in chapter 7.2).

To sum up, it appears that the type of fish resource more extensively exploited in Greece between the 5th century B.C. and the 7th century A.D. was that of the inshore fish. This consists of a variety of species, available collectively all year-round, easily accessible even from the shore with simple technological means. These are the fish which, given the accessibility and low technology required for their catching, can be targeted not only by professional fishermen, but even by amateurs. Marine schooling and marine migrating fish have also been exploited to a considerable degree. The first case is less visible than the second, perhaps due to taphonomic processes. The exploitation of the seasonal fish schools is much more evident, mostly through indirect evidence. Both these resources produce an abundance of catch, often requiring some subsequent processing. Catching them requires not only specialized knowledge on the part of the fishermen (of school movement, weather and water conditions etc.) but, at least in the case of migrating fish, it requires specialized fishing technology and communal effort. The euryaline fish, those moving in and out of the brackish waters of estuaries, deltas and lagoons was another exploited fish resource. These are only lightly visible archaeologically, and can be better traced by inference through the literary tradition. Its intensity, albeit strong in some cases (e.g. Krania) can not at the moment be safely deduced. Anadromous/catadromous fish and fish are an equally elusive, albeit certainly exploited fish resource. Again, few actual fish remains and somewhat more substantial literary data attest to their exploitation, but, as in the previous case, the intensity of such exploitation remains obscure. What can be certainly deduced from the available data is that all types of fish resources had been exploited to some degree in various parts of Greece. The range of exploited resources and the geographically specific data will be presented and discussed in the following section.

5.3 The geographical distribution of exploited fish resources.

Given the diversity of the character and richness of fish resources throughout Greece, it is perhaps instructive for the present study to explore the geographical distribution of their exploitation in the period in question. Also, it might be informative to infer from these data, as far as possible, the presence of fishing communities involved in their management. For the purposes of this particular quest, the database used in the previous section will be enriched with more data. These are the remains of the technology involved in the exploitation of the fish resource (Appendix 3). At this point, these remains, the fish hooks, net weights, netting needles, fish ponds and so on are taken as yet another type of indicator of fishing activity. Because these finds are in most cases reported summarily, in catalogues, with minimal contextual information, the presence of different types of fishing gear can not be used in any straightforward way to indicate emphasis on different types of fishing, such as line and hook versus net fishing (for a discussion of the problem of the fishing tools as representing the scale of fishing activity see Højte 2005, pp. 135-138). The information derived from fishing gear can be used however to add yet another dimension to the activities which were related to the exploitation of aquatic resources.

The tabulation of the fish and fishing-related data (Table 5.5) involves a bias in the geographical distribution of some aquatic resources, which should be kept in mind during the following discussion. Wherever fish-bone assemblages have been systematically collected (by water-flotation) and studied, the breadth of the resources exploited in the past increases. This is the case with Krania in northern Greece, Kalaureia in the Saronic gulf, Itanos and Kommos on Crete and to a lesser extent Mytilene, on the island of Lesvos (for the character and fish of the most important sites referred to in the following discussion see Appendix 4 with all related references).

Northern Aegean (Fig. 5.7a-b).

The review of the available fishing-related data from around the northern Aegean, shows, not surprisingly, the most complicated picture, despite the fact that only one site, Krania, provides a full, detailed fish bone record. Three types of sites can be discerned: islands, coastal sites on the mainland and inland waters.

The islands, Thassos, Lesvos, and Samos, give evidence for the exploitation of inshore fish and of the migrating schools of tuna. Being situated in one of the most eutrophic sea areas of Greece, and right on the migration route of the various Scombridae moving in and out of the Black sea through the Sea of Marmara (Economidis, pers. com.), this is to be expected. The frightening descriptions (Appendix 1: Ael. 13.20) of the large numbers of sea monsters and *ketoi* (probably sea mammals such as whales and dolphins or even very large tunas and sharks - Papadopoulos and Ruscillo 2002) in the coastal waters on the eastern coast of Mounth Athos are not irrelevant to this richness, since these animals typically follow large schools of migratory fish, feeding on them. Furthermore, lagoon fishing is attested for the Island of Lesvos by Aristotle, who performed much of his biological field research there (Lee 1948; 1985). It is reported that in the Pyrrha lagoon in his days the scallop

Table 5.5 Geographical distribution of all fishing-related data.

	Sites	Fish bones	Literary data on fishing	Fish and seafood in cult and myth	Fishing related technology
Northern Aegean	Avdera		+		+
	Olynthus		+		+
	Lake Volvi		+	+	
	Strymon river		+		
	Torone, Chalkidike		+		+
	Pella		+		
	Krania, Platamon	+			
	New Halos, Thessaly	+			
	Thassos		+		+
	Samos		+		
	Samothrace,			+	
	Chios			+	
	Lemnos			+	
	Mytilene	+	+		
Mainland west coast and Ionian Sea	Corcyra		+	+	
	Kassope	+			
	Amvrakia		+		
Corinthian gulf	Delphi	+			
	Bulis		+		
	Kalydon lake		+		
	Achaia			+	
	Corinth	+	+		+
	Sicyon		+		
Sterea mainland	Kopais lake		+		
	Tanagra			+	
	Akraephia		+		
	Boeotia indeterminate			+	
	Thebes		+		
	Delphi	+			
Euboic gulf Central Aegean	Anthedon		+		
	Karystos		+		
	Eretreia		+		
	Aedypsos		+		
Eastern coasts of Attica	Marathon		+		
Attica indeterminate	Agora, Athens	+			
	Athens		+		
Saronic gulf	Halai Aixonidae		+	+	+
	Phalero		+		
	Rheittoi River			+	
	Eleusis			+	
	Megara		+		
	Salamis		+		
	Isthmia				+
	Kenchreae				+
	Kalaureia	+			
	Troizen		+		
Peloponnese mainland	Arcadia			+	
	Pyrgoudi, Argolid	+			
	Nemea				+
	Elis				+
Argolic gulf	Halieis	+	+	+	+

Table 5.5 (Continued)

	Argolid indet.		+		
	Argos		+		
Laconian gulf	Aegiae			+	
Messenian gulf	Messene	+			
Cyclades	Thera	+			
	Tenos		+		
	Naxos		+		
	Serifos		+		
	Delos		+	+	
	Myconos			+	
Dodecanese	Cos		+	+	
	Rhodes		+		
	Castelorizo				+
Crete	Knossos	+			
	Chersonessos				+
	Eleutherna	+			
	Kommos	+			+
	Itanos	+		+	+
	Mochlos				+
	Siteia				+
	Ierapetra				+
	Stalis		+		

*The table is based on combined data from Appendix 1, 2, 3, Table 5.1a-g and Table 5.3.

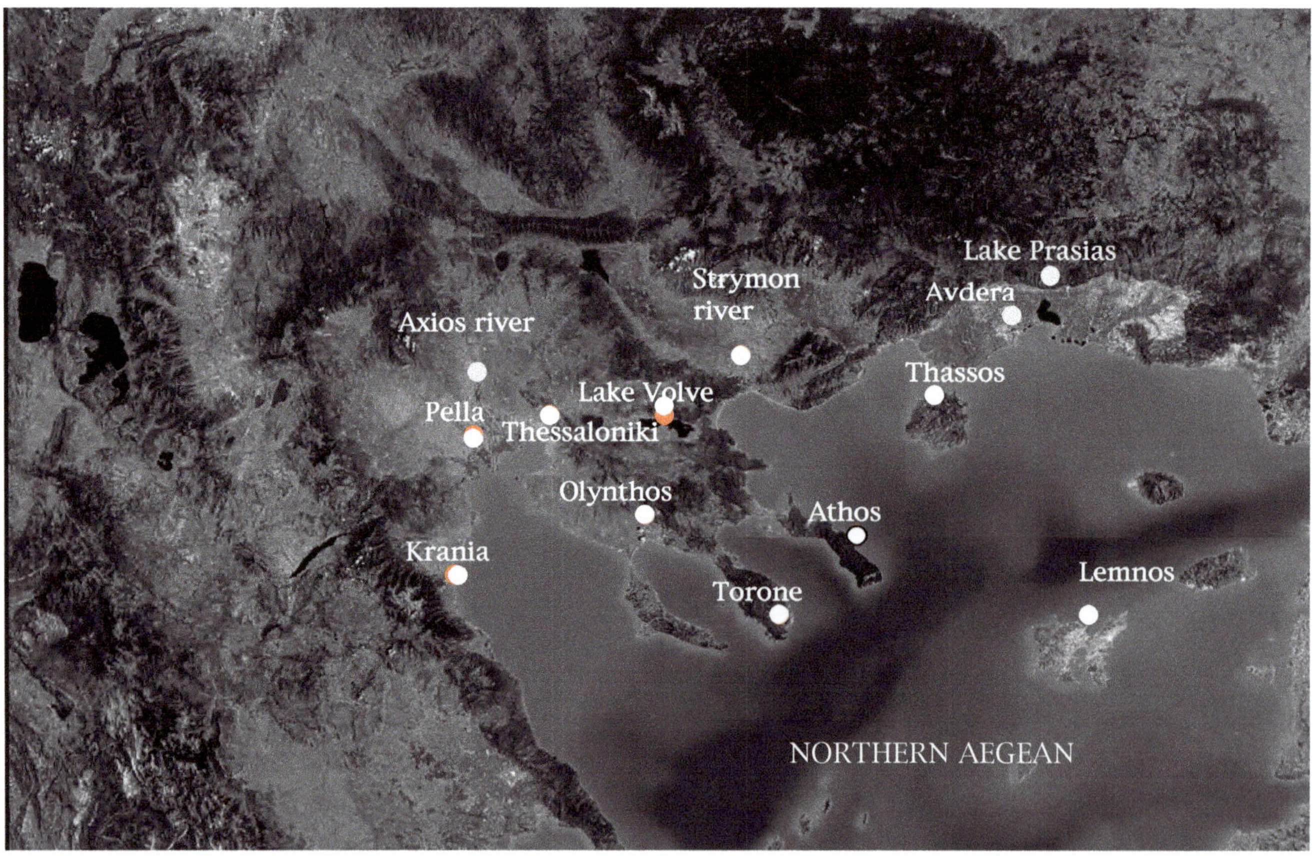

Fig. 5.7a Satellite image of northern Greece and part of northern Aegean (from TalentCruiser, www.cruiser.gr/en_main.html accessed 25 November 2006).

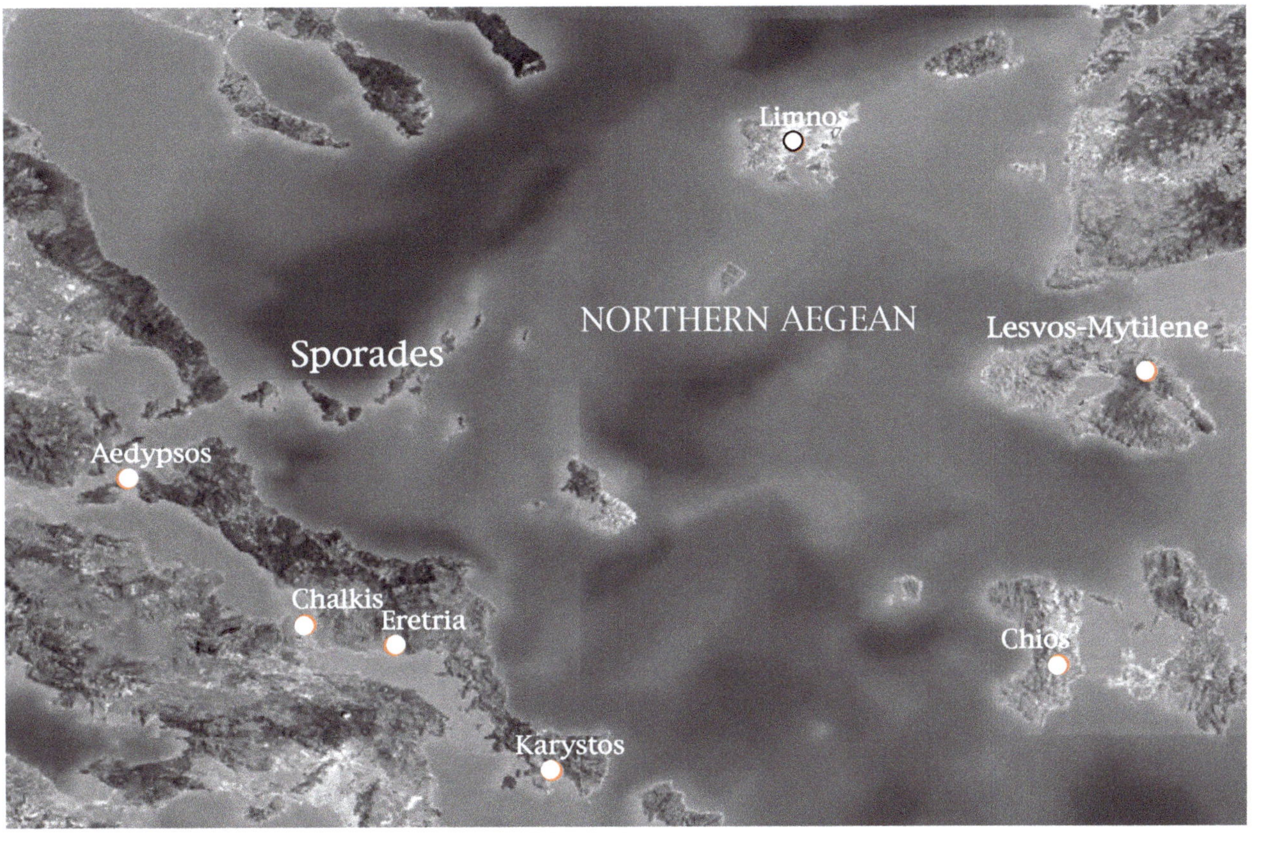

Fig. 5.7b Satellite image of part of northern Aegean and Euboea (from TalentCruiser, www.cruiser.gr/en_main.html accessed 25 November 2006).

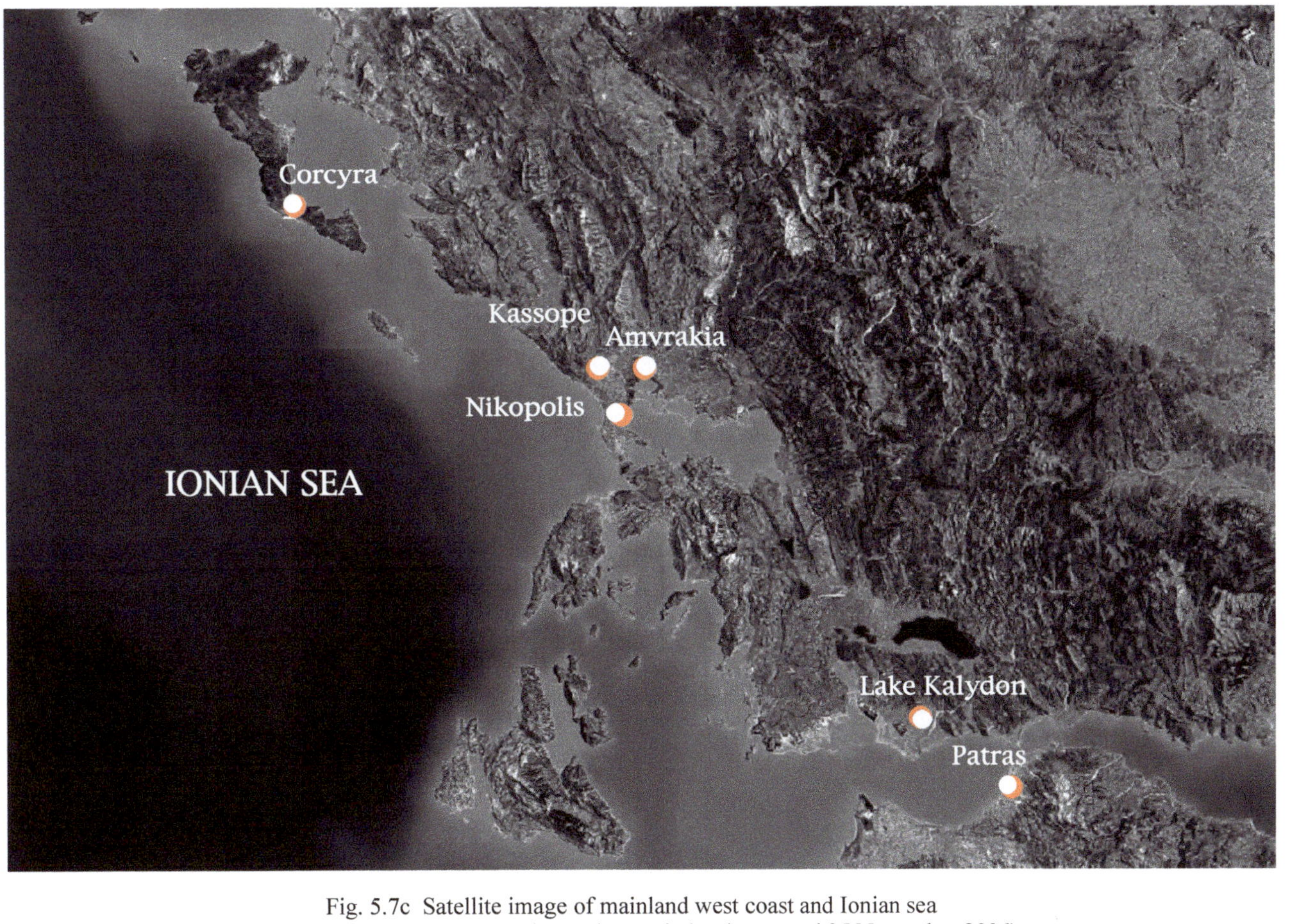

Fig. 5.7c Satellite image of mainland west coast and Ionian sea (from TalentCruiser, www.cruiser.gr/en_main.html accessed 25 November 2006).

Fig. 5.7d Satellite image of southern Thessaly and Sterea Ellada (from TalentCruiser, www.cruiser.gr/en_main.html accessed 25 November 2006

Fig. 5.7e Satellite image of Corinthian gulf (from TalentCruiser, www.cruiser.gr/en_main.html accessed 25 November 2006).

Fig. 5.7f Satellite image of Attica, Saronic gulf and Argolic gulf (from TalentCruiser, www.cruiser.gr/en_main.html accessed 25 November 2006).

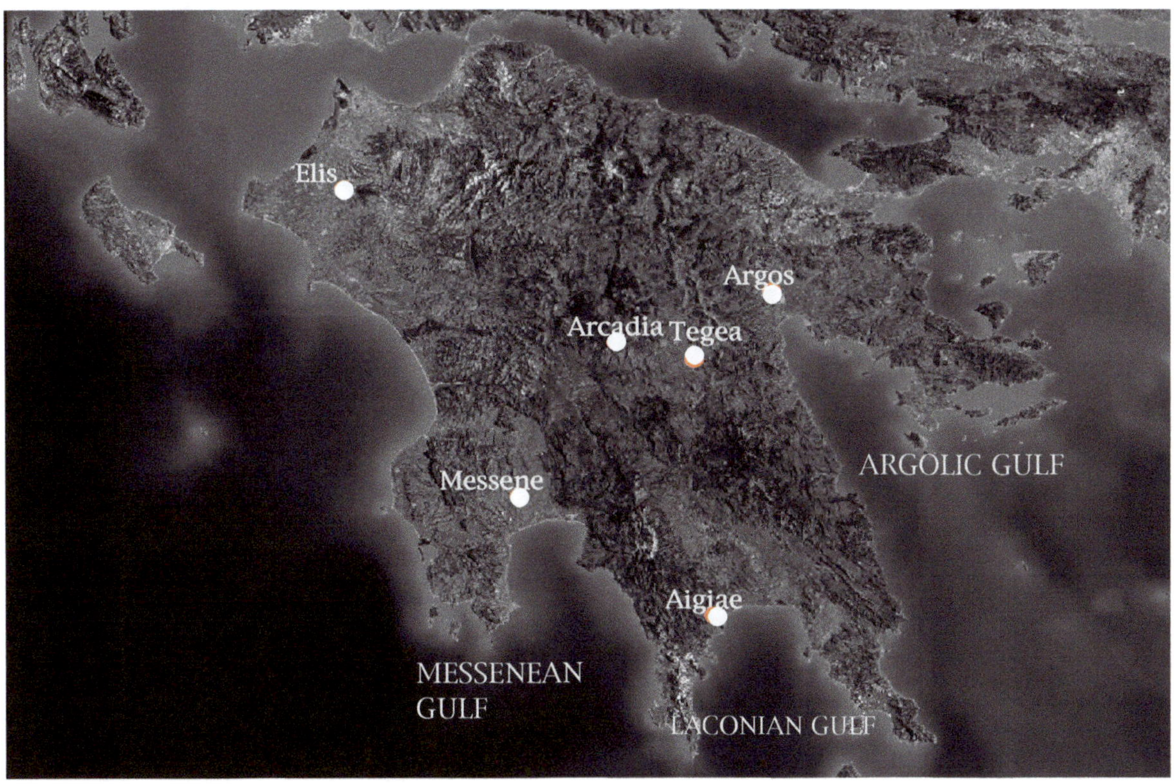

Fig. 5.7g Satellite image of Peloponesse (from TalentCruiser, www.cruiser.gr/en_main.html accessed 25 November 2006).

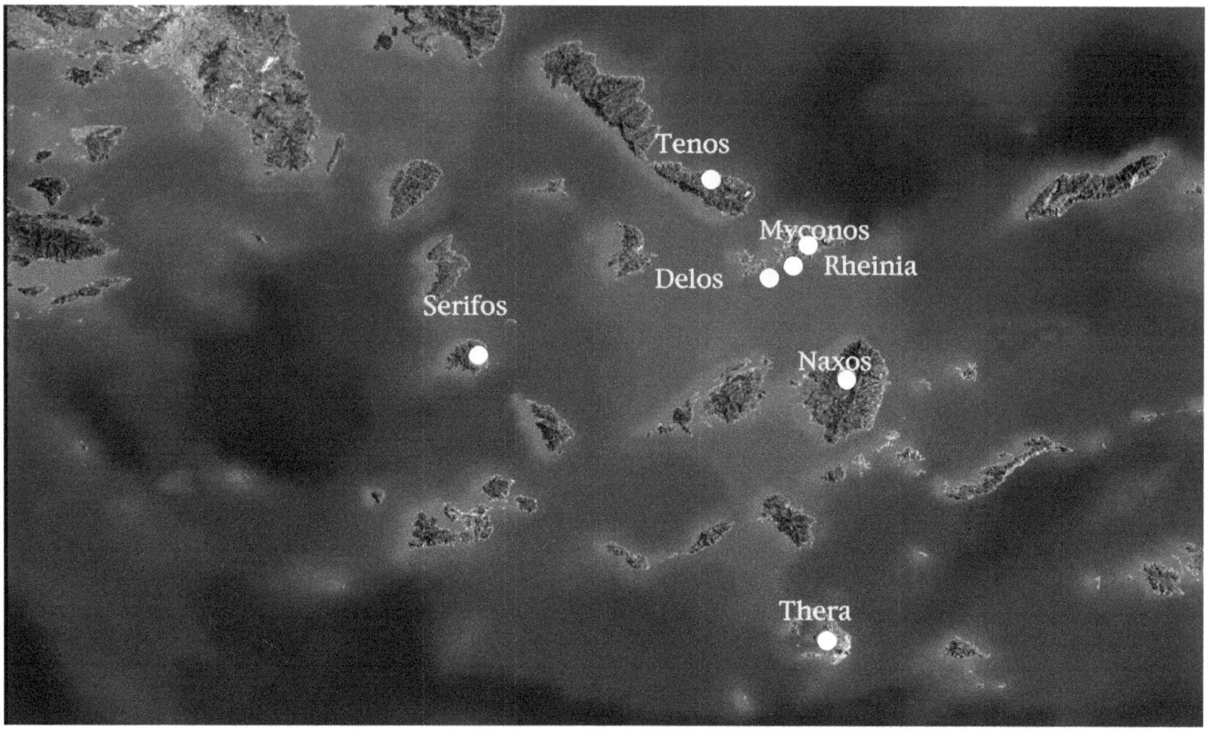

Fig. 5.7h Satellite image of Cyclades (from TalentCruiser, www.cruiser.gr/en_main.html accessed 25 November 2006).

Fig. 5.7i Satellite image of Crete (from TalentCruiser, www.cruiser.gr/en_main.html accessed 25 November 2006)

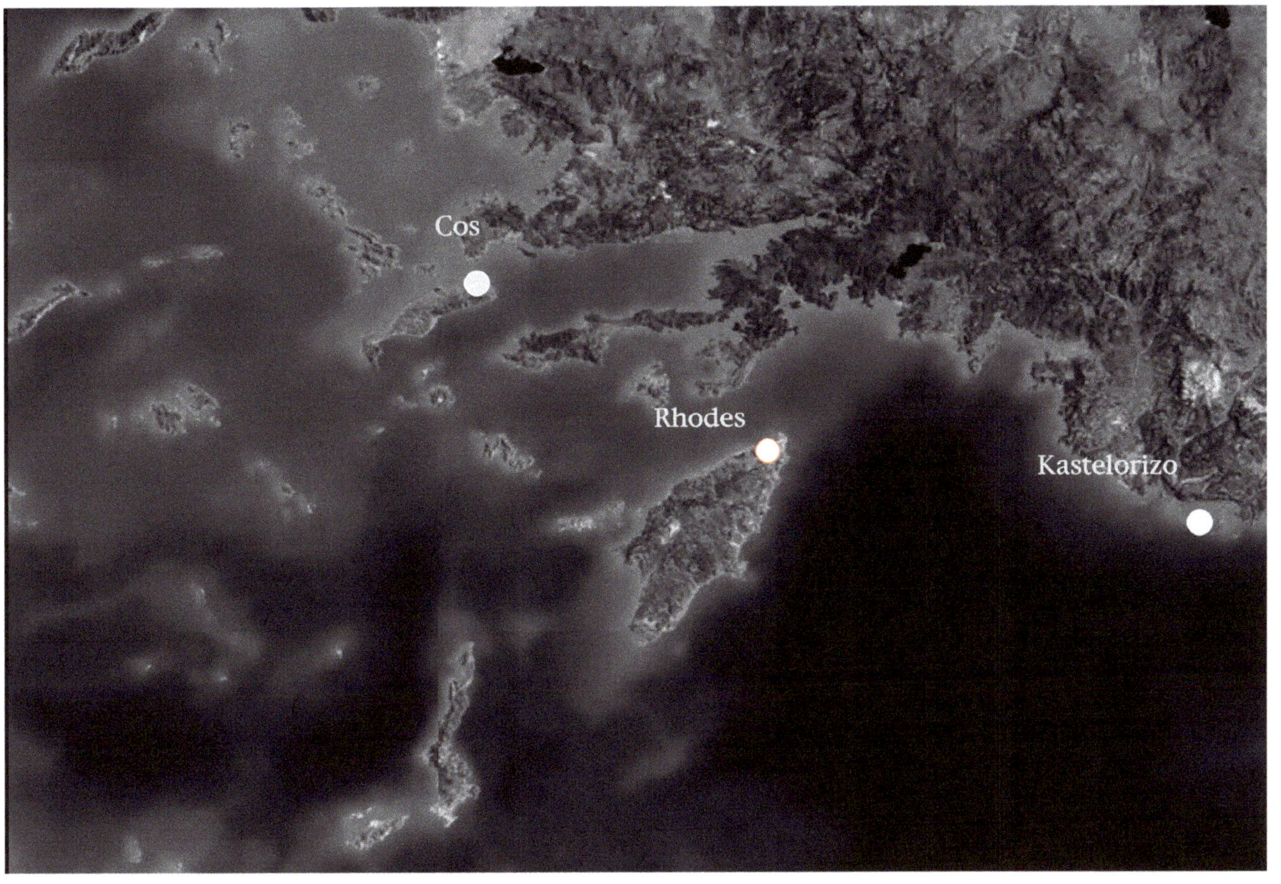

Fig. 5.7j Satellite image of Dodecanese (from TalentCruiser, www.cruiser.gr/en_main.html accessed 25 November 2006).

population had been diminished, due both to adverse climatological conditions and to over-fishing (Appendix 1: Arist. *Hist.an.* 603a-20-25).

A serious involvement with fishing on the islands of the northern Aegean is highlighted by the presence in various spots of fishing gear (Table 5.5 and Appendix 3a-b) and by the existence of legends involving deities (e.g. Appendix 1: Antikleides *Inversions* fr.7) or immense fish, as is the case of a story, recounted by Herodotus (end of 6th - beginning of 5th century B.C.). This was about a huge fish brought to king Polycrates of Samos as a present by a local fisherman, only for it to be discovered that it contained a valuable ring that the king had previously thrown in the sea in order to make it disappear (Appendix 1: Hdt. iii.41-43). In the case of Thassos Island the emphasis on the exploitation of the migrating fish is also echoed in the reputation attained by the island in Classical times, for its preserved fish products, which had been an object of trade (Appendix 1: Ar. *Ach.* 671). A reputation of fish connoisseurship of the Thassians is also attested by the fact that a special fish dish, which is known from sources several centuries apart, was called "Thassian pickle". This consisted of broiled small fish plunged into a pickle (Appendix 1: Ath. *Deipn.* viii 329b).

The coastal sites of Abdera, Olynthos, Torone, Pella, a coastal city in the 4th century when Archaestratos visited it (Struck 1908; Bintliff 1976; Astaras and Sotiriadis 1988) and Krania are those for which some published data are available. It is interesting that most of them are situated in highly eutrophic areas, such as large river deltas (Abdera, Pella, Krania) or recesses of rich bays with lagoonal conditions (Olynthos). For most of these sites, despite their extremely promising geographical position, the archaeological data are scant, but this is probably related to excavation and publication biases rather than the actual lack of data. Many north Aegean harbour cities were favoured by the gourmet traveller of the 4th century B.C. Archaestratos, who priced specific fish. According to him, Abdera had been an attraction for its grey mullet, obviously a product of the Vistonis lake or Lagos lagoon (Koutrakis 1999, p. 368), Olynthos for a mysterious, unidentified fish, named *glaukos*, Torone for its shark and Pella for is euryaline meager (*"chromis"*-Sciaenidae) (Table 5.3). The excavation of the first three sites has produced the remains of fishing gear (Fig. 5.8), which provide some material evidence to the cities' elusive ancient reputation (Appendix 3a). A Macedonian tomb at Foinika, one of the eastern suburbs of Thessalonike, which dates to the 4th century B.C., housed the remains of a wealthy aristocratic couple and at a later stage those of a less affluent individual. Among the prestigious objects common in such graves (remains of

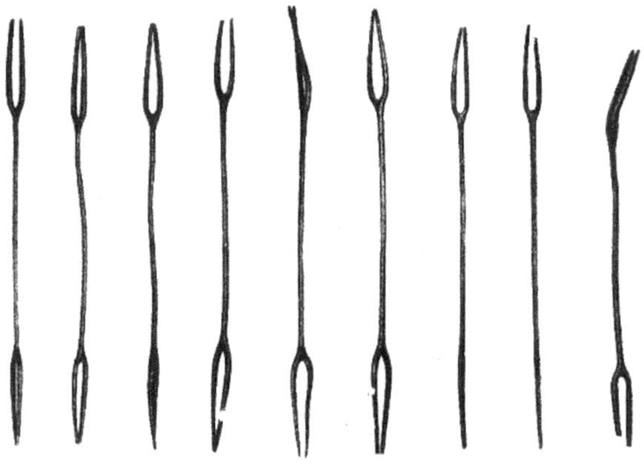

Fig. 5.8 Olynthos netting-needles (from Robinson 1941, pl. CXVI).

golden wreaths, wooden decorated *klinae* -narrow beds, etc. one fish-hook and a netting needle have been found. From the publication of the material it is not clear which burial they are associated with. Interestingly, however, in the same grave the mattresses of the *klinae* were filled with sea weed, common nowadays in the inner recesses of Thermaikos gulf. Two species have been identified, *Padina pavonia*, and *Ulva Lactuca* (Tsimbidou-Avloniti 2005, p. 42, 66).

Hellenistic Krania provides as yet the best case in which the exploitation of fish resources in such a northern Aegean coastal site can be understood. Krania, like many of the sites already mentioned is situated on the coast, with direct access to the rich fishing grounds of the Thermaikos gulf. Besides that, it is situated very near the old estuary of the Pineios River (Poulaki-Pantermali, pers. com.), one of the largest rivers of Central Greece, thus having access to the euryaline fish inhabiting the estuary, to the river fish of its lower flow and to the anadromous and catadromous fish moving to and from the river to spawn. The fish bone record from this site (Tables 5.1a and 5.6) indicates that all these resources were exploited. The large amount of fish bones found in relatively restricted areas and time span (see detailed discussion in chapter 8.2 and Appendix 4) indicates an intensive consumption of fish of various kinds. Small and large, seasonal and perennial, schooling or solitary, were all targeted. Judging by the fact that the vast majority of the Krania fish bones have been collected from water floated soil samples, and only a few by hand picking, Krania provides perhaps a good example of what the situation might have been like in other excavated sites around north Aegean, if appropriate collection methods had been applied during the excavation.

A third category of sites in this area of the Greek world are inland. These are situated near large rivers or lakes. No site provides any archaeological testimony of this inland fishing in the period under study. An extraordinary case is described by Herodotus and refers to the exploitation of lake fish resources by the Thracian lake dwellers of Lake Prasias, probably Lake Vistonis, near the estuaries of the River Nestos (Appendix 1: Hdt. v.16). This area is geographically close to the city of Avdera discussed above. According to Herodotus, Lake Prasias was so rich that fish were even used as fodder for the animals. Two species are reported by name, the *telones* and the *peprakes*, both unknown (see discussion of this case in chapter 6.3). The Strymon River, further to the west, was apparently famous for its large eel, which reached the Athenian market in a preserved form, and for its wels (Table 5.3). Another river, which in Aelian's time (end of 2nd-beginning of 3rd century A.D.) was called *Astreus*, probably the Axios River, was considered the place where the artificial fly had been invented for the catching of a specific fish (Appendix 1: Ael. 6.1).

Lake Volvi, at the northern borders of the Chalkidike peninsula, seems to have been exploited for at least one type of fish, the Danube bleak (*Chalcalburnus chalcoides*), which performed an annual small-scale spawning migration from the lake to its bordering rivers (Economidis and Sinis 1988). According to an anecdotal story by the 2nd century B.C. historian Hegesander, preserved in Athenaeus's *Deipnosophistae*, the inhabitants of the area, caught the fish in large quantities and preserved them for later consumption (Appendix 1: Hegesander, *Commentaries*). This appears to have been a small-scale exploitation, which only involved the locals. The practice of massive fishing was enmeshed in a ritual of ancestor- worship. Volvi, a lake goddess, granted rich catches of fish during two months of the year to those who worshipped her son, Olynthos. A late testimony by Strabo, however, refers to the *taricheia*, the fish preservation establishments of Lake Lycnydon, the present day Orchida Lake, in modern FYROM (Appendix 1: Strabo, 7.7.8), which testify to the possibility of a systematic exploitation of lake fish resources.

In conclusion, the north Aegean fisheries had the potential to be important in terms of both resource variability and richness. The available evidence suggests that these resources, both marine and fresh-water, were indeed exploited, often in a systematic manner, which was in some cases only of local importance (e.g. lake Volvi, possibly Krania). In other cases, the value of the resources was appreciated in distant markets. The export of preserved tuna from Thassos and eel from Srymon River, as well as the reputation for excellence of certain fish from specific sites in the area testify to this. However, much future systematic work is required for this evidence to be more precise and quantifiable.

Mainland West Coast and Ionian Sea (Fig. 5.7c).

This is one of the areas of Greece, where, despite the presence of very rich fish resources in certain areas, fishing is very poorly documented. The gulf of Amvrakia with its adjoining lagoons is one of the most productive marine ecosystems in Greece. Archaestratos of the 4th century B.C. is again our only literary source which directly testifies to any kind of systematic fishing in the area. He talks about the excellence of sea-bass, meager (*"chromis*", Sciaenidae) and the elusive *kapros* (Table 5.3). *Kapros* has been tentatively identified by Thompson (1947, pp. 101-2) as the wels (*Silurus* sp.), a fish inhabiting slow flowing rivers. If this identification is correct, then it was probably caught in a river emptying in the Amvrakikos gulf, perhaps River Arachthus, which passes by ancient Amvrakia, at the location of modern Arta. It should be noted here that the presence of this fish in a river of western Greece contradicts Aelian's statement that among European rivers (in his known world) wels can only be found in Strymon River (Ael. 12.14). Alternatively *kapros* could be the carp *(Cyrpinus* sp.*)*, perhaps named thus after its habit of feeding on the roots of water plants, in a manner similar to that of the wild boar (*kapros* in ancient Greek) that feeds on the roots of shrubs (Economidis, pers. com.).

The riches of the Amvrakikos gulf were exploited by Kassope, a city founded by Eleians just before the middle of the 4th century B.C., east of ancient Amvrakia. Excavations at Kassope have revealed remains of euryaline fish (sea-bass, gilthead sea bream, grey mullet) but also purely marine creatures, such as tuna, grouper and an indeterminate cartilaginous fish (Table 5.1c). Unfortunately, no contextual information is available, apart from the fact that the remains originate from domestic contexts (Appendix 4). It is interesting however, that although the remains of euryaline fish are relatively numerous, those of the marine fish are very few (one of each species). This might be an indication that they may belong to fish obtained through trade, and not through direct tapping of open sea fishing grounds. Furthermore, Plutarch in the 1st – 2nd century A.D. deems one of the symposiasts, in his *Table Talks*, Symmachus, to be a suitable defender of the superiority of sea food to land food, because he is from Nikopoli (Appendix 1: Plut. *Quest. conv.* 667-669). Apparently, the city, being near the Amvrakicos gulf, provided its inhabitants with ample opportunity to accumulate a thorough knowledge of the issue of sea-food and fish in particular.

The only other data available on fishing in the northern part of the Ionian Sea, during the centuries under study refer to the island of Corfu (Corcyra). According to a story cited by Pausanias, the Corcyreans, some time in the 5th c.entury B.C., dedicated a bronze bull statue to Apollo at Delphi and sacrificed a bull to Poseidon because an oracle helped them find a way to catch a multitude of tuna off their coast (Appendix 1: Paus. ix.9.3-4). This, and the single tuna bone from Kassope are the only evidence for the exploitation of the migrating fish in the area (for the Corcyrean case see Purcell 1995, p. 139).

Sterea Ellada (Fig. 5.7d).

Sterea Ellada, the part of mainland Greece, just north of the Corinthian gulf and south of the Thessalian plain, is characterized by its access to some of the richest seas of Greece, namely the Corinthian gulf and the gulf of Euboea and by the dominating presence, until the beginning of the 20th century of Lake Kopais, an extensive shallow lake, which has now been completely drained. The sites that provide evidence for the exploitation of fish resources in mainland Sterea are all located on the coasts of the famous Lake Kopais, the source of eel with an immense reputation, at least in Athens (Appendix 1: Antiphanes, *The sheep owner*, fr. 191; Archaestratos in Ath. various lines; Ar. *Ach.* 884cf, 894; Ar. *Pax* 1013-4; Paus. iv.24.1; Table 5.3). It appears that the exploitation of this resource had an acknowledged antiquity and a local importance. This was reflected in a peculiar Boeotian religious custom, according to which eel were sacrificed to the Unknown Gods (Appendix 1**:** Agatharchides, *European History*), when fish sacrifice to the gods was an uncommon phenomenon in Classical Greece (see chapter 8.1.2).

The lake was exploited not only for eels but also for other fish, some of which are named on an inscription featuring a Hellenistic fish price list at Akraephia (Feyel 1936; Vatin 1971), a small town on the eastern coast of the lake. These fish were part of a broader economic regime which also involved the exploitation of other wetland resources such as reeds (Appendix 1: Antiphanes *A Lady's maid*, fr. 127 and Theophr. *Hist. pl.* iv.x.1-2, vi.x.6; Strabo 4. 325). The establishment of coastal villas in the area, which is indirectly attested by a 4th century source (Appendix 1: Antiphanes *Adulterer*, fr. 159) probably reflects these interests. Along the same lines, the presence of two border stones, one of the 6th-5th and the second of the 3rd century B.C. (Appendix 2: IG VII, 2792; SEG XXX 440) which settle the borders between Kopais and Akraephia, both settlements on the coast of the lake, also highlight the interest of the locals in defining the exploited grounds (rich pastures on the coast of the lake, fish, other resources).

These inland towns also had access to marine fish, both coastal and migratory. The Akraephia price list refers to various such fish, which represent coastal fishing, fishing of migratory schools and in brackish waters (Table 9.1). Access to the sea for fish appears to have been also an option for Thebes of the end of 4th century (Appendix 1, no. 61) and probably for Delphi as well (Table 5.1c).

Euboikos and Pagasitikos gulf (Fig. 5.7b).

The sea in this area is among the most eutrophic of Greece. Again, the richness of the waters in fish is not reflected in the quantity of data, about fishing, available to us. These are mostly literary reference to local myths or comments on the excellence of local fish. The coastal waters of these seas were certainly exploited, as were the migratory fish, which even in recent times are particularly abundant in these waters (Athanassopoulos 1926; Zappas 1968; Guest-Papamanoli 1984). Plutarch, in the early 2nd century A.D. comments on the riches of the waters around Aedypsos, in northern Euboea

(Appendix 1: Plut. *Quest. conv.* 667) and Archaestratos talks about the abundance of fish around Karystos on the southern part of the island a few centuries earlier (Appendix 1: Archaestratos in Ath. vii. 304d). In Classical times Chalkis, Karystos and Eretreia, could all boast of certain fish of excellent quality (Table 5.3). Anthedon, on the Boeotian coast, a few kilometres west of Chalkis was a harbour town which was reputed to be the hometown of the sea-god Glaukos (Appendix 1: Aesch. *Glaukos Pontion*; Paus. ix.22.5). According to the 4th century B.C. geographer Heraklides Kritikos, the inhabitants of Anthedon turned to fishing because their land was too poor to support them (Appendix 1: Heraklides kritikos *Description of Greece*).

It would be expected that excavated sites in this area would produce large numbers of fish and selfish remains. The excavation at Hellenistic New Halos, on the western coast of Pagasitikos Gulf, is interesting in this respect (Table 5.1c). The town had a military character and was situated near a brackish marsh. The residential district of the town was probably occupied by solders and their families (Reinders 1988). The excavation of six of these dwellings revealed a modest number of molluscs (NISP: 504) and only one fish bone (Prummel 2003b, p. 200). Whether this is the result of the admittedly restricted water-sieving applied during the excavation or reflects a deliberate choice on the part of New Halos residents can not be determined at this stage.

Corinthian gulf (Fig. 5.7e)

The Corinthian Gulf is a closed sea, on the coasts of which several important ancient settlements of both Peloponnese and Greek mainland are located. It is a fairly eutrophic area (Fig. 5.2) which supports a wide range of fish species (Kaspiris 1973). Despite these facts, very little is actually known about its exploited fish resources. Marine inshore and offshore fish were certainly caught, as becomes evident from the fish bones found at Corinth (Table 5.1b and 5.6). The small sea bream found in Classical deposits at the Sanctuary of Demeter and Core, whether fresh or preserved, could be caught locally, while the sea-bream and tuna remains found in the Classical Punic Amphora building (Williams 1978; Zimmerman-Munn 2003), are apparently imports and will not be discussed here. The exploitation of the available fish resources are more substantially documented for Roman Corinth. Bones of coastal fish in a Roman cistern fill and from a 2nd century A.D. butcher's shop which also combined some cooking and selling of food, are the only ones available at present, but the importance of fishing in the daily economy of the city in Roman times is further attested by the existence of a specialized fish-market, in the agora (Appendix 3b), and the abundant remains of fishing gear (Appendix 3a), which are mostly late in date (Roman and Byzantine).

Euryaline and fresh-water fish resources were also exploited, in some cases systematically. The eels from Sicyon, on the northern coast of the Peloponnese, had a reputation in Classical and Hellenistic Athens (e.g. Appendix 1: Philemon *Soldier*, fr. 82). Also, the fish from Lake Kalydon, which was probably one of the coastal lagoons, which form the Messolongi lagoon complex, were systematically managed by fishermen, belonging to Roman Patras (Appendix 1: Strabo 10.2.21), after the forced abandonment of Kalydon and its territory by its inhabitants (Rizakis 1997).

No evidence is available for the exploitation of small schooling marine fish or of the migrating fish. This however, may be a result of poor research rather than a narrow focus of past fisheries to a few resources only.

Eastern coasts of Attica and Attica indeterminate (Fig. 5.7f)

The exploitation of fish resources in this area is little known. Our only data is a reference to the excellent quality of whitebait in the area of Marathon (Table 5.3). Some of the fish, the remains of which have been found in Athenian contexts (Table 5.1f), and the large number of fish in the Athenian market, could in theory have been caught in these waters.

Saronic gulf (Fig. 5.7f)

The Saronic Gulf, enclosed by the western coast of Attica and the north-eastern coast of the Peloponnese, is a relatively rich sea in terms of nutrients. Geographically it was also one of the more densely inhabited areas of ancient Greece, with major centres all along its coasts and its islands. It was also, in the period under consideration, heavily used by naval traffic of every sort.

Fishing in this area was probably quite intensive. The small schooling fish, such as the picarel, white-bait, but also small squid and other coastal creatures were apparently a major exploited resource, and Athenian dishes made of these fish had attained some reputation in the ancient word (Table 5.3, see chapter 9.2.2). Remains of fish belonging to this group have been found in the Hellenistic strata in the sanctuary of Poseidon in Kalaureia (Table 5.1a; Table 5.6).

Some other inshore fish, a flat fish (*psitta*) from Eleusis, the mysterious *glaukos* from Megara, and the red mullet, sting-ray and saddled bream from the coasts of Halae Aexonidae, near present-day Voula (Table 5.3), were also famous catches from this type of fishes. This reputation, which does not originate solely from Archaestratos, as in the case of most fish from elsewhere in Greece, but from a number of different sources, probably reflects the situation fairly accurately. Fish bones found in Classical, Hellenistic and Roman deposits in the Sanctuary of Poseidon on Kalaureia (Table 5.1a and d; Table 5.6) give some material evidence for the catching of a wide range of inshore species. It is, however, interesting that fish with more offshore character such as the sharks and some of the rays and skates are absent from this site. An emphasis on the exploitation of inshore fish can also be inferred from the existence and function of at least one complex of six fish tanks at the harbour at Kenchreai, the eastern port of Corinth (Appendix 3b, Fig.7.4; see chapter 7.1.2).

Further material evidence on the exploitation of coastal fish resources is provided by the remains of fishing gear recovered from excavations at the Sanctuary of Poseidon at Isthmia. Concentrations of fishing tools, such as lead net weights, fish hooks and one iron fishing spear (Appendix 3a, Table 5.5), all offerings to the god, testify to the presence of fishermen in the area. More such fishing implements have been recovered at the so-called "Priest's House" mentioned earlier, located on the coast, near the Temple of Apollo Zoster, in the deme of Halae Aixonidae in Attica (Appendix 3a, Table 5.5). There, there has been found a concentration of lead net weights of various types, fish hooks, netting needles, along with one obsidian blade and several clay loom-weights, which could also be part of the same fishing tool kit.

Halae Aexonidae is also a place which was involved in the fishing of migratory fish. The first tuna caught every year was offered to Poseidon (Appendix 1: Crates *Sacrifices at Athens*), apparently in the spirit of "first fruits" offerings, a ritual quite common in Ancient Greece (Rouse 1976, pp. 39-44; Burkert 1985, pp. 66-8). The formalization of this behaviour, through ritual, indicates a well established fishing activity. The systematic exploitation of tuna is also attested by a Hellenistic inscription (Appendix 2: IG 4[2] l. 76-77) which regulated the management of the income generated by the tuna catch from the common waters between Troizen, on the Peloponnesian coast, and a neighbouring, unspecified city (probably Arsinoe/Methana). Such a formal agreement, again testifies to a well-organized and sizeable activity. Some fruits of this type of activity, tuna and bonito bones have been found in the Sanctuary of Poseidon at the nearby Kalaureia (Table 5.1a; Table 5.6).

Euryaline fish were also targeted by fishermen in the Saronic Gulf. Pausanias recounts that the fishing rights in the sacred Rheittoi River were reserved for the priests of the Sanctuary of Demeter and Core at nearby Eleusis (Appendix 1: Paus. i.38.1). In the modern era the Rheittoi River is but a stream connecting the sea to a spring-fed lagoon. In the 1830's soldiers from a nearby army camp were fishing for grey mullets there (Levi 1971, vol. 1, p. 106, no 226, Rose 2000, p. 523-4). Given the europhization of lagoonal environments, the reservation of fishing rights for the priests makes more sense, as it might represent a considerable income. The Kalaureia fish-bone assemblage (Table 5.1a) also provides evidence of fishing in brackish waters in the form of grey mullet and eel remains. Such coastal environments abound today on the Peloponnesian coast across Kalaureia.

Argolic gulf (Fig. 5.7f)

The Argolic gulf appears to be fairly rich in nutrients. Little specific information is available for this geographical area. According to a playwright of the 3rd century B.C. the Argolid was famous for its boar-fish (Table 5.3). A runner, winner at the Olympic Games some time in the 5th or 4th century B.C. (Appendix 1: Arist. *Rh.* 6.55), is said to have transported on his shoulders fresh fish caught near Argos to inland Tegea (perhaps as a form of training). The large Serranidae vertebra found in the Late Roman farmhouse at Pyrgouthi, in the Berbati valley of the inland Argolid (Table 5.1d) is probably also a product of the inshore fisheries of the Argolic gulf. Migratory fish were also caught in the Argolic gulf. Here too, at Halieis, according to a 3rd century B.C. source, as in the deme of Halae Aixonidae in Attica, the first tuna caught every year was offered to Poseidon (Appendix 1, no. 66). The fishing activity at Halieis is further attested by hoards of several dozens of fish-hooks (Fig. 5.9) found in the lower town and several fish-hooks and lead objects, probably net weights, found on the Acropolis of the town (Appendix 3a; Table 5.5). These objects can be dated to the historical period but not more accurately. The picture is further enhanced by the recovery at the now submerged Sanctuary of Apollo at Halieis, of two fish bones, one belonging to a gilthead sea-bream (Table 5.1d; Table 5.6). The date of the unidentified bone is mixed (6th-5th century B.C. to Roman) while the sea bream is dated to the 5th century B.C.

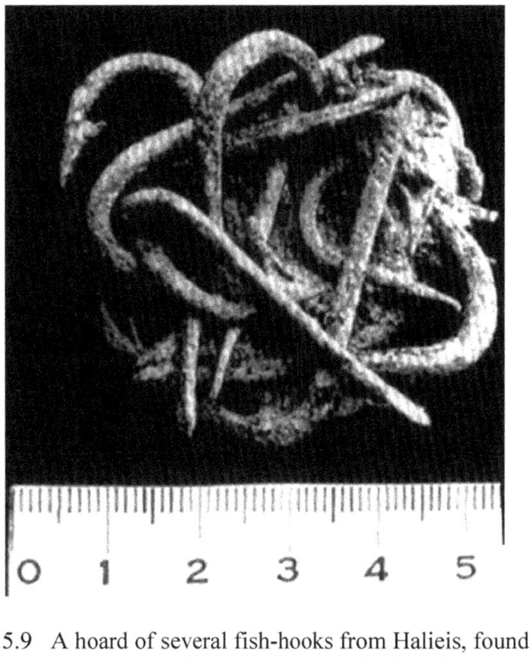

Fig. 5.9 A hoard of several fish-hooks from Halieis, found at the Lower Town (from Jameson *et al.* 1994, fig. 5).

The fishing activity in the Argolic gulf was variable. The harbour town of Hermione was already famous around 600 B.C. for the purple dye produced there (Jameson *et al.* 1994, pp. 316-319). The focus of the Hermionian culture upon the sea and its exploitation is also reflected in a local religious custom: a diving competition taking place during the celebration of the festival of *Dionysos Melanaigis* (Dionysos of the dark goat skin) (Appendix 1: Paus. ii.34.9. ff) (Burkert 1983, p. 211).

Table 5.6 Geographical distribution of archaeologically or literarily attested fish resources.

	Sites	Inshore-Offshore	Marine schooling	Marine migrating	Euryaline	Anadromus-Catadromus	Fresh-water	Exotic	Indeterminate or generic
Northern Aegean	Avdera				+				
	Olynthus								+
	Lake Volvi					+			
	Strymon river					+			
	Torone	+							
	Pella	+							
	Krania	+	+	+	+	+	+		
	New Halos, Thessaly								+
	Thassos	+		+					
	Samothrace	+							
	Mytilene	+		+					
	Chjos				+				
	Heraion, Samos	+							
	Samos indet.	+		+					+
Northern Ionian Sea and islands	Corkyra			+					
	Kassope	+		+	+				
	Amvrakia	+			+		+		
Corinthian Gulf	Delphi								
	Bulis	+							
	Calydon				+				
	Corinth	+		+				+	+
	Sicyon					+			
Sterea mainland	Kopais lake, Boeotia					+			
	Thebes		+						
	Akraephia, Boeotia	+							
Euboic Gulf Central Aegean	Anthedon								+
	Karystos		+	+					
	Eretreia	+							
	Chalkis								+
	Aedypsos								+
Eastern coasts of Attica	Brauron, Attica	+							
	Marathon, Attica		+						
Attica indet.	Agora, Athens	+							
	Athens, Attica		+						
Saronic gulf	Halai Aixonidae, Attica			+					
	Phalero		+						
	Rheitoi River						+		
	Eleusis	+							

Table 5.6 (Continued)

	Megara								+
	Salamis		+						
	Kalaureia, Poros	+	+	+	+	+			
	Troizen, Argolid			+					
Argolic gulf	Halieis, Argolid	+		+					
	Argolid indet.	+							+
	Hermione	+							
	Pyrgoudi, Argolid	+							
	Argos								+
Laconian gulf	Messene	+		+					
	Aigaiae				+				
Peloponnesian mainland	Arcadia						+		
	Pharae, Achaia						+		
Cyclades	Thera	+		+					
	Naxos								+
	Tenos								
	Delos	+			+				
	Myconos								+
Dodecanese	Rhodes indet.	+				+			
	Cos			+					
Crete	Knossos, Crete	+?							
	Eleutherna, Crete	+		+				+	
	Kommos, Crete	+	+		+				
	Itanos, Crete	+	+	+					
	Stalis, East Crete								+

This table is constructed on the basis of data included in Table 5.1a-g, Table 5.3 and Appendix 1 where the relevant literature is cited.

Messenian gulf (Fig. 5.7g)
This gulf appears to be quite rich in nutrients at its recesses. The only fishing evidence available originates from Messene, an inland city. The few fish bones, found in different spots across the town indicate both inshore fishing and the exploitation of migratory fish (Table 5.1e and 5.6). The paucity of data in the area is probably related to the low number of published excavations.

Cyclades (Fig. 5.7h)
The Cyclades are located in an oligotrophic area of the Aegean, but nevertheless have been known in recent times for their fishing tradition, of both the local resources and of the passing schools of migratory fish (Bintliff 1977, pp. 118-9, 594-595; Pennewise 2004). In Hellenistic Thera (Santorine) tunas and several large inshore sea-bream (Sparidae) were brought and consumed at the Sanctuary of Zeus, inland (Table 5.1f and 5.6).

Tenos and Delos are both mentioned in Archaestratus's list of sources of excellent fish such as the unidentified *levias*. Furthermore, the Delian coasts, and those of the neighbouring Rhenia and Myconos were systematically exploited for purple shells (Homolle 1890, pp. 455). An incident of hostility between fishermen and their urban customers which developed around the landing of a fish catch on the coast of Naxos, as described by Aristotle (Appendix 1: Arist. *The Constitution of Naxos* in Ath. viii 348b) is the only evidence of fishing on this island at present.

Inscriptions and literary references provide some evidence of the exploitation of euryaline and/or fresh-water resources in the Cyclades. Delian inscriptions refer to the taxing by the sanctuary of Apollo of the fish caught in the sacred lake (Appendix 2: Homolle 1882, 1890). The sacred lake referred to here is either a pond or a brackish-water one which provided breeding grounds for prolific fish resources such as grey mullet, eels etc. The Delian sacred lake will be discussed in detail in chapter 7.1.3.

Dodecanese (Fig. 5.7j)
Very little evidence for fishing exists from the area of the Dodecanese, in the period under study. The few cartilaginous fish vertebrae reported by Reese on Rhodes Island (1984) are either of uncertain date or Archaic. This lack of fish data is probably related to the situation of the archaeological research in the area rather than to a disinterest in fishing in antiquity. Archaestratos' list of excellent fish mentions the dog-fish (another cartilaginous creature) from Rhodes as well as the "*antakaios*", which is probably a sturgeon (Table 5.3). The fishing of this anadromous fish was probably done at sea, as the fish were moving to a river on the nearby Asia Minor coast. The exploitation of fish resources is further attested by the presence on Kalstelorizo Island (ancient Megisti) of two rectangular vats, possibly used for salting (Appendix 3b).

At Cos the migratory fish were exploited in a systematic manner and "*thynnia*", i.e. permanent net arrangement and watching towers, were formally rented (or taxed?) (Appendix 2: SIG[4] 1000; Table 5.6). The fish element on Cos is further accentuated by the evidence of purple shell trade, and probably production on the island which is connected to the silk industry, especially in the Roman (Imperial) period (Sherwin-White 1978, p.242).

Crete (Fig. 5.7i)
The exploitation of fish resources appears to be very well documented for Crete, an area, within the most oligotrophic waters of the Greek seas and having the fewest inland water bodies. This is perhaps a reflection of the intensive archaeological work, which has taken place on the island for over a century. Some of the more systematically collected fish-bone assemblages of a historical date (e.g. Kommos, Itanos) originate from here. Fishing related issues are much better documented for prehistoric sites, such as Kommos, Palaekastro, Mochlos and Pseira (Rose 1994; 1995a; 1995b; 1998; Mylona 2005; Mylona in press) thus providing some historical perspective on the issue of the fish exploitation. Most of the data at hand originate from the Eastern part of the island, where archaeological activity is more intensive.

Coastal fishing was predominant on Crete. At Late Roman Itanos, on the East coast of Crete, both inshore species and small coastal schooling fish such as the picarel are almost the only types of fish caught and consumed in a domestic context (Table 5.1a; Table 5.6). At Hellenistic Kommos the situation is similar (Table 5.1a; Table 5.6). Fish of this category have also been found in Late Roman Eleutherna, an inland site of Central Crete and at Knossos, at the Sanctuary of Demeter (Table 5.1g; Table 5.6).

The systematic targeting of inshore fish is also reflected in the presence, at several locations on the island, of a number of fish tanks, of a type known from Roman sources (Appendix 3b). Fish tanks at Mochlos, Siteia, in the Ierapetra region and probably at Zakros and at Phalasarna on the North Western coast of Crete are all considered, without any clear proof, to be Roman in date. These are constructed on the rocky shore, at wave line and are said to have been used to keep certain coastal fish alive, after they are caught in open sea. Alternatively they could be used as breeding places for certain fish. Judging by the location and physical characteristics of the Cretan fish tanks, their function was probably connected to the fishing of inshore rock fish from the vicinity of these locations (for further discussion on the character and function of these fish tanks see chapter 7.1.2).

It is interesting that the fish-bone assemblages from Crete are much poorer in variety of cartilaginous fish, with sharks and dog-fish, the large predators, being almost absent. In the Cretan cases, stings and rays (Dasyatidae and Rajidae) appear to be more common (see Table 5.1a and g). The migratory fish seem to have been exploited very lightly. Itanos provides some evidence for the fishing of the smaller Scombridae (mackerel) while at Hellenistic Kommos no remains of this group of fish have been found. The only other occurrence of migratory fish and more specifically of tuna is at Late Roman Eleutherna (Table 5.1g). In that case however, the fish could have reached this inland city in a preserved form originating elsewhere.

The euryaline fish, and specially the grey mullet (Mugilidae) and to a lesser degree the sea-bass (Moronidae) and the most versatile of the sea-breams (Sparidae) are also present on Cretan sites (Table 5.1a and g; Table 5.6), providing some indication that local brackish micro-environments were probably exploited also.

The importance of the exploitation of marine resources on Crete is further supported by the epigraphical evidence. A treaty which regulated the obligations of Stalis to Praessos, on the Eastern part of the island specified the taxation on fishing (Appendix 2: SIG[4] 427), thus indicating that the income generated by the exploitation of marine resources was considerable. This income could however not rest exclusively on the exploitation of fish but also of the purple shell, which appears to have had a long tradition in this area. Korobios, the purple shell fisherman from Itanos, led, according to legend, the colonists of Thera to Cyrene, on the northern coast of Africa, apparently exploiting his knowledge of marine routes (Appendix 1: Hdt. vi.151 and Mylona 2003b). The Leuke Island, off the south-eastern coast of Crete, is also said to be a place where purple dye was systematically extracted, and purple-shell beds have been recovered archaeologically along with a number of fishing implements (Appendix 3; Appendix 4 and references therein).

The above review aimed at establishing that the use of aquatic resources was widespread in Greece between 500 B.C. and the end of Roman period (ca 7th century A.D.). The same review however indicates, at least in some cases, that this exploitation was also intensive and of considerable scale. The inclusion of fish in formal arbitrations, the strict organization of the fishing of migratory fish as this is indicated by regulations inscribed on stone, and the connection of fishing to cult, point to systematic fishing, or at least to an activity which had some permanence. In almost all these cases we are not at present able to elaborate on the details or the scale of the organization of fishing. Such a task would require further focused field research.

5.4 Conclusions

This chapter brought together all available data, archaeological and literary, on the fish resources exploited in the Hellenic area in the centuries between about 500 B.C. and 7th century A.D. It also brought forth a large amount of related evidence, mostly technological, but also linguistic, along with myths and legends. Greece has been viewed not on the basis of modern geopolitical divisions, but through the divisions dictated by the nature of its aquatic resources. This novel perspective, along with the accumulation of data, revealed an interesting picture.

The first point to be made is that fishing took place almost everywhere in Greece, by the sea and inland. The exploitation of aquatic resources was not closely linked to the general richness of the water or the availability of great masses of fish. The example of Crete, where fishing appears to have been quite important despite the oligotrophic nature of its waters is illuminating. The uneven relevant archaeological research in areas with rich or poorer marine resources precludes any reliable conclusions regarding differences in the intensity of exploitation between areas of differing fish availability. Marine fish were not the only ones caught. All possible fishing grounds (sea, rivers, lakes, lagoons) were exploited, often intensively. It appears that the vast variability that is inherent in Greek land- and seascapes and in the fish resources was appreciated by local communities, and fishermen made the most of it.

The above review makes it clear that fish was part of the diet of people in all parts of Greece, both coastal and inland. The rich scatter of evidence for a wide variety of fishing modes employed in the area of Greece between 500 B.C. and 7th century A.D. cannot adequately illuminate the intensity and importance of this fishing locally. This is mostly because the majority of the available data are random, non quantifiable, and often out of context, thus precluding any meaningful comparisons and generalizations.

A systematic character of fishing in certain areas can however be assumed on the basis of indirect evidence. The data on the existence of the organized production and marketing of preserved fish presupposes organized fishing that would support this enterprise. The Thassian preserved fish for example (Appendix 1: Ar. *Ach.* 671), would not have reached Athens and recorded in sources as a distinctive product, had it not been systematically produced. The same importance can be inferred in the cases where fishing had been included in state treatises and public agreements. The tax on fishing for example would not have been included in the treatise between Praessos and Stalis on Crete (Appendix 2: SIG[4] 427), had it not been sizeable and systematic enough to represent a considerable income. Furthermore, the importance of fishing locally may be indirectly assumed by the fact that at certain sites there is evidence that suggests a percolation of fishing-related ideas into other aspects of life, besides food procurement.

Fishing-related data (fish bones and fish-related technology) in some places often co-exist with elements of a sea-oriented culture, i.e. special language, legends, cults, where fishing appears to be prominent. In cases where all these types of evidence co-exist, we might assume that we are dealing with the remains of maritime or fishing cultures, with a possibly distinct identity and way of life. The foundation story of Phaselis for example, a coastal town at the bay of Antalya on Asia Minor (Appendix 1: Hierophytos *Cronicles of Colophon;* Philostephanus of Cyrene *On the Asian Cities*), presupposes such distinct spheres. The pastoralist Kylabras is said to have parted with a large piece of his coastal lands (where the colony of Phaselis was established) in exchange for preserved fish. Preserved fish in this story appears as a typically marine and alien product that represented a different culture, familiar with the sea (Purcell 1995, p. 144). Such maritime cultures have not been identified and studied as such in Greek archaeology. The evidence on the exploitation of fish resources that have been presented here might provide a starting point for this type of research. The following chapter goes some way towards analyzing this issue.

CHAPTER 6

THE INVISIBLE FISHERMEN

A discussion of fish resources would perhaps be deficient, without reference to the fishermen (here the word is used in a generic sense with no reference to gender). Fishermen are the agents who transform a natural resource, the fish, into food to be consumed. An understanding of the fishermen, their role and their worldviews is crucial, because their abilities and choices determine, to some extent, which resources will be exploited and to what degree. Yet, the ancient Greek fishermen are very little known, both archaeologically and literarily, despite the prominence of their produce in literature. The fishermen, their craft and the materiality of their life are not central issues of this book. No detailed analysis of the relevant data therefore will be attempted here. This chapter focuses mostly on matters located at the interface between fishermen and fish consumers. It also focuses on the interface between land and sea (or other water bodies). As fishing in ancient Greece was mostly coastal (chapter 6), the coast (of sea, lakes and rivers) and the waters near it is perhaps the place where the world of the fishermen could be found.

In order to fill the gaps left by the poor representation of the fishermen in both the literary and archaeological record, anthropology will be employed to shed some light onto some crucial characteristics of the fishermen and their world. Subsequently, an attempt will be made to trace some of these features in the existing literary and archaeological record. By looking at issues such as the names given to fish by fishermen and consumers, or to the different ways in which fishermen and land-bound people viewed the aquatic environment and related to the fish folklore, I will attempt to explore some aspects of the fishing mentality. A treatment of issues such as the above may be useful in explaining the invisibility of the fishermen and its implication for the status of fish as food.

6.1. What is a fisherman?

The fisherman, in the tradition of Classics is a shadowy figure, destitute, weatherworn, always at the margins of society (Fig. 6.1; Richardson 1969, pp. 170-172; Laubscher 1982; Pollitt 1994, pp. 185-191); almost a total stranger. The fisherman has no voice. Educated elite speak for him or instead of him sometimes. In this context a literary *topos* is repeatedly employed, which spans several centuries of literary production. Inscriptional evidence is a more eloquent source of information about the social and economic status of fishermen (Appendix 2: Mentzou 1975; OGI 496; IEphesos, V 1503; IKyzicos, I 211; Frisch 1983; SIG[4] 1000). Archaeology on the other hand, despite its ability to locate the traces of the fishermen's existence through the remains of their living (settlements), craft (fishing instruments) and produce (fish) has done little to illuminate this shadowy picture.

Fig. 6.1 Marble statue of an old fisherman, Roman copy of a 2nd-1st c. B.C. original, Palazzo dei Conservatori, Rome (from Pollitt 1944, fig. 156).

It is a conviction of this book that this above described phenomenon is part of a more general lack of knowledge and understanding on the part of archaeologists and Classicists more generally, of the world of the fishermen, be it technical, social or ideological. Modern scholars followed the ancient stereotypes and imagined the ancient fisherman to be actually a "hungry farmer", someone who turns to the sea when there is no other source of living available. This generalized model takes several forms (see also discussion in chapter 2.2). Whole communities turn to fishing because their territory is too poor to support agriculture (e.g. Appendix 1: Hyp. *Against Arastagoras*; Strabo 6.1.5 and 43.3.5). Anecdotal stories such as these are repeatedly recounted in modern discourse (e.g. Dalby 1995, p. 409; Purcell 1995). Alternatively, individuals are thought to have become fishermen because they were too poor to be anything else. Finally, hungry farmers are said to have turned to fishing (and seafood gathering) in order to survive during hard times or to broaden their subsistence base thus achieving security (see section 2.2).

All these variations of the basic scenario assume that the fishermen possessed a basically land-bound farming mentality. Fishermen are viewed as isolated units. They are hardly ever perceived as parts of a fishing community with distinct and dynamic rhythms of life and worldviews. Furthermore, this scenario views fishing as a predicament, as a last resort, not a choice or something one might be content with or take pride in. Fishing is viewed as something one would wish to get out of. In all cases, the focus is placed on a supposed inadequate outcome of the fisherman's effort, which in any case is evaluated on the basis of an arbitrarily defined optimal standard (see chapter 2.2) and with minimal understanding of the physical and technical parameters of the act of fishing itself.

Fishermen all over the world, whatever the medium they work in, the type of technology they employ or the cultural context in which they act share some basic characteristics, some features that set them apart from people working on the land: farmers, herders, hunters or even industrial workers. These distinctive characteristics are best encapsulated by Acheson in a review article on the anthropology of fishing (1981) and by Blount in a review of several related current works (2005).

According to a large number of studies oriented towards fishing communities, what sets fishermen apart is the fact that they work in a highly dangerous environment, not compatible to human biology. They can only get into the water, where fish can be found, by help of technical means. Their catch is for the most part invisible beneath the water's surface. The building of a body of knowledge about their prey and its habitat can only be done indirectly. Fishermen need a whole set of very specialized knowledge to be able to practice their art, and this knowledge is unique to the mariners. Meteorological, ecological and biological knowledge, navigational skills, ability to manufacture and maintain their own tools and equipment, as well as the fishing skills are some broadly defined categories of this specialized knowledge (Morrill 1967; Acheson 1981; Pálsson 1994a). The fish world may serve as a vehicle for symbolic thought and traditional fishing knowledge may also reflect people's association and connections with the spiritual world (Pálsson 1994b). Demarcation of sacred sites in the sea and their connection to creation myths and important stories is one aspect of this association (Diegues 2002).

It is perhaps crucial, for our understanding of fishermen, to consider the manner in which this wealth of knowledge is usually transmitted. In traditional fishing communities the enskilment in fishing is not done by tutoring or by mere participation of the novice to the process of fishing itself. Instead, the novice is enmeshed in every aspect of life of such a community. Enskilment is a dynamic process, where knowledge is generated, confirmed and transmitted at the same time (Pálsson 1993). Thus, a solitary farmer who occasionally catches fish for fun or because of need may in no way be viewed in the same terms as a professional or part-time fisherman, who acts as a member of a fishing community.

The fishermen's work rhythm often precludes their presence in the places where political decisions are made. Fishermen's communities often function in ways which do not always conform to the standard rules of the related land-bound communities. Their lifestyle is not only distanced from that of the literate elite, as indeed are those of farmers, shepherds and hunters, but it appears to be distant enough to preclude even mere observation (Acheson 1981).

One of the most interesting points in anthropological work on fishing communities world-wide, and of particular relevance to the study of ancient fishermen, refers to the construction of the fishermen's identity. It is a common feature of many small-scale fishing communities all over the world to have multiple livelihoods, fishing being combined with other types of employment, farming, manual labour etc., depending largely on what is available locally. Yet in cases of combined occupations, fishing is the essential cultural context, the one which the fishermen see themselves as fundamentally belonging to, and to which they return again and again (Blount 2005, p. 10). This observation has important implications in the study of ancient fishermen, whether they are exclusively full time professionals or more mobile, in and out of fishing and farming activities. The farmer's mentality imposed on them both by the ancient discourse and modern preconceptions needs to be investigated and proved rather than taken for granted. Also, the materiality of the fishermen's mentality needs to be explored.

A notion which is very popular in modern anthropological thought, the seascape, could prove crucial in understanding the fishermen's mentality, and might be a useful notion by which to explore the issue in an archaeological context (for a discussion of various issues pertaining to the study of archaeological landscapes, but relevant to seascape as well, see Knapp and Ashmore 2000). Seascape is a term that encompasses the sea, the coastal land and the land viewed from the sea (Cooney 2003). It is inhabited on three different, yet interconnected levels, as natural, human and imagined space (after Cole 2004, p. 1). What defines a seascape is not only its natural characteristics (morphology, hydrology etc.), but also the way humans make use of it through activities such as habitation, food procurement, space use, demarcation of spaces as common or private, secular or sacred etc. A seascape, however, is also an imaginary place, a place inhabited by spiritual creatures, by gods, monsters, dead ancestors or heroes (McNiven 2003). It is a place where part of the world's creation may have taken place in the distant or mythological past. These parallel seascapes exist simultaneously and constantly shape each other (after Cole 2004). The seascapes, encompassing all the above become, like the landscape, a materialisation of memory, they express identity and offer a key to interpreting society (after Knapp and Ashmore 2000, pp. 13-17). Although current anthropological and archaeological literature focuses on landscapes (dry land) and to a lesser degree on seascapes, similar issues are relevant to the study of river- and lakescapes as well.

A fisherman is caught up in a constant negotiation of these different seascapes, in a way which is similar to other mariners, to people who live from the sea (or indeed from any water mass) (Lindenlauf 2003, pp. 416-419 and references therein). What perhaps differentiates fishermen from other mariners, for example those people involved in maritime travel, is the fact that their interests focus on making use of the resources of the sea. In the context of ancient Greece, another differentiating parameter is the fact that fishermen mostly operate near the coast (see chapter 6, but see also Appendix 1: Men. *Fishermen*, fr. 13), while mariners often used the more open sea (e.g. Morton 2001, pp. 162-172, 215-228). Those fishermen's seascape therefore was not one of open water, but a seascape tied to the coast and its various features (natural, cultural, spiritual; see Morton 2001, pp. 185-206 for a discussion of various such landmarks pertinent to coastal navigation). This close association with the coast brings fishermen, at least partly, into the realm of the land-bound coastal inhabitants. Because of this contact, the ways in which people of dry land perceived the sea (but also lakes and rivers) are relevant to the fisherman's world (for a review of the literature on such perceptions see Lindenlauf 2003, pp. 416-419 and references therein; Vermeule 1979, pp. 179-209). In the following two subchapters aspects of these types of beliefs are explored, and both the fishermen and the land people's perceptions of the aquatic environment and of the fish in particular are investigated.

6.2 Looking for the fishermen in the fish names.

It seems that in ancient Greece there was a great variety of terms to describe fish and marine life in general. Less is known about the language used for fresh-water creatures. The following observations are mainly based on data collected by D'Arcy Thompson (1947), and to a lesser degree on older or more focused works such as the fish name lists by Wood (1927, 1928a, 1928b) or a paper by Lacroix (1937b). Other relevant works such as Hoffman and Jordan 1892, Palombi and Santorelli 1960 and Campbell 1982 were not accessible and thus have not been consulted.

D'Arcy Thompson has compiled a corpus of terms from ancient Greek and to a lesser degree Latin literature, which are related to fish, molluscs and marine invertebrates. Secure identification has been possible for only a small number of them, based either on the presence of a characteristic anatomical or behavioural feature, which is made explicit in ancient sources, or on the similarity of the ancient name to a modern one still in use in one of the circum-Mediterranean countries.

The following discussion is based on the idea, which is amply documented anthropologically, that the patterns of naming fish, fish behaviour, fishing practices and technology are part of a complex ecology which includes both the fish and their environment and the fishermen and their society (Nettle and Romaine 2000, p. 77). Fish names in other words could be treated as important and very informative remains of fishing cultures of the past.

The etymology of the fish-related terms indicates that many of these words had been constructed using words connected to land creatures (see various entries in Wood, 1927, 1928a and 1928b and Lacroix 1937b), on the basis of morphological or behavioural analogies. The name for *Akanthias*, the spurdog, for example, originates from the word *akantha*, a generic name for thistle like plants (Wood 1927, p. 300; Baumann 1993, p. 209). The choice of this name was apparently inspired by the hard, well-developed spines of the first and second dorsal fins, which are characteristic of this genus.

Despite a number of problems concerning the identification of specific fish species with their ancient names (e.g. Andrews 1949; Dalby 1996, p. 67, note 61), this corpus illustrates a wealth of terms related to water creatures. Considering that the data available to us had been collected in antiquity by scholars and not fishermen, sometimes indirectly, through older texts of unknown or obscure character, we could perhaps, consider them, as only a faint reflection of the variety of terms initially in use. Ethnographic observations concerning the language of maritime communities in Northern Europe reveal another possible problem concerning the representativeness of the fish names written in texts. It appears that in those communities, there is often a language barrier between land and sea. Some words-names are considered taboo and are not spoken when fishermen are at sea or on land (depending on the case) (Lockwood 1955; Westerdahl 2005). We should therefore, consider the possibility that even in ancient Greece, some fish names, or fish names under certain conditions, could not have been revealed to outsiders, thus never entering the written record. No proof of this can be provided at the moment, but the possibility remains open.

Thompson's fish names list indicates the existence of a localized terminology for certain fish. Some terms, such as *anosteos* and *orthagoriskos* for example were encountered in Lakonia, while others, such as *thraitta*, *varakos*, *opisthotila* appear to be Boeotian. The numerous known synonyms for certain fish may also be words of geographically distinct fishing vocabularies.

A factor which seems to have influenced the survival of certain fish names is their relevance to urban life. It is perhaps instructive that among the surviving fish-related terms, many form groups (synonyms or names designating specific fish age, size, part) which refer to fish popular in the urban fish-market (Table 6.1). The largest such groups are those which describe the sharks, rays and the rest of the large cartilaginous marine creatures, the tuna and other migratory fish, the euryaline fish, mostly the varieties of grey mullet and the sea bass, and finally words that describe the small fry, the cheapest variety of fish found in the market.

The fresh-water fish, especially those caught in lakes, are scarce within Thompson's corpus. Nile fish names predominate, usually found in accounts of exotic strange phenomena. Other river fish, such as the vocal fish of various rivers, are also treated as curiosities. From this scarcity, and in light of the fish bone and literary evidence discussed in the previous chapter, we could probably infer that fresh-water fish rarely found their way into large urban markets, especially Athens, missing the chance to be known literarily. Taking the argument further, we could perhaps also infer a cultural distance between the urban fish consumers (and their educated representatives) and that part of the rural population which made a living near inland water bodies such as lakes and rivers. If such a distance existed, then the fact that only few words related to river and lake fish and fishing entered the literary vocabulary would not be surprising.

Even if this situation existed, there were apparently other ways in which these relations could be organized. The mixed assemblage of both fresh-water and marine fish bones in the same consumption contexts (Appendix 4, Table 5.1a), at Krania tavern (see discussion of this context in chapter 8.2) could probably be seen as such an alternative. Marine and riverine fish had been consumed in the same establishment, if not by the same consumers, at least contemporarily. They were apparently included in the same menu. This fact in itself implies a familiarization of the consumers with both the marine and riverine fishing sphere. In the Krania case however, we lack the literary evidence which would permit any comparative, in depth exploration of the issue.

A cultural distance such as the one presumed above is not observed in the case of the euryaline fish, and especially the grey mullets, which are described with a large variety of terms (Koutrakis 1999 for a collection of these terms and their linking to modern Mugilidae taxonomy). This

Table 6.1 Ancient fish names of popular fish varieties.

Cartilaginous	Migratory	Euryaline	Small schooling
Aetos	*Amia*	*Adonis*	*Atherini*
Akanthias	*Auxis*	*Akarnax*	*Afritis*
Alopix	*Aulopias*	*Daktyleus*	*Afyi*
Amia	*Chelidonias*	*Gomfarion*	*Aktar*
Asterias	*Eleginos*	*Gylarion*	*Alosa*
Darta	*Kitos*	*Kephalinos*	*Box*
Galeos	*Kolias*	*Kestreus*	*Egkrasicholos*
Karcharias	*Kybion*	*Kestrinos*	*Eritimos*
Kentritis	*Melandrys*	*Leukiskos*	*Lykostomos*
Kentritis	*Orkynos*	*Myxinos*	*Thrissa*
Kentrophrys	*Pilamis*	*Nistis*	*Thritta*
Kestra	*Primades*	*Peraias*	*Trichis*
Kyon	*Saperdi*	*Pordias*	*Vemvras*
Lamia	*Sarda*	*Porias*	
Lamna	*Skepanos*	*Vakchos*	
Leios	*Skomvros*	*Voreus*	
Leovatos	*Skordyli*		
Narki	*Skordyli*		
Nevrias	*Synodontis*		
Notidianos	*Thorineus*		
Poikilos	*Thranis*		
Poikilos	*Thynnos*		
Rini	*Xiphias*		
Skyla			
Skylion			
Skylion			
Skymnos			
Troktis			
Vasiliskos			
Vatis			
Vous			
Zygaina			

is perhaps related to the fact that these fish were intensively exploited and important in the fish-market. Also, the grey mullets seem to have been considered marine species, despite the fact that they were caught in brackish and inland waters also (see discussion in chapter 9.1). The consumers' heightened interest in grey mullet may have triggered a familiarization with the words related to these fish. Furthermore, grey mullet was widely consumed in urban contexts in preserved form (Wilkins 2005). Preserved fish, being processed and transformed, are much more distant from the fishermen than fresh fish (see discussion in chapter 7.2). Such a condition could have permitted a familiarity with the euryaline fish which did not include a familiarity with their producers.

6.3 Looking for the fishermen in the fish myths.

Anthropology has contributed towards defining the characteristics of the fishermen and their communities in a rough way, and the analysis of the ancient fish names provides some hints on the existence of such identifiable Greek fishing communities. What remains to be explored, however, are further ways of identifying the ancient Greek fishermen and their culture in more detail and elaboration. Detailed excavation and contextual analysis of features such as coastal settlements with emphasis on individual houses, cemeteries of coastal populations, sacred places in the coastal zone, features on the wave line, etc., could contribute towards an archaeology of the fishing communities. Also, a re-examination of existing data, with specific research questions in mind, would be particularly valuable (e.g. Lindenlauf 2003, pp. 425-427). The scant available archaeological evidence will not be discussed in this chapter but it is presented throughout this book according to the requirements of the argument.

Literary sources offer a complementary, eloquent source of information in the form of fish-related myths (Buxton 1994, pp. 97-104) and anecdotes concerning fish-related phenomena, customs and superstitions. Any consideration of these data, however, must begin with a realization of their inherent bias. Their source is a large number of authors and works which, despite their diverse interests, style, context of production, etc., all relate to an educated elite, by position being thus distant from the world of fishermen as well as from any other sector of the population involved in manual labour (Finley 1979, pp. 40-41; Mossé 1969, pp. 25-30). Even those authors who acquired their knowledge about the fishing world from direct, personal experience, such as Aristotle who spent time doing fieldwork at Lemnos and Assos in the Northern Aegean (Lee 1948; 1985), were mere observers. They were probably recounting stories told. The element of participation in the whole set of actions and relationships of the fishing community, which is so crucial to learning and understanding (see chapter 6.1) was completely absent.

In order to use the information provided by the literary sources, in our quest for traces of the fishermen's mentality, some distinctions and clarifications are required. First there are elements of this discourse about the aquatic, fish-related supernatural world, which are very widespread, geographically, socially and chronologically, while others are specific to particular places or population groups. Although both cases may reflect the way fishermen viewed their world, the second case probably bears more weight in this respect. Some of the supernatural ideas about fish and the sea are presented in our sources as literary *topoi* which appear to be more relevant to those producing and consuming this literature (urban educated elite and land-bound people) than to those producing the fish (see below). On the contrary, situations and ideas which are presented in the literature as bizarre, alien phenomena, encountered at the periphery of the Greek polis world (social or geographical) could perhaps be seen as local traditions, reflecting ideas of the fishing communities of the place.

In the following discussion myths will not be analyzed, and their theological or cosmological significance will not be discussed, even though these dimensions are obviously important. For the purposes of the present analysis, the mere presence of such stories is of interest, because they may be taken as markers of a distinct aquatic ideology; also because, paraphrasing Schama (1995, p. 15), myths and memories inscribed on landscapes (or seascapes in our case) reflect the complexity and antiquity of local traditions of landscape. Myths in this sense *are* memory and the seascape as memory is linked to the identity of its inhabitants (Knapp and Ashmore 2000, pp. 13-4).

Poseidon, one of the major Olympians, was the god of the water (he also had other personae which linked him to the earth). His cult was connected both to the sea and to fresh-water. It was widespread across the Greek world and some aspects of it were formal, with typical sacrifices, of sheep, goats and bulls offered to him (Farnell 1977, pp. 1-55; Burkert 1985, pp. 136-9; Romero Recio 2000; Mylonopoulos 2003). Interestingly, horses and bulls were also sacrificed to Poseidon in special ceremonies that involved the submersion of the animals in water, being it the sea, spring or river (Robertson 2005). His symbol was the trident, a fishing instrument adapted to the catching of large fish (Opp. *Hallieutica*, iv 253, v. 151, 255), and in literature and iconography he was closely linked to fish and fishermen (Shapiro 1989). Occasionally, he was the recipient of fish sacrifices (Appendix 1: Antigonos of Carystos *On Diction*, fr. 56A-B; Crates *Sacrifices at Athens*). The seas were inhabited by numerous other minor gods, anthropomorphic or monstrous. Some had been born as gods while others, according to myth, were humans who later became gods. Amphitrite, Ino, Glaukos, Enalos, Palaemon, Halios Geron (Old man of the sea), the Nereids, are some of the more popular of these deities (Shepard 1940; Price and Kearns 2003). Springs were also home for such spirits, the Nymphs, and the rivers were often considered as gods (Brewster 1997; Burkert 1985, pp. 174-6; Larson 2001, pp. 8-11).

It is interesting that some of the sea gods have a very close affiliation to fish. Glaukos and Triton (probably local names for the same divinity – Shepard 1940, p. 39) were gods whose lower body was that of a fish and the upper one of a man. Triton(s) were considered semi-imaginary. Although they were venerated in certain parts of Greece as minor gods, there are stories of eye-witnesses of their presence (Appendix 1: Ael. 13.21). Certain areas, such as Tanagra and Anthedon in Boeotia (Appendix 1: Aesch. *Glaukos Pontion*; Euphoros *Geryones*, fr. 3; Paus. ix.20.4) and Itanos on Crete (Fig. 6.2) had close links to Triton or Glaukos being related either to his birth or death. There are cases, however, where a god who is not usually connected to the aquatic environment may have taken on such a character. Apollo, as a protector god of Elis, near Olympia in the Peloponnese, was called *ichthyophagos* (fish-eater) (Appendix 1: Polemon *Letter to Attalus*), apparently due to some special importance of fish or fishing in the area. Also Hekate, the goddess of the cross-roads and of a strongly chthonic character was venerated by fishermen, together with Poseidon at Eleusis, to grant them rich

Fig. 6.2 Silver stater from Itanos. On the one side is illustrated Glaukos holding a trident and fish, on the reverse two sea horses. 380/330 B.C.
(from http://www.coinarchives.com/a/lotviewer.php?LotID =82924&AucID=89&Lot=899 accessed 23 December 2006).

catches (Hes. *Theog.* 440-443; von Rudloff, 1999, pp. 85-86).

Certain areas were renowned for seemingly strange customs and ceremonies (Parker 1983; Simoons 1994). The diving competition of the cult of "Dionysus of the Black Skin" in Hermione (Appendix 1: Paus. ii.34.9ff), and the *Thynnaion* festival at Halae Aexonidae in Attica and at Halieis in the Peloponnese (Appendix 1: Antigonos of Carystos *On Dictions*, fr. 56A-B; Crates *Sacrifices at Athens*), are some such ceremonies, clearly connected to the local population who lived from the sea. It is apparently no coincidence that a strong fishing character of these areas emerges clearly from our admittedly deficient data base (see chapter 5.3). The abundance of fish sent by Volvi to those who venerated her son, the dead hero Olynthus, on the coast of Lake Volvi, and the sacrifice of eel to the Unknown Gods in Boeotia, on the coasts of Lake Kopais (Appendix 1: Agatharchides *European History*; Hegesander *Commentaries*) give two examples of this type of custom in an inland aquatic setting.

Local traditions involving a special relationship to fish appear in various places throughout Greece. In the sea off the island of Serifos a kind of lobster (*tetyx enalios*) was not caught by fishermen because they were considered the playmates of Perseus (Appendix 1: Ael. 13.26). In Arethusa spring, near Chalkis, the eels were tame and adorned with jewellery, and so were those in the natural pool by the Shrine of the Old Men on Chios island (Appendix 1: Ath. *Deipnosophistae* viii 331e-f; Plin.(E) *HN* xxxii, vii.16). On Lemnos, crabs were deified (Appendix 1: Hesych. *Lexicon*) and in the waters of northern Aegean the escort fish was especially honoured among the Samothracian Gods (Appendix 1: Pancrates, *Occupations at Sea*). At Pharae in Achaia, sacred fish lived in a stream which belonged to Hermes (Appendix 1: Paus. vii.2.4).

Fish however were involved in several less formal aspects of the fishermen's life (for the following see Thompson 1947). *Gnafeus* for example, an unidentified fish, was supposed to produce a broth which could be used to remove stains. The weaver could only be taken safely out of the water using the left hand, while the shark (*karcharias*) was considered one of those fish that it was unlucky to dismember. The *kallichthys*, another unidentified fish, was beloved by the sponge fishermen, because when it appeared, no evil creature would approach, and *glaukos* was considered by fishermen an omen (e.g. Appendix 1: Philarchos Histories; Polycharmos *History of Lycia*; Semos *History of Delos*). The extensive use of fish in medicine (see Thompson 1947) apparently had its origin in coastal populations and the fishermen. A story about the cure of mad dog bite aptly illustrates the issue. A fisherman's two sons from Rethymno on Crete were bitten by a mad dog. Their father, in desperation, discovered the medication for the disease, making a potion of powdered sea-horses (Appendix 1: Ael. 14.20 see also Plin.(E) xxxii, vii.16).

While there is clearly a domain of folk beliefs around fishing, which we may assume to be closely related to fishing communities of the Greek coasts, there is a parallel set of beliefs which probably reflects the ideas of the land-bound population about the sea (probably lakes and rivers as well) and its creatures. Even a cursory review of the relevant literary fragments highlights one issue as particularly prominent. For those consumers who get their fish in the market, fish are dangerous; essentially man-eaters. This idea is presented in various forms. Often it is expressed metaphorically - fish are so expensive that they ruin their consumers financially and morally (Appendix1: Alexis *The Greek* Woman, fr. 76; Dem. *On False Embassy* 9.229; Diphilos *The Merchant*, fr. 31; Paus. x.9.3-4, Wilkins, 2000, p. 296, Davidson 1997, pp. 186-190). But fish, especially the larger ones, were considered man eaters in the literal sense (Appendix 1: Archaestratos in Ath. vii 310e; Fig. 6.3). Any shipwrecked human being was in danger of being eaten by fish or sea monsters (Appendix 1: Alexis *The Greek Woman*, fr. 76; Archaestratos in Ath. vii 327e; Xanthos of Lydia *Lydiaca*, fr. 17; Vermeule 1979, p. 184ff; Purcell 1995, pp. 133-4; for a collection of relevant epigrams see Douglas, 1974, pp. 94-114).

This sense of fear of fish is part of a more generalized fear of the sea (Linder 1996). Although the sea is a domain intensively occupied by the Greeks (trade, warfare, etc.), it remained mysterious and threatening. Lindenlauf (2003) has demonstrated how the sea has been considered a place of no-return, a place where Greeks disposed not only of ordinary unwanted things (e.g. urban litter) but also polluted items and bodies. According to Lindenlauf (2003, p. 427), the sea was considered by ancient Greeks as something good when people were making use of it, but as something dangerous and potentially evil, when they were subjected to its forces (for an ethnographic example of a similar idea see Pálsson 1994b, p. 120). Perhaps the discourse around sea monsters, which included myths and an extensive

Fig. 6.3 Marble statuette of a man eaten by sea monster, 3rd-4th c. A.D. Clevelant Museum (from Farkas *et al.* 1987, pl. XXI).

treatment in art (Boardman 1987; Papadopoulos and Ruscillo 2002), was part of this mentality that focused on fear of the sea seen as an ambiguous element.

There was however another idea among Greeks which, by its persistence over time, shows another aspect of this fear. A recurring theme in geographical treatises was that of the strange *Ichthyophagoi* – the fish-eaters (e.g. Appendix 1: Arr. *Indike* 29-30; Diod. Sic. iii.15-17; Plin.(E) *HN* vi.30.5.32; Strabo 15.2.2). Fish-eaters were people who mostly lived in far away places and their lifestyle was extremely bizarre. The description of the fish-eaters of the Red Sea, given by a Roman writer, Diodorus Sicilus, is perhaps the most elaborate (Appendix 1: Diod. Sic. iii.15-17). The *Ichthyophagoi* had been so named because they subsisted on fish and even made bread of it. Their eating was uncontrolled, and people gorged themselves, led by pure desire. They did not use fire to cook but exposed the fish to the sun's warmth instead. They even used fish as fodder for their flocks. The fish eaters made minimal efforts to catch fish, since most of them simply gathered fish, small and large, left behind by the subsiding tide. When bad weather did not permit the fish gathering, they ate the leftover fish bones just like scavengers. Some built their houses using the bones of very large fish and seaweed. When one of them died the body was left to the tide to be eaten by the fish. The same story, with fewer details was recounted by various authors before and after the time of Diodorus Sicilus and of different cultural backgrounds (e.g. Appendix 1: Arr. *Indike* 29-30; Strabo 15.2.2). Some times the name *Ichthyophagoi* was used to describe the people of a specific geographical area, with no elaboration on their habits, as if those were well known already (e.g. Hdt. iii.19; Paus. i.33.4).

The fish-eaters' life-style was described as the direct opposite of what was familiar to Greeks (subsistence on grain, cooking of food, rules governing food consumption, burying of the dead). Although this is a typical rhetorical scheme to describe the "other" in comprehensible terms (Hartog 1988), it is interesting to note that a life so closely connected to fish and the sea in their natural state (fish almost placing themselves in the hands of people, eaten raw or sun-baked, the dead eaten by fish as a natural end of life) was presumed to be the inverted image of the cultured way of life (Marchiori 2000; Jacob 1997, p. 246ff). In this scheme, eating in connection to fish (eating fish or eaten by fish) looms as a theme of major importance (fish-eating and identity is an issue that is discussed in chapter 8). One can not but observe in this scheme the relevance of Levi-Stauss's ideas about the significance of the raw and the cooked in the arrangement of the famous "culinary triangle", where raw food signifies the natural state, and the cooked signifies the cultural transformation of it (Levi-Strauss 1969; 1997).

The *Ichthyophagoi* however, was not a cultural category which was only to be found in such distant, exotic places as the Red Sea. Some fish-eaters lived much closer. The Thracian lake-dwellers of Lake Prasias, for example, used to feed their beasts with fish just like the fish-eaters of the Red Sea, and the catching of fish was for them little more that mere fish gathering (Appendix 1: Hdt. v.16). These Thracians however lived, in the days of Herodotus, just a few kilometers away from such a civilized Greek place as the city of Abdera, which had been a colony of Clazomenai as early as the 7th c. B.C. and of Teos in 545 B.C. (Lazarides 1971; Loukopoulou 2004, pp. 872-5).

The theme which forms the ideational framework, within which such a varied discourse around fishing developed, was a juxtaposition of land and sea (Wilkins 1993, pp. 191-2 and note 3; Purcell, 1995, p. 133). The land is considered familiar, while the sea is mysterious. It is perhaps not irrelevant that, in some ancient Greek philosophical schemes, life was thought to have begun in the water and fish were considered as the progenitors of humans (Osborne 1990 on Herakleitos; Kahn 1974 on Anaximandros, Appendix 1: Empedocles *Katharmoi,* frg.117; Plut. *Quest. conv.* 730e).

In Classical Greece, wealth and status had been traditionally based on land, its ownership and produce (Purcell 1995, p. 134; Austin and Vidal-Naquet 1998, pp 136-141). Wealth originating from other sources, maritime trade being a large part of these, had a tint of un-respectability (e.g. *Pl. Leg. vi. 705a*; Morris 1994, p. 68; Cartledge 2002). The sea was considered risky and corruptive (Purcell 1995, Marshall 2000). Plato, in his

Laws (Appendix 1: Pl. *Leg.* vii. 823) grouped fishing with piracy under the heading of "sea-related activities" which he considered negative. Comparable land based activities, such as hunting, were on the contrary characterized as noble. In a different, contemporary approach to the issue (Appendix 1: Theopomp. *Philipica*), fishing was presented as a source of luxurious behaviour which, in contrast to agriculture, entailed gain without hard work. Here we obviously deal with a discourse that defended the social, political and ideological values constructed around land ownership. The sea and its wealth was apparently a completely different world, almost totally incompatible with the world of the city state and land property. The Epidaurus inscription which regulated the sharing of the fish profits that were generated from commonly held waters (Appendix 2: IG IV2[1], 123) provides an indication that attempts were made to apply to the sea rules of ownership similar to those regulating property affairs on land. Here again we have traces of the imposition of land-based values and arrangements upon the sea.

In this antithetical worldview, the coast, the domain of fishermen, was perhaps the most ambiguous space, it was a liminal zone, an area of transition (Helskog 1999, p. 79-81). Such places are often considered by traditional societies as places of danger and of unfamiliar social and ideological arrangements (Buxton 1994, p. 102). That is probably why, in ancient Greek culture, this was a place where polluted humans and objects were purified (Osborne 1990, pp. 19-20; Lindelauf 2003). Conspicuous features of the coast, such as rock carvings, rock cuttings etc. are perhaps traces of an effort to manage the liminal nature of the coast (Marangou 2002). The natural positioning of the fishermen in this ambiguous, dangerous zone is perhaps a reason why fishermen are left outside any discourse generated by city dwellers, thus becoming almost invisible.

6.4 Conclusions.

The discussion in this chapter has worked towards illuminating the ancient fishermen and their worldviews. This has been done in a rather indirect manner. This chapter began by building a framework of understanding based on anthropology and comparative ethnology, because issues related to the identity of fishermen and their communities, have been so little explored within the ancient literary tradition and modern archaeological/historical discourse. Literary and archaeological data have consequently been reviewed within this framework. This anthropologically-informed approach to the issue has shown that being a fisherman is not only about having the technology to catch fish, but it is mostly a way of living. It involves a wide range of specialized knowledge and most importantly the participation in a distinct set of social relations. Also, being a fisherman means viewing the world from a distinctive perspective.

The existence of distinct fishing cultures is implied by two different types of evidence. On the one hand, the corpus of the extant fish names indicates the use of local fish vocabularies. A preservation bias for certain groups of fish names in the written sources could be taken to suggest the existence of a cultural distance between certain kinds of fishing communities and urban fish consumers. Lake and river fishermen appear to have been the least visible. On the other hand, the existence of a small but highly indicative corpus of fish-related myths, beliefs and superstitions, which are geographically specific and culturally peripheral, could also be taken as an indication of the existence of distinct, localized fishing cultures. These beliefs are best understood if viewed against the dominant culture and ideology, which was basically land-bound. This includes a number of more widespread, recurrent ideas about the sea and the fish which are mostly characterized by fear or awe of the strangeness of the sea and the fish world.

The material, social and ideological differences between communities living from the land and from the sea (but also lakes and rivers) was expressed by several antithetical schemes. The distance between fishermen, as primary producers of fish, and urban populations, as consumers, is just one of them. The juxtaposition of the educated elite, as producers of the discourse around fishermen (and other manual labourers) versus the uneducated sectors of the population, is another dimension of this image. A moral distinction is also relevant. Respectability, morality and a whole range of social, political and ideological values in a typical city state society were based, on land ownership or lack of it. Fishermen worked in a regime outside this scheme, and had thus no distinct place in the social order of a land-bound society. This liminality of the fishermen's position might account for the invisibility of fishermen in the ancient literary discourse.

The above conclusions are only preliminary and serve to show that the issue of fishermen, their communities and worldviews demands further exploration. Future focused work is required on this field. The discussion of the issue in this book however, even in this generalized form, is necessary because some of the ideas pertaining to fishermen and their worldviews as well as their perception by modern research are crucial to the understanding of the discourse, ancient and modern, around fish-eating. Some of the issues highlighted here will be discussed and further elaborated upon again in later chapters.

CHAPTER 7

FISH BECOMING FOOD - DISTRIBUTION AND MARKETING

To understand the importance of fish as food in Greece from the Classical period to the end of the Roman period (ca 7th century A.D.) however, it is necessary to explore one crucial aspect in the fishing — fish-eating circle: the distribution and marketing of fish, the ways in which fish became available to the consumers. A review of the evidence indicates that two distinct mechanisms were in place, one dealing with fresh fish and the other with preserved fish in its various forms. These mechanisms set in gear a transformative process in which fish turned from a harvested resource into a commodity and ultimately into a dish, a menu item. Such a transformation involved a number of technologies, social structures and ideas, which were tightly enmeshed. The following discussion illustrates how the supply of and demand for fish, in its various forms, was articulated by exploring certain aspects of the distribution and marketing of fresh and preserved fish. In doing so it begins to clarify how fish as a foodstuff acquired multiple meanings during its "journey" from the hands of the fishermen to the tables of the consumers. The issue that looms prominent in this discussion is that demand for fish played, in places, a much more important role than previously thought. The actual accessibility and supply of fish appears, in some cases, less important than the demand and the desire for fish consumption. A range of processes had been set in place in order to overcome problems of accessibility.

Fish was a food item that was sold and bought in a systematic structured manner. It should be noted here, that in the following discussion the term "market" is not used in its restricted form, i.e. a space where goods are sold and bought, but more as a geographical and social space where people meet, exchange news, ideas and goods, with these transactions taking a variety of forms (Millett 1990; Dilley 1992). The extant literary references to fish make it clear that fish, in this context, was not simply a foodstuff to be sold or bought. It also provided an excuse for different types of people to interact and define themselves and others. It has been claimed that the fish stalls were places where politics were practiced (e.g. Davidson 1993; 1997, Wilkins 1993; 2000; Fisher 2000, pp. 367-369).

In literary sources fish appear to stand out as an exceptional food item in the market. But its uniqueness as a marketable object or its relation to other foodstuffs such as red meat is not so clear-cut. Food and drink items such as wine, some spices and perhaps bread and cakes are also prominent in the ancient discourse around the market (for a review of such items see Dalby 1996) but the analysis of their role and standing in this context is as yet minimal. The role of red meat as a marketable object, which is typically contrasted to fish as an alternative protein source, is in my opinion obscure and far from clarified (the few existing focused works such as Isenberg 1975; Steinhauer 1994 and Frayn 1995 do not adequately cover the issue). The multidimensional discourse developed around the selling and buying of fish in the Classical market is usually attributed, in modern scholarship, to the fact that fish is a source of protein which, unlike meat, is free from religious overtones, that might place restrictions upon the manner, scale and elaboration of its consumption (Davidson 1997, pp. 15-16). This however is just a hypothesis. Alternative explanations might be developed based on the analysis of the role of fish as a consumable food item (see chapter 11).

It is clear from the above that no straightforward answer can be given to the issue of fish prominence in the market. Its importance as a food stuff acquired from the market would require a consideration of its function and meaning in a broader context. Here I will proceed to explore the mechanisms of fish marketing and distribution or, in other words, the processes of fish commodification (see discussion in chapter 3.2). Again, as in several points in this book, my approach will be selective, focusing on those aspects of the issue which find a robust archaeological documentation. Some issues are thoroughly examined while others, such as for example the role of fish as a luxury food, are only superficially touched.

Modern research on the ancient economy and on fish in particular, has often striven to extract precise data on economic issues such as prices of goods, market regulations and so on, based mostly on literary sources, despite their acknowledged deficiencies (see chapter 8.1; Heichelheim 1930; Ehrenberg 1951, pp. 219-226; Frost 1971, pp. 68-70; Millet 1990, pp. 172, 192, notes 13, 53). Although these are interesting and perhaps crucial issues in some respects, here they will be downplayed in favour of an approach based on relative values, on archaeological documentation and on inferences from these sources rather than direct literary illustration. The aim of this chapter is to explore the mechanisms which governed the distribution and marketing of fish and their practical and social implications.

7.1 Fresh fish.

The management of fresh fish poses certain insurmountable restrictions upon both fishermen and consumers. Fresh fish spoils quickly. In temperate areas such as Greece, and for most of the year, fish cannot last longer that one or two days at the most, depending on the type of fish, its size and the temperature of the environment.

7.1.1 *In situ consumption or transportation over short distances.*

The most obvious way to deal with the problem is to restrict fish consumption to near its source, so that no considerable time elapses from the landing of fish to its cooking. An archaeological reflection of this necessity is that the sites which produce the most sizeable fish-bone assemblages are coastal ones. The literary tradition gives a vivid picture of the direct passing of fish from the hands of fishermen to those of the consumers. Strabo, in the Roman period, commenting on the fish craze of certain people, described the landing of the boats loaded with a night's catch at Iassos, a small town on Asia Minor. The landing of fish was marked with the ringing of bells which invited the buyers (Appendix 1: Strabo 14.2.21). The customers were apparently buying the fish directly from the fishermen on the beach. A Hellenistic epideictic epigram elaborates on the same topic. The speaker on the shore hails a fisherman, standing on a rock with a view of buying from him, if his catch is suitable (Appendix 1: Phanios *AP* 6.304). The even more straightforward process of the fisherman and his family consuming fish from their own catch was apparently too obvious to warrant literary comment.

Transportation over short or medium long distances, from the coast to the consumers is somewhat better documented. Aristotle in the *Rhetoric*, in a discussion on the value of exceeding one's abilities, recites an epigram inscribed on the base of the statue of a winner in the Olympic Games (a runner?) (Fig.7.1). "Formerly, with a yoke on my shoulders from which baskets were hung, I was bringing fish from Argos to Tegea" (Appendix 1: Arist. *Rh.* 6.55). Argos is close to the coast, on the Argolic gulf, while Tegea is in the mountains at a straight distance of about 40 kilometres. Although this case has been cited to illustrate the speed of the runner, it does however, reflect an apparently real transporting activity. Surprisingly, the existence of an established trade route for fish between the coast and Tegea is confirmed by another source. A Classical inscription found in the Asklepieion at Epidauros, which recounts the names of patients, their illnesses and treatments applied, refers to a certain Amphimnastos, a merchant who also transported fish from Argos to Tegea (Appendix 2: IG IV2 [1], 123). As a payment for his treatment Amphimnastos promised to offer one fourth of his earnings from the fish trade. Transporting fresh fish to the mainland was apparently part of a more generalized flow of goods from the coasts inland (Paus. viii.5.8). Transportation of fish from inland sources was not unheard of either. The celebrated eels from Lake Kopais, were regularly brought to Athens to be sold fresh and occasionally they were specially and characteristically packaged (Appendix 1: Antiphanes *The Sheep Owner*, fr. 191; Archaestratos in Ath. various lines; Ar. *Ach.* 884, 894; Pax 1013-4). Comic references to travelling fishmongers are fairly common (Appendix 1, no. 28), but the case of consumers travelling for fresh fish to the coast is also attested. In *The Constitution of Naxos*, a work which is dubiously attributed to Aristotle of Stagera and has been preserved by Athenaeus (Ath. viii.348b), a relevant episode is described. The inhabitants of an inland town on Naxos used to go to the coast to buy fish. Animosity towards fishermen arose when they failed to get their supplies.

Archaeologically, the short distance transportation of fish is attested through the recovery of marine fish remains at inland sites. The remains of a grouper, a common sea bream and a tuna have been recovered in the destruction levels of a Roman villa in Eleutherna, Crete, while shark, sting ray, grouper and grey mullet remains have been found in Late Roman deposits at the same site (Table 5.1g). Eleutherna is located at the foothills of mount Psiloritis, at a straight distance of 9 km from the sea being however quite difficult to approach through the meandering paths of the mountainous terrain. The fish remains found there could have been brought on site either as preserved or as fresh fish (Mylona 2003b). Judging by the species represented, both possibilities appear plausible. The same is true for the case of Pyrgouthi, a Late Roman farmhouse located at the inland valley of Berbati in the Argolid. The large Serranidae vertebra could have been brought on site either as a fresh slice of fish or in a preserved form (Table 5.1d). Further archaeological evidence of this type of transportation is probably provided by a single large fish vertebra (unidentified) found within the *cella* of the Temple of Apollo at Delphi (Table 5.1c).

A different type of archaeological evidence for the transportation of fresh fish some distance from the coast is provided by the Akraephia fish price list inscription (Appendix 2: Salviat et Vatin 1971); for details on the inscription see bellow chapter 9.1). This list includes not only fish which were caught locally, in Lake Kopais, but also a wide selection of marine fish, apparently caught in the rich waters of the Euboikos gulf some 20 km to the East (chapter 9.1).

The consumption of fresh fish near the coast either by the fishermen themselves and their families or by the inhabitants of small-scale settlements on the coast or by lakes and rivers could be considered the most obvious way to make use of the fish resources available. This however is the regime for which we have the least information. Not only are these cases seldom considered as worth writing about by ancient writers, but even modern classical scholarship, and excavation more specifically, has ignored this type of settlement or this type of remains. Some possible reasons for this treatment have been discussed in the previous chapter. More information about fish consumption is available when fish become the objects of a more formalized distribution

Fig. 7.1 Representations of two men carrying fish on a pole. One is carrying two large fish (tunas?) and the other two baskets of smaller fish. Black-figure amphora by Ambrosios painter, 6th c. B.C., now in Berlin (Beazley 1963, p. 109, no 30; photograph from Sparkes 1995, fig. 11.5 and 11.6).

and trade. In few words, when they are associated with established structures of the Classical, Hellenistic and Roman society. Fish become increasingly visible as the level of their enmeshing in urban life becomes more intense.

7.1.2 *Fresh fish kept alive for later consumption.*

A solution to the problem of fish spoiling was adopted, at least in certain parts of Greece, some time at the beginning of the Roman period. This was the construction of fish-tanks, known in Latin contexts as "*piscinae*". These are artificial enclosures, either inland, or on the coast, which served for the breeding of certain types of fish, or the preservation of live fish as a ready supply for consumption, and probably for the market. Their function is closely connected to the largest or more desirable of the inshore, or euryaline fish (for the range of fish that were commonly kept in fish-tanks see Higginbotham 1997, pp.41-53). Latin sources, such as Varro (*Rust.*), Columella (*Rust.*) and others (see Higginbotham 1997, pp. 4-5), describe the specifications for their construction and use. Our knowledge of the fish-tanks is vastly enriched by the study of a large number of surviving *piscinae* on the Italian peninsula (Higginbotham 1997) and elsewhere (e.g. Linder and Laenhardt 1964; Yorke and Dallas 1968).

The combined analysis by Higginbotham of the literary prescriptions for a functional fish tank and of the data produced by the archaeological investigation of real *piscinae* in Italy highlights some issues which are of interest to the present study (Higginbotham 1997, pp. 9-40). The fish-tanks could be designed to accommodate rock fish, fish preferring sandy bottoms, euryaline fish, or even varieties. The critical factor in each case was the medium in which the fish-tanks were constructed and the quality of the available water. The water in the tanks could be sea-water, entering through channels, or brackish water, created either naturally from a coastal spring or artificially by adding specially brought to a marine fish tank. The fish-tanks could even be filled with if situated inland near a spring or lake. The Italian examples show that the fish-tanks could be of commercial use, or decorative, domestic features, which supplied pleasure and prestige as well as fish (Higginbotham 1997, pp. 55-64).

In Greece, the coastal fish-tanks which have been identified to date are all of a similar nature. They are basins, partly or wholly dug in the rock of the sea-coast, probably constructed to accommodate rock fish. The water in them constantly circulated through channels. Specially made grills permitted the water circulation keeping the fish trapped. In some cases, traces of a super-structure still survive. Those were either shades, or walkways which regulated the environmental conditions within the tanks and facilitated the handling of the fish (Hood and Leatham 1958-9; Davaras 1974; 1975; Shaw 1978, pp. 25-34; Pirazzoli 1987).

The distribution of known coastal fish-tanks in Greece exhibits a distinct geographical clustering (Appendix 3 for all the fish-tanks mentioned below), which, however, may reflect the state of relevant research on coastal areas rather than their actual presence (for similar biases on the Italian peninsula see Higginbotham, 1997, p. 6, note 25). Fish-tanks have been located mostly on Crete (Appendix 3b). Two more cases have been reported from Corinth on the Saronic coast, and from Kastelorizo Island in Dodecanese (Appendix 3b). Almost all of the Cretan cases have a rectangular or rectilinear shape. A cluster of three fish-tanks have been found at Chersonessos, near Heraklion, on the Northern coast of the island. Another, bi-partite fish tank is located at Mochlos, on the north-east coast and a cluster of ten tanks have been dug on the shore of Siteia, again on the north-east coast. A possible fish tank has been reported from Zakros on the east coast of Crete, but no further information is available on it. On the Southern coast a large fish tank has been located at Ferma, near Ierapetra and a possible one has been reported at Phalasarna, at the north-west corner of the island. In that case the fish tank was situated near the entrance to the enclosed port of the city, in an area used extensively for stone quarrying.

None of these tanks has been properly excavated. Furthermore, in none of these cases, with the possible exception of the, as yet, unpublished Phalasarna case, has the surrounding settlement been located or investigated in any comprehensive manner. Therefore, these fish-tanks stand as isolated finds, completely devoid of context. Even their dating is based only on conjecture. Their dating is assumed to be Roman based on Roman, sources that describe similar fish-tanks.

The Corinthian fish-tanks consist of a cluster of six basins cut in the rock at the harbour of Kenchriae, the Eastern port of Corinth (Fig. 7.2). In this case the dating of the fish-tanks is secure. They had been constructed sometime in the 1st century B.C. - 1st century A.D. and were in use until the end of the 2nd century A.D. or the beginning of the 3rd. This fish-tanks complex had been constructed on the rocky shore near the warehouses at the port. They seem to have been part of the harbour complex (Scranton 1978; Shaw 1978). The size, and consequently the capacity of these tanks vary. Because only the architecture of the site has been published, no more detail of their actual function is available at the moment. The report on the cluster of two fish-tanks from Kastelorizo is lacking in contextual information, much like the Cretan cases.

What could be deduced from the known data on these fish-tank complexes is that almost all (excepting the Siteia case) represent small-scale ventures. They consist of a few tanks each, some of them small in dimensions. We do not know what the productivity of such establishments was and one may surmise that it depended on various factors. The exact function of the fish-tank (breeding or conservation), the nature of its water and the suitability of its environment to achieve high survival rates for fish, the rate of replenishing it with fish from the sea are some of these. It is clear that the Siteia fish-tanks were of a far larger scale than the rest. This is perhaps related to the presence of a sizeable city (Papadakis 1989) with high demand for a certain type of fish. Such a large market however existed also near the Kenchreai fish-tanks (Scranton *et al.* 1978). The fact that that complex is relatively small may be accidental. For the rest of the known fish-tanks we can not offer any judgment due to the recovery and documentation problems connected with them.

The lack of contextual information in almost all cases of marine fish-tanks in the Hellenic area, and their total absence from the written records, precludes the drawing of any conclusion regarding issues of ownership, size of production and connection to local or more distant markets. It would however, be instructive to note that the concentration of the marine fish-tanks in the southern part of Greece, is probably related to the different nature of the fish resources, which were less abundant (than those

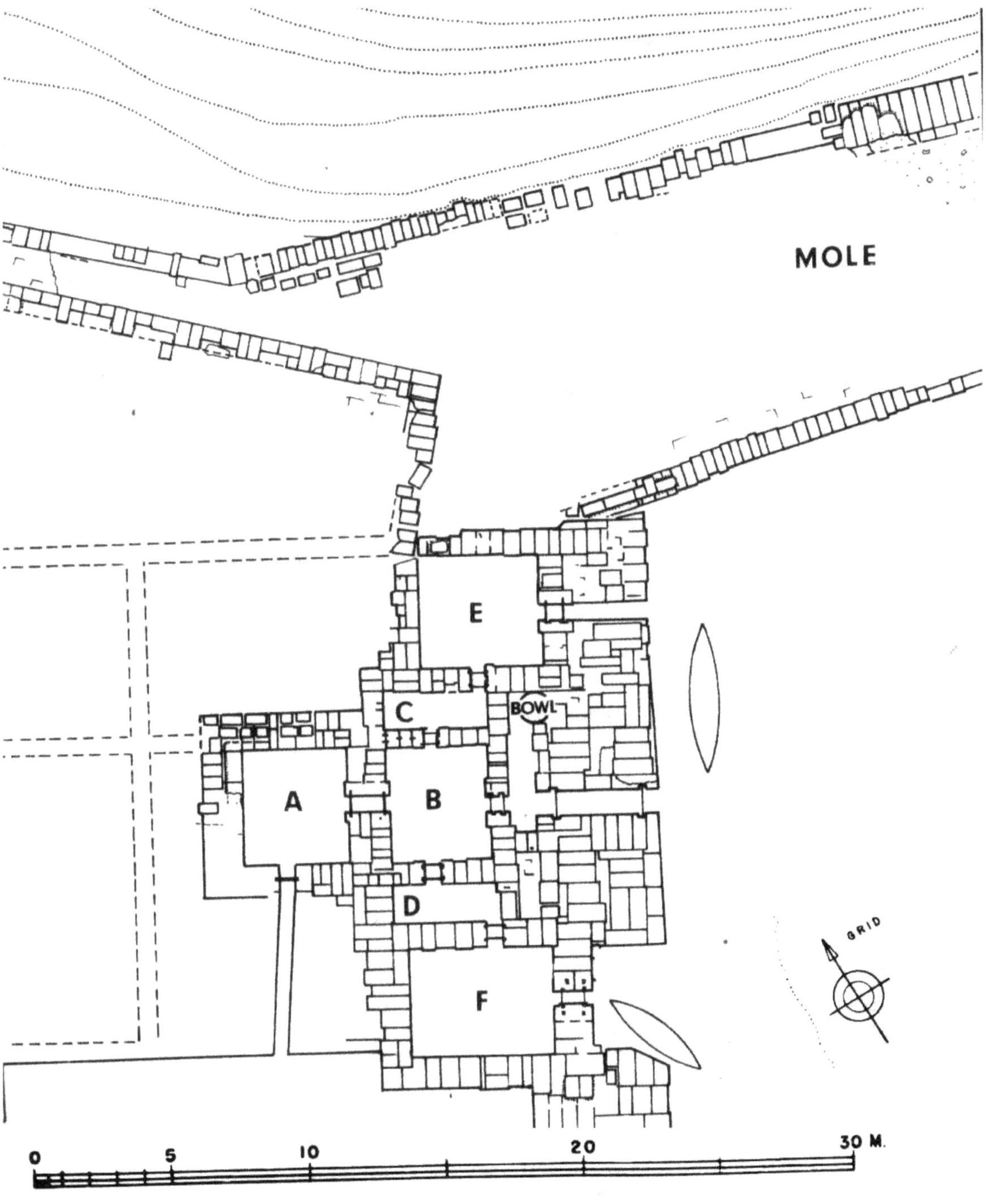

Fig. 7.2 Site plan of the Kenchreae fish-tanks. Spaces A-F are the interconnected vats (from Shaw 1978, pl. 11).

in other parts of Greece) but provided more greatly esteemed fish (for the evaluation of the taste of various fish see chapter 9).

7.1.2 Sacred lakes or fresh-water fish ponds?

Artificial or semi-artificial controlled environments for fish keeping and fish breeding are also probably represented in Greece by inland pools or coastal brackish water bodies. Because the material spatial arrangements needed for such ventures are usually minimal and perishable (e.g. reed fencing), most of the information available to us about these features are some references in ancient literature. They are known under the rubrics "sacred lake" (Appendix 1: Paus. iii.21.5; Appendix 2: SIG[4] 584; SIG[3] 997; Homolle 1882, 1890) or "sacred river" (Appendix 1: Paus. i.38.1 and vii.22.4). Some of these features do not seem to have had any relation to fish but they are referred to in relation to a range of cultic practices. Such cases are, for example, the sacred spring at the Sanctuary of Artemis at Brauron (Themelis 1971), or the pool at "Epidaurus Limera" in Laconia which,

Fig. 7.3 Sacred Lake on Delos (from Bruneau and Ducat 2005, p. 170, fig. 42).

according to Pausanias, was sacred to Ino (3.23.6). In other cases, fish were central, but their importance varied. They were ornamental, divinatory or simply curiosities (Appendix 1: Artemidoros *Geographies*; Ath. viii 331e-f; Paus. i.38.1, iii.21.5 and vii.22.4; Appendix 2: SIG[4] 584; Homolle 1882, 1890, also see chapter 7.1.3). Sometimes sacred lakes are referred to as sources of fish, the access to which could be restricted or taxed by a religious authority.

This is the case of the Sacred Lake on Delos. A number of inscriptions, which are actually the accounts and inventories of the Temple of Apollo, refer to the taxation of the fish caught around Delos and in some cases of the fish caught in the Lake (Appendix 2: Homolle 1882, 1890). These references span several decades in the 3rd century B.C. The word *Limni* (lake) has been taken to refer to the Sacred Lake, an inland pond, which was a central feature of the topography of the city of Delos (fig. 7.3). It is therefore assumed that this lake, drained since 1925 (Leguilloux 2003), was the source of large quantities of fish. In some cases the income from this resource was reported as tax or as profit from the selling of sacred objects by the Temple. In other cases they were reported in the list of incomes from rents of cultivated land or pastures. This might support the assumption that the Lake was in a rural area (Homolle 1890, pp. 454-5).

In recent years a geo-archaeological investigation has been undertaken to explore the availability and management of the water resources on the island (Brunet *et al.* 2001, Brunet *et al.* 2002). The morphology of the Sacred Lake in antiquity, the source of its waters and its placement in the architectural plan of the city were thoroughly investigated (Brunet *et al.* 2001). This exploration led to the conclusion that the sacred lake was formed in a natural depression, the outline of which was at some point framed by walls and terraces, following the construction needs of the city. The lake was mostly fed by spring water which was issued naturally at its depth. Channels had been constructed to remove any overflow, which might result from a particularly rainy season. There did not seem to be any permanent connection of the lake to the sea.

The question which arises is whether this is the lake referred to in the sacred inventories, which was so rich in fish that it provided a considerable income for the sanctuary. Obviously, this, due to its restricted size, is not a feature which could sustain a naturally renewable fish population which could furthermore be systematically harvested. In theory however, it could be a fish pond, which might serve as a breeding area for fish such as the grey mullet, which can grow in large numbers even in fresh-water. For such a fish-breeding programme to be feasible, a renewal of fish spawn in the lake is necessary, therefore, a source of it must have been available. The aforementioned geo-archaeological project suggested that the waters of Inopos, the main ravine of the island which was flowing seasonally, at least before the 2nd century B.C., reached the sea in the Sacred Port area, after passing through a coastal marsh which was sustained by spring waters and the Inopos seasonal input. Such an environment could perhaps be the source of the fish larvae or fry which would then be placed in the Sacred Lake.

Alternative possibilities for the identity of the "Lake", however, should also be explored. One of the definitions of the word "*limni*" given by the Liddell-Scott dictionary of Greek language (Liddell and Scott 1929, v. 3) is that *limni* is a mass of stagnant marshy water. Delos and the nearby Rheneia are, unlike the rest of the Cyclades, rich in underground waters which spring near their coasts (Paris 1916, p. 11) creating minuscule wetlands. Records of the Temple estates refer to the incomes generated as rent for an estate called "*Limnai*" (Lakes) on Rhenia, a nearby island. At the middle of the 20th century there existed, on Rhenia, an inland pond, which turned into a marsh during the drier periods of the year (Kent 1948, p. 252, 305). Kent claims that, by the time he did his research on the temple estates of Delos, this pond was not open to the sea. However, as no detailed examination of the area has been conducted, the issue remains open. An alternative wetland, which could perhaps be called "Limni" (lake), is the coastal marsh, at the estuaries of Inopos, mentioned above.

Despite the problems in the exact identification of the Delian "*Limni*", as a possible case of a temple-managed fish-breeding establishment, epigraphical and literary records refer to the sacred waters as a fairly common element in the landscape of sanctuaries. From the variety of types of attested water features it becomes clear that "sacred waters", be they lakes, rivers or springs, obviously had functions other than the utilitarian (sources of fish). What dictated the inclusion of sacred water bodies in a sanctuary layout? In chapter 6 the importance of the sea as an imaginary place has been discussed briefly; the sea as a place with its own distinct but not always visible geography, where gods and mythical creatures lived, and where mysterious and transitional things could happen. Inland sacred waters could be viewed in the same manner. Their nature, their contribution to cult and their relation to the gods and to local mythologies are perhaps issues to look into. In such a framework, the breeding of fish, but also the contribution of fish to cultic processes is particularly relevant.

7.1.3 Fresh fish for the urban market.

Fresh fish sold in the urban market has been quite a stereotypical image as far as our perceptions of life in an ancient Greek polis are concerned (Davidson 1996, p. 62). A wealth of passages in ancient texts, mostly comedies, Old, Middle and New have amply illustrated the issue (e.g. Appendix 1: Alexis *The Woman who Smacks*, fr. 57; *The Greek Woman*, fr. 76; *Heiress*, fr. 78; *Cauldron*, fr. 130-1; *The Meeting at Pylae*, fr. 204; *Phaedon*, fr. 249; Antiphanes *Adulterer*, fr. 159; *Rich Men*, fr. 188).

Most of the information available to us on fresh fish in a market context refers to Athens. It appears that the Athenian market, at least in the Classical and Hellenistic periods, was stocked with fish which could be caught in the surrounding seas, mostly the Saronic gulf. When Xenophon in the 4th century B.C. recounted an episode of the Peloponnesian war (Piraeus raid by Spartan Teleutias in 388 B.C.) he vividly described the Athenian fleet of fishing boats returning to Piraeus at dawn, apparently ready to supply the Athenian market with the bounty of the Saronic gulf (Appendix 1: Xen. *Hell.* 5.1.23). The whitebait from Phalero and Megara, and a variety of other species were sold at the fish stalls in the Agora (see discussion in 5.3.7). Fish originating in more distant areas were also sold there (see the case of the Kopaic eels mentioned earlier). Very little of this wealth has been uncovered archaeologically (Table 5.1f).

The different fish species that have been recovered at the Roman Akropolis on Mytilene (Table 5.1g) and in the Late Roman Merchant's house at Itanos (Table 5.1 a) appear to have been purchased at the market. The Itanos case is quite interesting. There, 24 fish taxa have been identified, almost all coastal. Many of the fish consumed in the house had been large, and their catch probably required professional effort. Although the smaller of the identified fish in the Itanos assemblage were represented with all parts of their skeleton, the largest of them, were mostly represented by head bones (Table 7.1). This anatomical selectivity could be viewed as intentional and could be taken as an indication that only that part of the fish were bought and brought on site. Most of the fish remains found in other urban contexts (Table 5.1a-g) could also be viewed as purchases from the market, but no tangible indication of this is available.

Based on the literary and inscriptional evidence, ancient historians agree that the "*agora*", the market place, was organized in a coherent manner, with different types of goods sold in different parts of it (Ehrenberg 1968, pp. 138-9). One, therefore, would expect the fish stalls to be concentrated in one place. Linguistic evidence, such as the worlds *almaiopolis* (the seller of *almi,* a mixture or salt and water used in fish pickling), *garopolis* (the seller of garum), *tarichopratissa* (the female seller of preserved fish), *temachopolis* (the seller of *temachi*, a type of preserved fish) which denote a narrowly specialized trade (Curtis 1991, index), suggests that fresh and preserved fish were sold in different places, thus supporting the idea of fish-market segregation.

Table 7.1 Itanos. Fish anatomical part representation.

	Cranial and branchial bones	Pectoral and pelvic bones	Vertebrae	TOTAL NISP
Dasyatis sp.			1	1
Engraulidae			1	1
Congridae	8		2	10
Gadidae			2	2
Serranidae	22		20	44
Epinephelus sp.	3		2	5
Serranus cabrilla/scriba	3		3	6
Sciaena umbra	1			1
Mullidae			1	1
Sparidae indet. sp.	34		87	121
Boop boops	18		4	22
Dentex dentex	4		1	5
Diplodus sp.	2		2	4
Lithognathus mormyris	1			1
Pagrus pagrus	8			8
Pagellus sp.	3			3
Pagellus erythrinus	1		9	10
Sparus aurata	1			1
Spondyliosoma cantharus	3		1	4
Sparidae/Centracanthidae	2		21	23
Centracanthidae	14		10	24
Chromis chromis	3			3
Labridae indet. sp.	2			2
Sparisoma cretense	53		20	73
Uranoscopus scaber	1			1
Scombridae			1	1
Sphyraena sphyraena			1	1
Scorpaenidae	1			1
Indeterminate small	6		43	49
Indeterminate medium	3	1	54	58
Indeterminate large	2		4	6

The agora in Roman Corinth provides an archaeological testimony of where fresh fish were sold in an urban market. There, the *tholos* building, situated near the Peirene Fountain, on the road that connected the harbour to the forum, apparently housed the Roman fish and meat market (Appendix 3b and 4). Architecturally it is very similar to a securely identified fish and meat market establishment in Pompeii (Mau 1904, pp. 94-101). Furthermore, at Corinth, two almost identical Latin inscriptions, which have been found broken in many fragments in the vicinity of the tholos building (Appendix 2: Kent 1966, West 1931), advertise this trade. They are probably the signs hung above the two entrances to the establishment and they identified both the function of the building and the owners of the place. The *tholos* building found in the area of modern day Glyphada, a modern suburb of Athens (ancient deme *Aexonae*) could be another case of a fish market (Keramopoulos 1919; Konsolaki-Giannopoulou 1990, pp. 28-29, fig. 5).

What is perhaps most interesting about the *tholos* building in Corinth, is the fact that at least in that case, in the Roman period, fresh fish and meat were closely associated. The fish and meat segregation which appears so prominent in the Athenian market of an earlier era (see above) does not seem to apply here. It would be interesting to know whether this is coincidental or if it reflects a different organization of the market. The material bridging (within this building) of the conceptual gap between fish and red meat, which was so fundamental in Greek culture, might reflect a different attitude in Roman times towards the issue. Here, archaeology, as it stands at the moment, can be of no help. The literary sources are also problematic. The division between fish and red meat, which corresponded to a segregation of the market stalls, is mostly documented by literary texts which had been produced in Classical and Hellenistic period but, interestingly enough, were preserved and re-circulated in Roman times (mostly by Athenaeus). Such a reuse of the older literary tradition was a result of broader intellectual developments in the Roman world (Bowie 1970; Swain 1996) and it was very much an affair of the elite, be it Roman or Greek (Alcock 1993, pp. 16-17). Obviously for a certain social group the conceptual distance between fish and red meat was still meaningful in Roman times. But did it bear any significance to widespread sectors of the population? This issue can not be explored here, but it is possibly worth pursuing further.

References to fish sold in the urban market are abundant, albeit somewhat restricted in focus. The high prices of fish and the slyness of the fishmongers are usually the issues around which the story develops (e.g. Davidson 1997). Fresh fish were sold by fishmongers, both male and female (Ehrenberg 1968, pp. 138-9 based on evidence from Old Comedy), with the latter having an even worse reputation than the former (Appendix 1: Pherecrates *The Kitchen or Midnight Festival*, fr. 70; Wilkins 2000, p. 168 and chapter 6; Arnott 1996, pp.361-393). The mechanisms by which the fish became available to consumers in the market are fairly obscure. The relation between fishmongers and fishermen in Athens (and perhaps other cities on the Hellenic peninsula) and in the period under consideration is not quite clear.

In Hellenistic and Roman Egypt, fishermen and fishmongers formed different occupational groups often organized in professional guilds. Fishermen provided fish to the fishmongers. These acted as middle-persons making the fish available to the customers. Some overlapping of activity was not unheard of (Chouliara-Raiou 2003, pp. 361-362). The same was true in Rome during the Republic, where however, the professional boundaries were somewhat more fluid (Chouliara-Raiou 2003, p. 361, note 104).

The situation in Greece with regard to a distinct role of fisherman and fish mongers is unclear. The numerous references to fishmonger, as opposed to the almost complete silence about fishermen, in the written sources could perhaps support the idea that a similar separation was followed in this case and was taken for granted. Various inscriptions from Ephesos and Kyzicos on the Asia Minor coast, in the Roman period attest to the organization of fishermen into professional associations (Purcell, 1995; Appendix 2: OGI 496; IKyzicos I, 211; Frisch 1983). Most of them (e.g. Frisch 1983) were probably related to the exploitation of migratory fish (Dumont 1976/77). In some cases, like the Late Roman tombstones from Korykos, some specialization is implied (Appendix 2: Mentzou 1975), as we find not only the generic term *alieus* (fisherman) but also terms like *sagineus* (net fisher). Whether this was also happening on the Greek mainland and in earlier periods remains uncertain but quite plausible, judging by certain terms used casually in literature (e.g. Appendix 1: Aesch. *Net Drawers;* Ar. *Nub.* 559). It also remains uncertain whether fishermen of a different type of fish resources (inshore or euryaline) were also organized in some way.

Whatever the organizational status of the fishermen might have been, the separation of fishermen from fishmongers, which is implied in the available data, bears some interesting implications. If the fishermen had no direct contact with the urban consumers but only with the middlemen, the fishmongers, then there was very little opportunity for the two worlds to meet. Perhaps this division of activities, fishing and fish selling was the mechanism which set in gear the intellectual and conceptual gap suggested in the previous chapter (chapter 6.2). It should be noted however that this gap can only be inferred in relation to Athens. What the situation was in other parts of Greece can not be assumed on the basis of the available evidence.

One of the most conspicuous and repeated motifs found among the extant fish-related fragments of ancient literature is the need to regulate fish prices and the proper manner of trade conduct, not only on the part of the fishmongers but also of the customers. Compulsive fish buyers are criticized in comedies and oratory as severely as ruthless fishmongers (Appendix 1: *The Woman who Smacks*, fr. 57; *The Greek Woman*, fr. 76; *Heiress*, fr. 78; *The Meeting at Pylae*, fr. 204; *Phaedon*, fr. 249; *Rich Men*, fr. 188; Archippos *Fishes*, fr. 14-34; Chrysippos of Soli *Inscriptiones Insanos Esse et Impios*, fr. 667; Diphilos *The Busybody,* fr. 67; Eupolis *Kolakes*, fr. 160 and 174; Lynceus of Samos *How to Buy Fish in the Market*; Pl. *Chrm. 163b; Resp.* 372e-373; Plut. *Quest. conv.* 668a; also see Wilkins 1993, p. 195; Davidson 1993). No consensus has been reached among scholars as to whether prices of goods, especially foodstuffs, were actually controlled or not and under which circumstances (Millet 1990, p. 192, note 53; Arnott 1996, pp. 361-393). For the present discussion, the issue of the actual existence of such controlling mechanisms for fish prices is not crucial. A point to be noted is the fact that fish in the context of an urban market (especially Athenian) are stereotypically characterized as expensive. When more details are given, it becomes evident that some fish, certain species or size ranges, were the most expensive, but there was also a range of fish, mostly of small size, which were very cheap (Davidson 1993, p. 56; for fish prices in Classical Athens as given in comedy see Boeckh 1886, for an alternative view of the fish prices see chapter 9).

More instructive in our quest is the fact that imaginary or real control mechanisms found a place in the most public form of literature, the comedies. Fish in comedies are typically referred to in connection with constantly heightened covetousness and gluttony. Such an approach gives an impression of vivid market transactions around fish. The ability to buy fish despite the obstacles (high prices, fishmonger's slyness), is presented in the sources, especially comedies, as an achievement. It implies high status, education and a cosmopolitan life style. In the literary discourse around fish in the market, which covers several centuries, fish appears as a luxury food (e.g. Appendix 1: Alexis *Crateia or the Drugist*, fr. 115; *Principal Dancer*, fr. 200; Ar. *Ran.* 701-3; Diphilos, *The Merchant*, fr. 31), the acquisition of which requires not only money but also the mastering of high taste (Appendix 1: Alexis *Principal Dancer*, fr. 200; Antiphanes *Butalion*, fr. 69; Lyncaeus of Samos *How to Buy Fish in the Market*; *Centaur*, fr. 1). These issues have been extensively discussed in recent historical works (e.g. Davidson 1993; 1997, Wilkins 1993; 2000).

What needs to be emphasized at this point is the fact that all these ideas which connect fish to luxury are only part of the issue. Shopping for fish by ordinary people and for purposes other than elite luxury display was obviously a parallel reality, even if it remains largely invisible. The presence of fish bones in all sorts of non-luxury contexts (see Table 5.1a-g) and the negative, but quite detailed description of poor men buying cheap fish (e.g. Appendix 1: Alexis *Odysseus at the Loom*, fr. 159) imply that non-elite fish consumption was taking place as well and was perhaps quite widespread.

7.2 Preserved fish.

The preservation of fish has been widely used historically, and in many cultures it was a way to overcome the problem of quick spoiling. Preservation offers the opportunity to make fish storable and transform a seasonal abundance into food available throughout the year. While these strictly economic consequences of fish preservation are true, other less obvious forces are also at play. When fish is preserved its taste and texture change. As a consequence, the way it is prepared for consumption, the combination of foods it is eaten with and the very way it is consumed change drastically (see discussion in chapter 9). Considering the variety of fish, preservation methods and added ingredients available, it becomes obvious that, starting from a given range of fish, preservation offers the possibility of a great variety of different food items and tastes.

Consumption of preserved fish in various parts of the Greek world during the period under study is amply documented in the literary sources (e.g. Curtis 1991, pp. 6-15; Dalby 1996, pp. 75-76). Also, archaeological finds of fish salting establishments at the periphery of the Greek world, mostly the Black sea and the Western Mediterranean, and of transport vessels for such products are eloquent witnesses of how widespread these practices were, despite the fact that most of the details remain elusive (for a detailed discussion of the issue and extensive references see Curtis 1991; 2001, pp. 316-321, 402-417; 2005; Højte 2005; Lund and Gabrielsen 2005; Trakadas 2005). The identification, in recent years, of the actual remains of preserved fish in various types of containers, further illuminates the issue (e.g. Wheeler and Locker 1985; Struder 1994; Bruschi and Wilkens, 1996; Desse-Berset and Desse 2000).

7.2.1 The technical parameters of fish preservation.

The preservation of fish, especially on a commercial scale, requires repeated large catches, restricted to a short time period and geographical location. Thus, an abundance of fish is created, even after covering the needs of the nearby communities for fresh fish. Preservation can be achieved by various means, all of which aim to prevent spoiling for short or longer periods. Salting, drying, smoking and fermentation are the most widespread, with combinations being regularly applied (Fig. 7.4; Curtis 1991, pp. 15-18). Some of these methods are fairly simple, both in terms of materials and technology used (e.g. sun drying), while others require extra ingredients (e.g. salt, vinegar, wood/smoke) and a complex technology (e.g. for garum and other fish sauces) (Curtis 1991, pp. 6-26).

Fig. 7.4 Salted and sun-dried chub mackerels (*Scomber japonicus*), called in Greek "*tsiros*" (photograph by D. Mylona).

Some fish are susceptible to some preservation methods and not to others. Fish such as the Scombridae for example, are unsuitable for drying due to their high oil content. Fish oils are prone to oxidation when exposed to the air, so quick salting or/and smoking are more suitable preserving procedures (Cutting 1955; 1962, p. 175; Pyke 1964, p. 75). Different methods of preservation result in products which are susceptible to spoiling to varying degrees. As a general rule, the longer the exposure of the fish to the preserving medium the more durable the final product is, although it may loose the characteristic fish taste and odour and succulence of flesh (Curtis 1991, pp. 17-8). One example of such heavily cured fish in the Classical world were the *melandria*, a type of processed fish of hard texture that were said to be shaped and coloured like oak wood, or to have the form of darkened roots (references in Curtis 1991, p. 10, note 16). Salting is the most efficient way to preserve fish due to the chemical properties of salt (Curtis 1991, pp. 16-18; 2000, p. 407). Often areas famous for their "*taricheia*", the fish preservation establishments, were also famous for their salt pans (for sites on the Black Sea coast see Højte 2005 and references to relevant ancient sources therein).

All fish can be preserved. Ancient literary sources often refer to preserved fish with no elaboration on the exact fish species, using words which designate preservation method, body part or origin rather than species (see a collection of such terms in Curtis 1991, index). When the species is determined, these are usually the grey mullet, eel, cat fish, and some small fish such as the picarel and sprat (Curtis 1991). By far the most typical preservable fish however are the various members of the Scombridae family, with tuna being the most valued. The main reason for the focus of large scale preservation activities on a few kinds of fish, mostly the Scombridae, the Clupaeidae/Engraulidae and the Mugilidae is two-fold. Firstly, these fish have a seasonal behavioural pattern and they tend to be massively caught at certain times of the year. Secondly, being fatty fish with high blood content, they spoil quickly. Preservation makes this seasonally abundant but vulnerable resource available for longer time periods and to more distant places.

The preserved fish, may, as has been noted earlier, take a variety of forms, depending on factors such as the preserving medium, the type of fish, the length of preserving process, the additional flavouring ingredients, such as oil, wine and vinegar, the anatomical part of the fish chosen to be preserved and the shape of the fillet, if the fish is not whole. The preserved fish could be available in a dry form, as in dry salted fish or in a semi-liquid form, as in the case of fish in brine, or of "*hallec*", a paste, made of fermented assorted small fish and/or viscera of larger fish. Alternatively fish could be turned into a fish sauce, widely known during Greek and Roman antiquity as "*garum*" (Corcoran 1963; Curtis 1983; 1991). The basic principle in the production of garum was the prolonged exposure and fermentation of a mixture of fish and salt in the sun or alternatively by artificial heating. Garum could later be re-processed with the addition of flavours (Curtis 1979). The variety of preserved fish forms is reflected in the wealth of related terminology (Table 7.2). Taking into account that the relevant terminology known to us is partial, in the sense that it only reflects mass produced marketable varieties of preserve fish and only includes those terms that happened to survive to date, we may assume the actual existence of an even richer vocabulary and thus variety of end products.

7.2.2 *Evidence for fish preservation in Greece.*

Little concrete information on locally preserved fish is available from the Hellenic area. Classical Thassos in the northern Aegean was known for its preserved fish, which had been the object of trade and for the Thassian pickle (references in Curtis 1991, pp. 117-8, 139; see also discussion in chapter 5.3, Table 5.3). Eels from the Strymon River were apparently preserved, as they found their way to such distant markets as Athens (chapter 5.3, Table 5.3). We must consider, however, that fish that were preserved in various places around the Aegean may remain invisible archaeologically. Fish remains found at inland sites, such as the Hellenistic and Late Roman Eleutherna on Crete, or Pyrgouthi in the Berbati Valley in the Argolid (Table 5.1d and g) are probably the remains of such an elusive and certainly not renowned trade. More visible are the cases of exotic fish such as the cat fish found in Kommos (see discussion in chapter 9.2.3 and Table 5.1a)

The story of the massive catches of river fish at Volvi, in Northern Greece, and their subsequent preservation for local consumption (Appendix 1: Hegesander *Commentaries*) attests that such cases might have been common but are unknown to us. The Volvi case found its way into literature as anecdotal material, probably because of the connection of the particular practice to a local cult.

Additional, albeit indirect, evidence for the local production of preserved fish products may be provided by the inscriptions which refer to the systematic exploitation of certain fish resources. The inscription from Epidauros (Appendix 2: IG 4[2] l. 76-77) which regulates the revenues from the exploitation of the common tuna fishing grounds and salt pans between Troizen and probably Arsinoe in the Argolid, is one such example. There, the two basic ingredients of the fish preservation business are found together. An additional example originates from the island of Cos in the Dodecanese. An inscription dating from the beginning of the 2nd century B.C. makes reference to the leasing of public watch-towers, immobile structures used in tuna fishing, to private citizens (Appendix 2: SIG[4] 1000). Assuming that the investment in these towers indicates a systematic and profitable exploitation of the migratory shoals, the existence of some type of preservation establishments in the area may be considered certain.

The identification of structures related to fish preservation could be illuminating for the local production of preserved fish. Such establishments, which are often quite distinctive architecturally, have only been identified at one spot in Greece. This is the case of a group of two, now submerged vats, on Kastelorizo island (ancient Megisti), near Rhodes (Appendix 3), but as no elaboration is offered in the publication, their identification as fish preservation vats rather than fish-tanks is far from secure. Given the lack of research interest in this type of maritime issues, and the conviction among archaeologists and ancient historians that fishing (and related activities) in Ancient Greece was unimportant (see discussion in chapter 2.1), such structures might have gone unrecognized. There is one case which could fall into this category of miss-interpreted features.

Table 7.2 Ancient Greek terms related to the variety of preserved fish products.

Ancient Greek term	Translation	Quality
Trigonon	Triangular	Shape
Tetragonon	Square	Shape
Kyvion	Cube	Shape
Akropastos	Lightly salted	Type of salting
Imitarichos	Half salted	Type of salting
Iminiron	Half salted	Type of salting
Teleios	Fully salted	Type of salting
Ypogastrion	Stomach portion	Anatomy
Melandrya	Dorsal slice	Anatomy
Omotarichos	Pickled flesh of tuna's shoulder	Anatomy and type of fish
Tarichos tilton	Slated without scales	Processing of raw material
Tarichos lepidoton	Salted with scales	Processing of raw material
Kerameia	Fish pickled in earthen pots	Type of packing
Oraion	Pickled young tuna	Age of fish
Tarixos pontikos	Salted fish form the Black Sea	Origin
Sardika tarichi	Salted fish from Sardis	Origin
Garos	Fish sauce	Texture of product
Almi	Type of fish sauce	Texture of product
Tarichos	Pickled or salted meat or fish	Related term
Tarichiros	Pickled or smoked	Related term
Tarichion	Piece of salted meat or fish	Related term
Tarichopleos	Rich in salt fish	Related term
Zomotarichos	Stewed salt fish	Related term
Tyrotarichos	A dish made of salt fish, eggs, chicken livers and various other ingredients	Related term

The table is based on terms taken from Georgacas 1977 and Curtis 1991.

The fish-tanks located at Siteia, on the NE coast of Crete comprise a cluster of 10 rectilinear basins (Appendix 3b; see also chapter 7.1.2). In relation to them, there have been observed two more basins, one apsidal and the other trapezoid communicating via a narrow channel (0.10x0.7m) dug near the floor of the apsidal compartment. The trapezoid compartment has a direct exit to the sea through two more channels. The apsidal compartment, which had a floor sloping towards the sea and a narrow, low bench around its sides, unlike the rest of the fish-tanks, was plastered with hydraulic cement. The archaeologist who recorded the site recognized the different nature of these two basins, and interpreted them as part of a bath complex, which might have lain near the fish- tanks (Davaras 1974, p 93, pl. 15b, 16, 17).

While this may be true, and these unusually shaped depressions might have been a swimming pool with its feeding water reservoir, their alternative interpretation as fish preservation vats is equally plausible. The plastering of the depression with hydraulic cement, the sloping of the floor, the existence of a low ledge around the perimeter of the tank, and the availability of an opening for cleaning are standard or common features of fish preservation vats (Curtis 1990, pp. 53-55, Tracadas 2005, pp. 70-72). Furthermore, their association with fish-tanks is also common elsewhere, where such establishments have been properly investigated (e.g. the coast of Alicante in Hispania Terraconensis, Curtis 1991, p. 53, 180 and references therein). As it has been noted previously however (chapter 6.1.2), the Siteia complex has not been excavated, and for this reason we can only speculate on the exact nature of its features.

7.2.2 Preserved fish as an object of trade.

Preserved fish could be expensive, luxury products (Appendix 1: Diocles of Carystos *Hygiena*; Euthydemos *On Salt Meats*), often imported from long distances (e.g. Appendix 1: Alexis *The Epidaurian*, fr. 77; Ar. *Ach.* 671), or could be cheap fare, food for the poor (Appendix 1: Ar. *Vesp.* 490-5; Chionides *Beggars*, fr. 4; Pherecrates, *The Kitchen or Midnight Festival*, fr. 70) (Edmondson 1987, p.102, Curtis 1991). The archaeological detection of this type of fish is quite difficult. At the moment it is almost totally restricted, at least in Greece, to the indirect evidence of the vessels, which contained this type of product. The vessels, mostly amphorae, are identified as garum containers, on the basis of their shape and in some cases, the "*tituli*", a type of short inscription serving as a title or sign on the amphora's handles (Curtis 2005, pp.40-41). No chemical analysis has been used to confirm their contents. Furthermore, packaging in perishable containers, though plausible, would be archaeologically invisible. Examples such as the remains

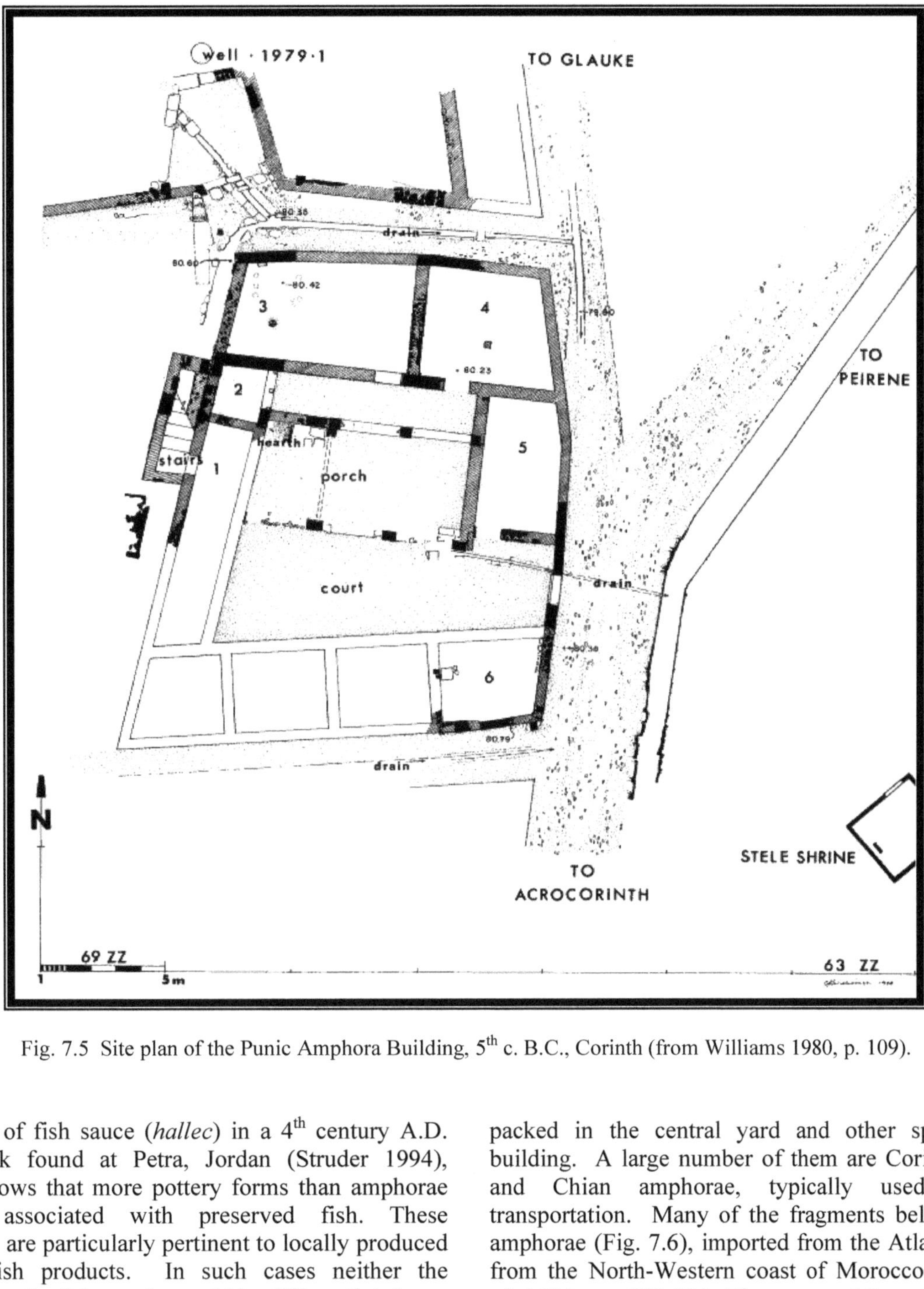

Fig. 7.5 Site plan of the Punic Amphora Building, 5th c. B.C., Corinth (from Williams 1980, p. 109).

of one type of fish sauce (*hallec*) in a 4th century A.D. pilgrim flask found at Petra, Jordan (Struder 1994), however, shows that more pottery forms than amphorae could be associated with preserved fish. These observations are particularly pertinent to locally produced preserved fish products. In such cases neither the containers nor the fish species could be differentiated.

Imported equivalents are in some cases more visible, mostly again due to the distinctiveness of the containers. The most renowned case of preserved fish imported from afar comes from Corinth. Remains of such imported fish have been found in the Punic Amphora Building in Corinth, which dates to the 5th c B.C. (Fig. 7.5). The building has been interpreted as the commercial establishment of a merchant, who specialized in imported goods (Appendix 4). The composition of the pottery assemblage in this building is peculiar. Very little domestic pottery is present. On the contrary, a great abundance of amphorae fragments have been found packed in the central yard and other spaces of the building. A large number of them are Corinthian, Coan and Chian amphorae, typically used for wine transportation. Many of the fragments belong to Punic amphorae (Fig. 7.6), imported from the Atlantic, perhaps from the North-Western coast of Morocco (Maniatis *et al*. 1984, pp. 208-221; Zimmerman-Munn 1983, p. 264; 2003, pp. 200-6). These were the typical containers of preserved fish. Fish-scale segments have actually been found still adhering to some of the fragments (Fig. 7.7). Furthermore, loose, sizeable fish bones have been found among the amphora fragments (Williams 1978, p. 20; 1979, pp. 113, 117-8).

The collection of these bones during the excavation has been done by hand, and there is no way to ascertain whether other, smaller fish remains were also deposited among the amphora fragments. The available fish remains, which have been identified only from photographs, appear to be from tuna (indeterminate

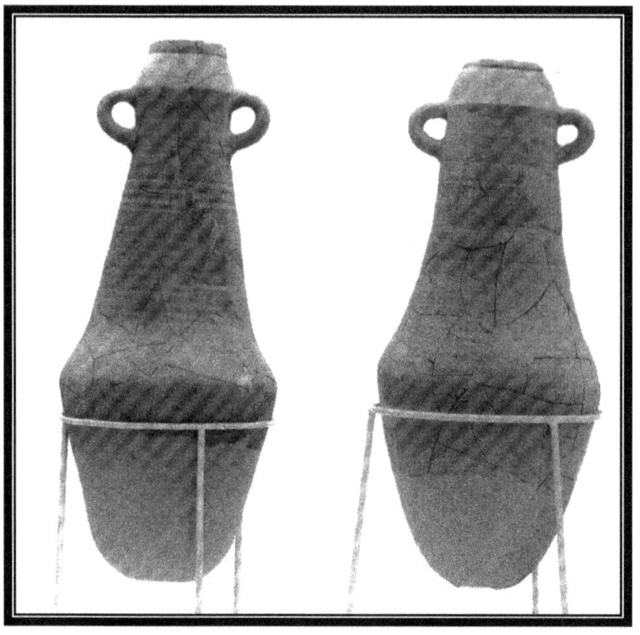

Fig.7.6 Punic amphorae of the Maña-Pascual A4 type (from Zimmerman-Munn 2003, p. 200, fig. 12.3).

variety) and sea bream, perhaps a gilthead sea bream (*Sparus aurata*) (Zimmerman-Munn 2003, p. 201, note 47). They had apparently been cut into slices, or rectangular fillet pieces, and packed in their containers. We do not know whether the fish were preserved in a dry or wet medium, but the fact that some of the Punic amphorae have a porous interior surface while others are water-tight, suggests that both versions existed (Maniatis *et al.* 1984, p. 221).

Apart from the preserved fish, the excavation at the Punic Amphora Building produced abundant evidence for the importation and sale of wine from sources such as Cos, Chios and Mende, all renowned in antiquity for their wines (Dalby 1996, p. 99-101; Papadopoulos and Paspalas 1999). The combination in the same trading venture of imported fish and drink implies perhaps a type of specialization which is oriented towards the origin and association of the goods rather than their nature. Such a phenomenon of trading ventures along lines of associated goods has been observed in other cases of various dates around the Mediterranean (Millett 1993; Laurence, 1994, p. 52; Kerschner 2004) In the Corinthian case, however, it might be instructive to consider that even in literature, such as Atheneus' *Deipnosophistae*, wine and fish, much of it imported, are two foodstuffs which are regularly associated together. Furthermore, both often represent luxury and even decadence.

A collection of terms related to the trade in preserved fish however (as these are found in Curtis 1991, see index) indicates that a commercial specialization based on one type of foodstuff, i.e. preserved fish, was common practice. The sources for these terms range in date from classical to late Roman, thus indicating a continuous tradition in this field. "*Opsopolis*", "*tarichopolis*", "*almaiopolis*" "*temachopolis*", "*garopolis*", "*oraiopolis*", "*tarichigos*" and "*tarichas*" are some such terms, which denote the profession of selling distinct preserved fish products and imply a specialization even within the field of preserved fish. An inscription testifies to the existence of at least one "*tarichopoleion*" i.e. a shop selling preserved fish on 4th century B.C. Delos (Appendix 2: Henning 1983).

In conclusion, the following points could be made in relation to preserved fish. The discussion of the economic importance of preserved fish in antiquity usually stops at the realization that preservation makes a temporary sensitive resource, storable and available for longer time periods, thus enabling its efficient exploitation. There is however one basic quality of the preserved fish, which adds to its importance, especially in certain contexts, such as that of the sophisticated, complex worlds of a Greek city. This quality is related to the transformation of the plain fish, or often of its discarded parts, into something else; into a tradable, highly identifiable good. Although such products are often invested with great economic and symbolic value, their distinctiveness applies both to cheap and expensive fair. Preserved fish often acquired a specific identity related to their place of origin. Extra processing and the addition of both labour and extra ingredients added to economic value, not unlike the case of elaborate textiles or decorated pottery.

7. 3 Conclusions.

Access to fish can not be viewed as uniform. The ease with which people could access fish depended, to high degree, on the proximity to the resource, with people living in coastal settlements or near rivers and lakes being able to eat fish more regularly and perhaps in greater quantities. Quite a lot of this type of transaction remains archaeologically but also literarily invisible.

Although the above observation is true to a large degree, our combined evidence suggest, that there was in place a parallel, complex way to access fish. This was through market mechanisms, structured ways by which fish became available to the consumers. Fish stalls in the market, travelling fishmongers, fish kept alive after their capture, fish breeding and marketing of preserved fish are some examples. When fish became available in this manner, the proximity of the resource and the type and amount of landed fish appear to be of lesser importance. Some of these mechanisms seem to have been set up because of the demand for fish, despite adverse conditions of availability. The twice documented, by independent sources, desire for fresh marine fish in mountainous inland Tegea in the Classical era, is a good example of this phenomenon.

Fish as a foodstuff became available to the consumers in two broad forms, fresh or preserved. These categories are quite distinct. Fish as food in each case requires a totally

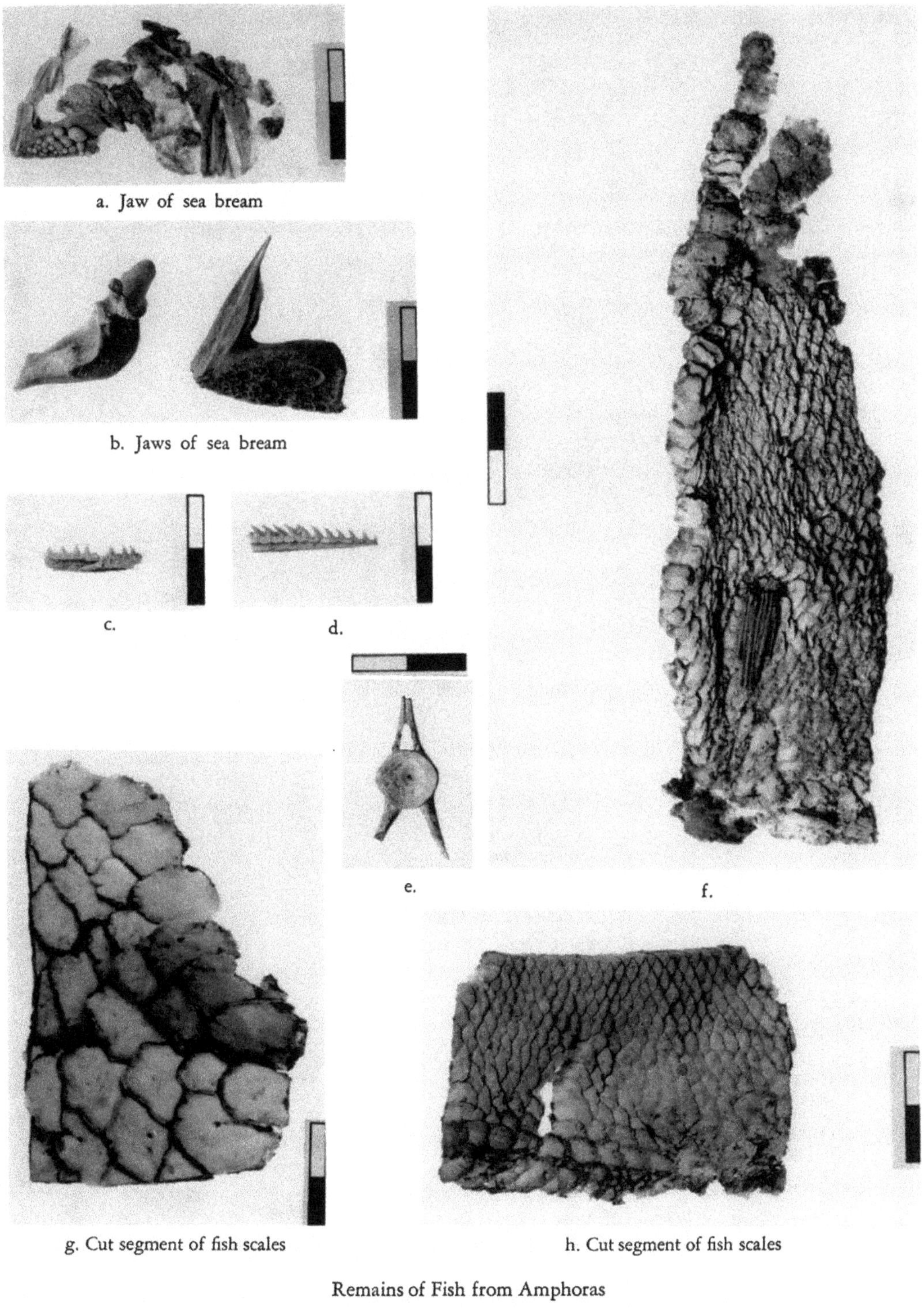

Fig. 7.7 Sections of fish scales found adhering to Punic amphorae fragments and other fish remains found in the Punic Amphora Building (from Zimmerman-Munn 2003).

different treatment. Furthermore, a totally different set of acquisition conditions was formulated for fresh and for preserved fish. Fishmongers, middlemen, market places, processing facilities, packaging and, in some cases, even advertising, were all crucial elements in this process. The cornerstones of these transactions, the fishermen and the consumers could be linked in different sets of relations. Sometimes they were close, even in direct contact. The fishermen and their families consuming their own catch or the fishermen directly supplying the consumers are two expressions of such a direct relationship. Although this appears to have been a simple and perhaps widespread pattern along coasts in Greece, little documentation for it exists to date. On the contrary, through the acquisition mechanisms described above, some fish (as material objects and/or as ideas) left the world of the primary producers, the fishermen, and became incorporated in a different world, the world of the consumers.

The discourse on fish-eating, which is such a prominent feature of ancient Greek literature, is based exactly on this dichotomy. It refers to the consumption of fish, fresh or preserved, which have been divorced from its natural state and turned into an element of urban culture, into something which lives in the city.

CHATER 8

WHEN, WHERE AND WITH WHOM TO EAT FISH - THE SOCIAL CONTEXT OF FISH CONSUMPTION

The literary sources insist on placing the consumption of fish and other sea food in the banqueting halls, with tables set somewhere in the city or its surroundings (on the location of food consumption in general see Dalby 1996, pp. 12-16). It is presented as an almost exclusively secular affair (Davidson 1993; 1995; 1996; 1997; Wilkins 1993; 1996; 2000; 2003; 2005), for the elites (Davidson 1997) but also for middling citizens (Fisher 2000, Wilkins 2000, p. 65-66). Archaeology on the other hand, especially in recent years, through the discovery and analysis of inscriptions and fish bones and shells, seems to have more to say about the consumption of marine food within the precincts of sanctuaries (e.g. Rose 2000). Being offerings or plain menu items, eaten by mortals or sacrificed to gods, heroes and venerated dead, fish appear to be involved in the cultic sphere, much more commonly than what was previously thought. Ordinary, day to day fish-eating is barely visible in either discourse.

This chapter will attempt to explore the social context of fish-eating focusing on the occasions where fish was considered suitable or unsuitable for consumption, on the localities where such consumption took place and on the individuals or groups that were entitled to participate in fish-eating. The treatment of the subject is by no means exhaustive. The cases discussed have been selected mostly because they provide archaeological evidence relevant to the issues under discussion. The distinct images created by the different types of data and the biases of our sources are crucial in the following discussion.

8.1 Fish-eating with the gods and the dead.

One of the most widely-held ideas about fish in contemporary historical and archaeological literature is that Greeks believed that fish did not belong to cult. It was not a suitable victim for sacrifice to the Gods. The tuna and the eel are considered exceptions to the rule (Fig. 8.1). The explanation given by scholars is that these are not like other fish because they are bloody, thus conforming to one of the demands of sacrifice, which is to shed blood on the altar (Burkert 1983, pp. 204-212; Durand 1989, pp. 127-128; Sparkes 1995).

The anatomy of this modern discourse is of some interest to the investigation of the role of fish-eating in cultic contexts. The above idea is actually based upon a few literary references to actual fish sacrifice that span in time from the 4th to the 2nd century B.C. (Appendix 1: Agatharchides *European History*; Antigonos of Carystos *On Dictions*, fr. 56A-B; Crates *Sacrifices at Athens*; see Durand 1989, p. 127, note 37). The discussion is of course about what is considered a formal sacrifice to the Olympian gods (as it is described in Burkert 1985 and Koinski and Olsen 2001), and "cult" is viewed in a strict manner, as a set of formal actions which aim at the veneration of a god. Fish involved in more profane activities within sanctuaries, such as dinning on food other than that of the sacrificial meat, fish breeding and possibly the selling of fish, were thus not considered as pertinent, despite the relatively early discovery of relevant data in the form of inscriptions and literary testimonies. The minimal interest paid by most ancient historians to the archaeological evidence in support of a more prominent role of fish in cult, perpetuated this idea. The body of zoo-archaeological, inscriptional and literary evidence which forms the raw data for the present book (Appendices 1, 2 and Table 5.1a-g) points to a more complex situation.

Fig. 8.1 Tuna catch (from http://www.travelblog.org/Asia/Philippines/Mindanao/blog-50042.html accessed 10 January 2007).

The following discussion is based on the assumption that sanctuaries were multifunctional places, where a range of activities took place, both purely cultic and more secular, with the clear distinction between them not always discernible. The worship might be formal or mystic (for various such cases see papers in Hägg and Alroth 2005). Most importantly, sanctuaries were places where people gathered and ate (Marinatos 1993). Whether eating in the sanctuary was always linked to sacrifice or could take several forms is open to debate. Assuming the latter, we may expect to find that fish-eating was an integral part of a variety of processes within cultic contexts.

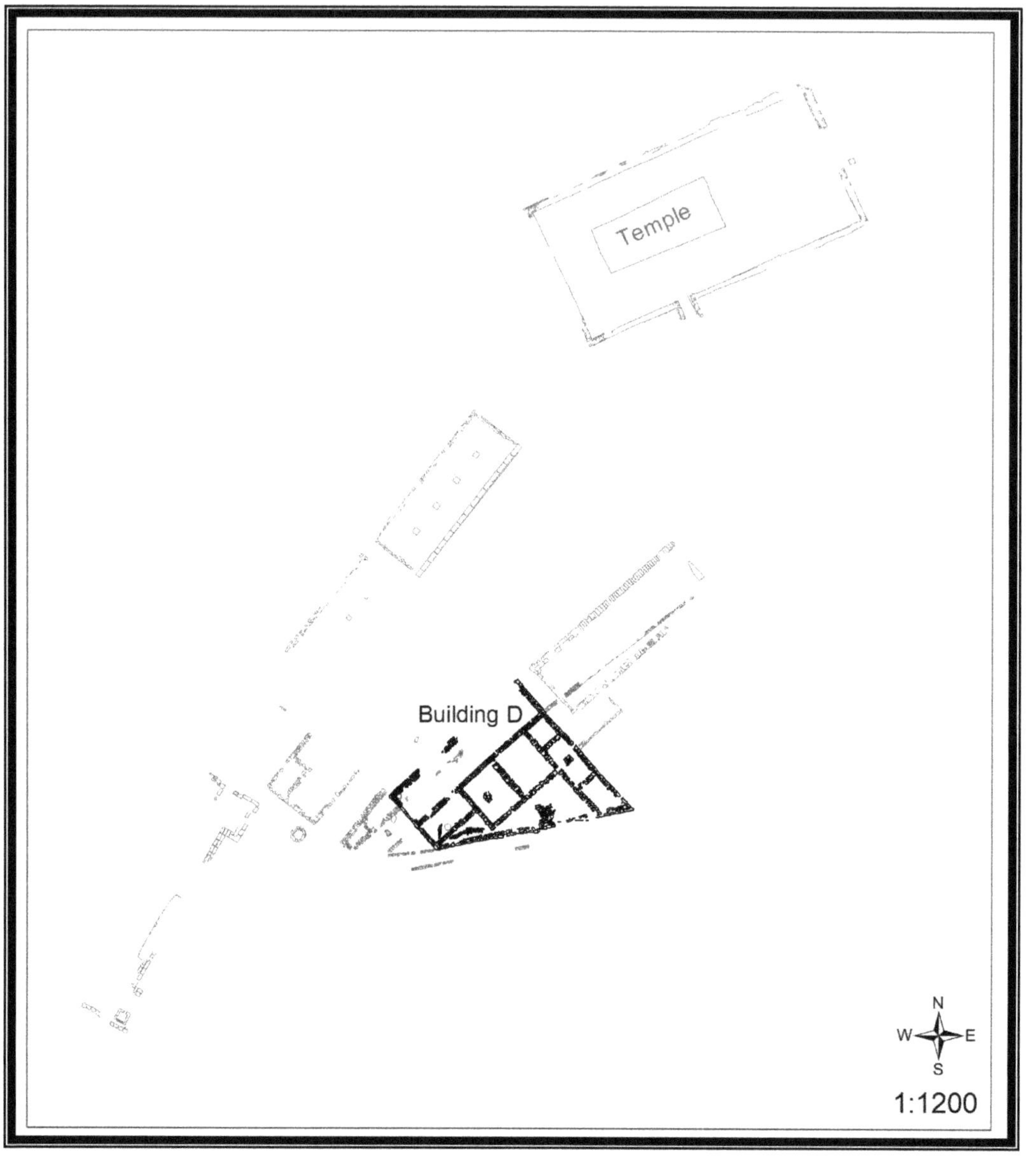

Fig. 8.2 Kalaureia site plan (The Kalaureia Excavation Project. Plan by E. Savini).

8.1.1 The dining deposit in the sanctuary of Poseidon at Kalaureia.

The recent excavations at the Sanctuary of Poseidon at Kalaureia, on Poros Island (Saronic gulf), conducted by the Swedish Institute in Athens, under the directorship of Berit Wells, aimed at the investigation of precisely this issue, i.e. the everyday life in a sanctuary and its physical environment (Wells *et al.* 2003; Wells *et al.* 2005; Wells *et al.* forthcoming; Wells and Penttinen, forthcoming). Excavation took place away from the most central feature of the sanctuary, the temple, in the area of a building of unusual architectural shape (Fig. 8.2). The building is conventionally named "Building D" and the excavation in and around it went through deposits and features ranging from the Early Iron Age (EIA) to the Roman period. The nature of the finds indicates that the area of Building D was connected to cult during the whole history of its use.

The excavation placed emphasis upon the detailed recording of contextual information for every find and, also importantly, the systematic sampling and collection of as wide a range of remains as possible. Water floatation was an integral aspect of the project and chemical analysis of soil and pottery was also used. As a result a great variety of bio-archaeological remains was collected, all tied to precise contexts (the information briefly described below is partly published in Wells *et al.* 2005; much of it is based on work, still in progress, conducted by several researchers - see Appendix 4). Fish bones were abundant, as indeed were sea shells. Mammals, birds, bird's eggs, small mammals, snakes, frogs and lizards were all present with one to several individuals each. Many of their bones bear traces of human modification (e.g. burning or cutting). Furthermore, charcoal and carbonized seeds of a variety of plants were also collected. The fish bones were present in almost every level and feature, sometimes in connection with clearly defined closed contexts and often in secondary fills. One of the most interesting features revealed by the excavation, particularly relevant to the present discussion is the so-called "dining deposit", dating to circa 165 B.C.

The "dining deposit" is an undisturbed accumulation of table debris consisting mostly of drinking as well as cooking and serving vessels, animal bones, sea shells, charcoal, carbonized seeds and small objects, including lamps and datable coins (Wells *et al.* 2005). Chemical analysis of a soil sample from this deposit revealed the presence of sea plankton on the spot, which obviously implies the presence of sea water, transported to the sanctuary from some distance. Lipid analysis of one of the cooking pots provides tantalizing evidence of what had actually been cooked in it. This consisted of non-ruminant meat, perhaps pork, and also leafy greens. The analysis can not clarify whether these had been cooked together or separately at different times (Isaksson, forthcoming). No vessel appears to be complete (Wells *et al.* 2005) and the mammal bones, despite the fact that they represent every part of the skeleton, do not comprise articulated carcass cuts (Mylona forthcoming). It appears that only a part of the dining refuse had been deposited at this spot.

The deposit had accumulated in a triangular space, specially constructed for the purpose, west of Building D (Fig. 8.3). Internal features of both the animal bone and pottery assemblages suggest a one-time deposition. The excellent preservation of the most fragile animal remains, which is quite unlike the situation almost everywhere else in the sanctuary, emphasizes this. Here follows a summary of the characteristics of this fish-bone assemblage (for an extended version see Mylona forthcoming) and a discussion of its possible significance in the archaeological exploration of the social parameters of fish-eating in a religious context.

The fish remains from the "dining deposit" consist a moderate number of identifiable bones (111) and many more non identifiable fragments (1092). These were not only well preserved in the soil but they had also been minimally affected by pre-depositional destructive factors such as scavenging or trampling. This assemblage is particularly rich in taxa (Table 5.1a). It appears that at least 18 different species of fish had been consumed. Among those, very few had been small fish (<15 cm. in

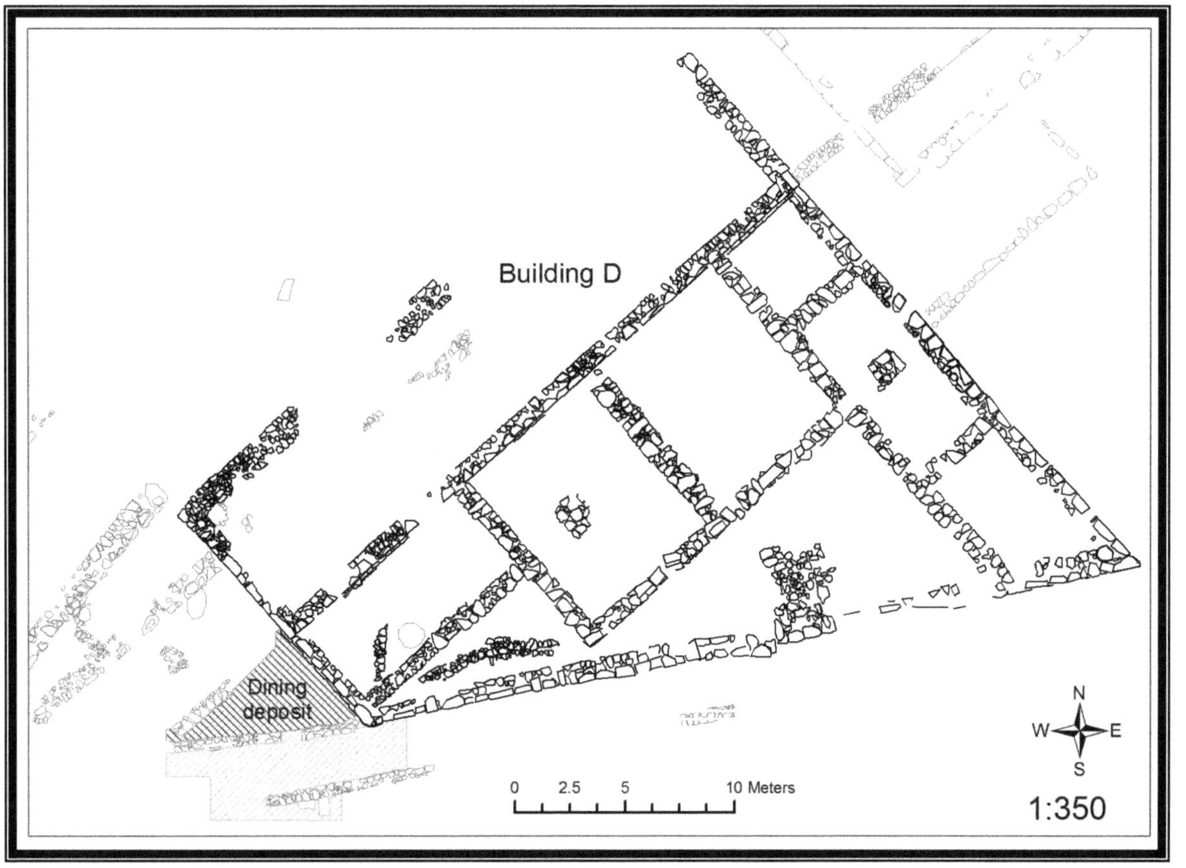

Fig. 8.3 Triangular area, west of Building D, Kalaureia. The dark shaded area denotes the main body of the "dining deposit", while the light shaded area spilled over material of the same deposit plan (The Kalaureia Excavation Project. Plan by E. Savini).

Table 8.1 Kalaureia – "Dining deposit". Fish taxonomic and body-size representation.

	Small size	**Medium size**	**Large size**	**Total**
Scombridae indeterminate			3	3
Thynnus sp.			6	6
Euthynnus aleteratus			1	1
Xiphias gladius			1	1
Seriola dumerili			2	2
Migratory total	(0)	(0)	(13)	(13)
Anguilla anguilla			1	1
Mugilidae			2	2
Brackish waters fish	(0)	(0)	(3)	(3)
Sparidae indeterminate	4	26	4	34
Dentex dentex			4	4
Pagrus pagrus		1		1
Pagellus erythrinus			1	1
Pagellus sp.		4	1	5
Sparus aurata		1		1
(Sparidae total)	(4)	(32)	(10)	(46)
Serranidae indeterminate	5	2	1	8
Epinephelus sp.		2	9	11
Serranidae total	(5)	(4)	(10)	(19)
Muraenidae/Congridae			1	1
Sciaenidae		1		1
Centracanthidae	2			2
Centracanthidae/Sparidae	5	1		6
Labrus sp.		1		1
Trachinidae		1		1
Scorpaenidae		1	2	3
Inshore various	(7)	(5)	(3)	(15)
Indeterminate large			6	6
Indeterminate medium		6		6
Indeterminate small	3			3
Indeterminate total	(3)	(6)	(6)	(15)
Total identifiable	**19**	**47**	**45**	**111**

Size ranges: small: <15 cm; medium: 15-30 cm; large: >30 cm.

length) and those are either picarels (Centracanthidae) or small combers (Serranidae) and sea-breams (Table 8.1). Most of the fish remains originate from medium size specimens (15-30 cm. in length) and the majority of them are sea breams. The common sea bream (*Pagrus pagrus*) and the gilthead sea bream (*Sparus aurata*), are those which could be identified with certainty. Groupers (*Epinephelus* sp.) and combers (Serranidae) of medium size had also been eaten along with one meagre (Sciaenidae), one wrasse (*Labrus* sp.), one weaver (Trachinidae), one scorpion fish (Scorpaenidae) and one very large picarel (Centracanthidae). It is interesting that most of these fish are species which could grow to even larger sizes (see Table 5.4). Their young age, which is reflected in their relatively small size, might be an indication that the fish had been caught near the shore (Wheeler and Jones 1989, p. 163) or alternatively might speak of intensively fished waters (Haedrich and Barnes 1997; Højte 2005, p. 140).

Finally, an almost equally large number of remains belong to large individuals (>30 cm. in length). This category of remains is the most varied. 13 of the 47 specimens are from large migratory species. Some of them are from tuna, about 1 m in length, one is from a little tunny (*Euthynnus alleteteratus*), some are from indeterminate Scombridae, one is from a sword fish (*Xiphias gladius*) and two from an amber jack (*Seriola dumerili*). Another category of large fish remains includes bones of euryaline fish which inhabit brackish and s, such as lagoons, estuaries and rivers. One eel and two grey mullet bones have been identified. Finally, several of the bones in this category belong to inshore fish, which inhabit coastal shallow or medium deep waters. Many of the remains in the large fish category are from sea breams. Those identified are the common dentex (*Dentex dentex*) and the pandora (*Pagellus erythrinus*). Several of the large fish bones belong to groupers (*Epinephelus* sp.); two of them to the scorpion fish (Scorpaenidae) and one to either a conger eel or a moray (Congridae/Muraenidae).

These fish represent various marine habitats and their catch requires different fishing techniques. The Scombridae are open sea, migratory creatures, which are caught seasonally by communal effort. The almost contemporary inscription which regulated the use of the tuna catch in communally owned waters between Troizen, across Kalaureia, on the Peloponnesian coast and another neighbouring town (Appendix 2: IG 4[2] I.76-77 no. 22) is certainly relevant. Other fish, such as the groupers or the larger sea-breams are coastal animals and are caught individually by hook and line and/or harpoon probably with the aid of a boat. Finally, the smaller fish such as the picarel or the small sea-breams, also coastal creatures, could be caught by net, even from the coast. The euryaline fish, offer another set of technological possibilities. The present configuration of the coasts across Poros, on the Peloponnesian mainland, offers ample habitat for such fish. Lagoonal environments were possibly present during Hellenistic times as well, and the reference to the shared profits of the common saltpans in the above mentioned inscription points in this direction. The nature and location of these habitats remains, however, to be proved. The analysis of the sea-shells which is in progress (Syrides forthcoming) will probably shed more light on this issue.

All the fish represented in the assemblage, apart from the migratory, have been brought on site and consumed whole. This becomes evident by the fact that all parts of the skeleton are represented among the fish bones found (Table 8.2). The migratory species, on the contrary, are only represented by vertebrae. No head bone, pectoral or pelvic bone has been located. This discrepancy in the anatomical part representation between the different groups of fish is perhaps an indication that the seasonal large fish were brought on site already processed to some degree, certainly beheaded, and perhaps already cut in slices or chunks. It is also possible that they were even preserved in some way. Apparently the two categories of fish went through a different process on their way to the consumers. The Troizen inscription mentioned above indicates that in the vicinity of Kalaureia there was an organized business of catching and probably processing of seasonal fish. That might be the origin of those remains found in the deposit.

No cut marks have been observed on any of the fish bones. Chewing/crushing is almost absent as well. Only one vertebra of a small picarel or sea bream shows chewing/crushing traces. Eating small fish whole, as common a practice today as it apparently was in the past, may account for the relatively low number of bones of the smaller fish in the assemblage (Jones 1986; Wheeler and Jones 1989, pp. 69-78).

Burning traces on the other hand are more common. About 20% of the identifiable bones are burned uniformly brown or burned brown on spots as if they had been thrown on a dying fire. Only the bones of the largest fish show traces of burning (grouper, sea-breams, grey mullet, scorpion fish, tuna and an indeterminate Scombridae). Burned bones are both cranial and post cranial. The bones give no clue to the way the fish had been cooked, apart from the fact that no burning traces usually associated to cooking on the spit (Colley 1987) have been identified. The same is true for the mammal bones also. The burning on all bones is more consistent with a scenario of burning on the fire, perhaps as a

Table 8.2 Kalaureia – "Dining deposit". Fish anatomical part representation.

		Migratory	**Euryaline**	**Sparidae**	**Serranidae**	**Inshore various**	**Indeterminate**
Neurocranium	Vomer				1	1	
	Otolith		1	4		3	
Jaw related bones and branchial skeleton	Articular				2		
	Premaxilla			2			
	Quadratum				1		
	Hyomandibular			1			
	Postemporal				1		
	Opercular					2	
	Pharyngeal bones			1	4	1	
	Teeth			1	1		
Pectoral and pelvic fin skeleton	Anterior abdominal vert. Scapula	2		5 1	2	1	2
Vertebral column	Posterior abdominal vert.			8	1	3	
	Caudal vert.	8		23	6	2	5
	Indeterminate vert.	3	2			2	8
TOTAL		13	3	46	19	15	15

*The table is based on the Number of Identifiable Specimens (NISP)

strategy of garbage disposal (Hayden and Cannon 1983; Walters 1988; Gifford-Gonzalez 1989, p. 187). On the basis of the above we may assume that cooking of fish most probably involved boiling and stewing. This option is supported by the fact that cooking pots, of various sizes and shapes, were among the most common pottery shapes in the deposit, with *chytrae*, the deep bodied cooking pots being more common than other, more open shapes (Wells *et al.* 2005; Penttinen pers. com.)

The dining deposit also produced various plant remains in the form of carbonized seeds. These are mostly from fruits: olives, grapes, figs, almonds, very few cereals and pulses and a few weed seeds. Their analysis suggests that these remains represent storage or consumption activities, rather than crop processing (Sarpaki forthcoming). This observation corresponds well with the assumption that the finds in the dining deposit are all remains of a single meal. We might speculate on possible culinary combinations or dishes of seeds, vegetables, fruits, meat and fish in the menu, such as the combination of leafy greens and non ruminant meat mentioned above (Isaksson forthcoming). I believe, however, that the mere coincidence of recovery spot can not lead to such conclusions. This has mostly to do with the varying taphonomic history of the different classes of remains. Most consumed animals leave remains (bones, shells) in the form of food refuse. These are relatively durable. Plant consumption on the other hand leaves in many cases no visible remains. And even if something is left (e.g. seeds), that can only be preserved if it happens to be burned (other preservation methods are not common in the area under study). In essence, archaeological plant remains, unlike animal bones, represent what has not been consumed. Considering that the materials in the dining deposit are leftovers from a single meal, the preserved plants and the fish (also the mammals) remains could not belong to the same dish.

To sum up the information gained by the analysis of fish bones from the "dining deposit", it appears that at least 18 different kinds of fish had been consumed in the occasion which led to the formation of this deposit. Target species with different habits and food preferences, different fishing grounds (in terms of sea-bed physiology and depth) and different technology required in their catch, make it clear that the fish represented in the dining assemblage do not form part of one single catch from one coastal locality. They were rather the fruits of the labour of different fishermen at different spots around the island of Kalaureia or the mainland coast of Troizen and Methana. These fish varied in size from over a meter long for tuna to about 15 cm for the picarel, the combers and some sea-bream. The seasonal migratory fish, regardless of size, were brought on site already beheaded and gutted, also probably preserved to some degree. On the contrary, all other fish were probably brought on site whole. The fish were most likely cooked in ways other than grilling.

On the basis of the above, and the corresponding analysis of other classes of data (see references above) it has been suggested that some feasting took place around 165 B.C. at a peripheral spot within the sanctuary of Poseidon, quite some distance from the central Poseidon Temple. The banqueting involved perhaps over 200 people (Wells *et al.* 2005). The occasion for the gathering must have been extraordinary, so that the arrangement of a special spot to deposit part of the dining debris had been considered essential. The menu involved a large variety of ingredients, perhaps in a large number of combinations (distinct dishes). How closely connected were these foodstuffs to the formal sacrifice to a god (as this is graphically described by Koinsky and Olsen 2001), can not be accurately assessed. Several of the animals eaten in the gathering are typical sacrificial victims. Whether the consumed meat originated from a previously performed sacrifice or not remains open to debate but quite plausible. What can be securely asserted is the following. The banqueting tables were loaded with a variety of dishes which had been cooked in different cooking pots, either in situ or elsewhere. The variety of fish, the high number of individual edible items, the variety of types and sizes of cooking and serving pots makes it plausible that individual diners or groups of diners contributed to the event with their own provisions.

A similar situation is perhaps aptly illustrated by a 19th c. literary description of a religious festival, which took place on the island of Skiathos, at Christmas. Such festivals, annual affairs which take place in honour of particular saints on their celebration day, are an integral part of the celebratory calendar of the Orthodox Church. The whole community participates, according to the prescriptions for each such festival, which vary from place to place, or from Saint to Saint (Megas 2004). In the following example different participants contributed to the celebratory meal with whatever food they possessed or could afford.

> Papa-Fragoulis [the priest] asked for the bread offerings to be placed in sacks, along with some biscuits; he also asked for the olives and fish-roe to be placed on two large platters. He filled two flasks of seven okades (weight unit) each with his own wine. He wrapped two or three sun dried octopuses in a piece of paper and he filled a small box with dried figs and large raisins. [...]
>
> At dawn of the festive day the goat shepherds slaughtered and roasted two tender kids, while two woodcutters brought from the mountain several dozens of salted black birds. And captain Constandis brought from his schooner, [....] two bladders of wine and a basket filled with eggs and kaskavali (a type of cheese) from Ainos and half a dozen of chicken and a small cask of salted mackerels.

(Papadiamantis 1892, translated by D. Mylona)

Alternatively, the menu might have been centrally regulated in a manner comparable to that known from several sacred laws. An inscription of the 2nd c. B.C. from the island of Delos offers a good example (Appendix 2: ID 440, l. 60-71). The inscription regulates the exact menu at the festival of Eilithyaia, in honour of Goddess Eilithya. Food items and the money available for their purchase are referred to in detail (Linders 1994). The variety and fragmentation of the record observed in Kalaureia might just be a result of the size of the gathering. Feeding so many people would require large quantities of foodstuffs which would have to be procured from various sources and localities.

8.1.2 Fish for the gods and fish for the supplicants.

Any discussion of fish in cult had until recently begun with two testimonies. An idiosyncratic case of eel sacrifice by the Boeotians to the Unknown Deities, apparently peripheral to contemporary practice, was already in antiquity attributed to an ancestral custom (Appendix 1: Agatharchides *European History*; Radcliffe 1974, p. 215; Burkert 1993, p. 136, note 4; Simoons 1994, p. 275). The sacrifice of the first tuna caught every season to Poseidon at the deme of Halae Aexonidae in Attica, in a special rite called *thynnaion,* also has the flavour of an unusual practice in the 4th century B.C. Athens (Appendix 1: Crates *Sacrifices at Athens*; Burkert 1993, p. 136, note 4; Simoons 1994, p. 275). Fish offerings in the form of artefacts such as fish miniatures, figurines or metal engraved plates were, however, fairly common (Bevan 1986, p. 133f). Dedication of cooked fish (Appendix 1: Apollonides, *AP* 6.105) and fishing gear by fishermen to certain gods was apparently also common over several centuries (e.g. Deonna 1938, pp. 200). Such fishing implements, fish hooks, netting needles etc. have been found in sanctuaries (Appendix 3) and Greek epigrams of various dates poetically emphasize the widespread and diachronic nature of such gestures (e.g. Appendix 1: Anonymus *AP* 6.23; more related epigrams in *Anthologia Palatina*).

Zooarchaeological work in various sanctuary sites (see list in Mylona 2003a) and the analysis of a number of inscriptions (Appendix 2: SIG[4] 1024; Henning 1983; Homolle 1882, 1890; SIG[3] 1106, l. 177.42, 62; SIG[4] 1000) reveal a situation in which fish were more actively involved in cult processes than just as small offerings or unusual, peripheral sacrifices. A review of various zooarchaeological finds and a discussion of relevant data and their association with particular deities by Rose (2000) demonstrate a potentially more complex role of fish in cult than was previously accepted.

The available evidence suggests that, on certain occasions, fish stood as a prominent item in dining events within sanctuaries. At Kommos, a coastal site on Southern Crete, 509 fish remains have been found in Hellenistic layers in Temple C, mostly from a floor deposit but also a few from hearths and other fills (Rose 2000, pp. 556-559, Table 6.23). These fish bones originate from a variety of species, all coastal. An exception to this is the single bone of a catfish (*Clarias gariepinus*), a fresh-water fish imported from either Syria-Phoenicia or Egypt (Rose 2000, p. 499 and pp. 498-506, Table 6.14). Most of the fish were small or medium in size, but larger fish, such as large groupers or cartilaginous fish, were also present. All the bones, with the exception of five specimens from the interior of a hearth, are unburned (Rose 2000, pp. 536-537) in contrast to the fish assemblage from the Geometric Temple B, where many of the fish remains, especially those found in association with altars, had been burned. According to Rose, the fish remains from Temple C probably represent formal dining refuse (Rose 2000, pp. 536-537). Other foods which might have been consumed along with the fish in Temple C are the usual domestic mammals (Reese and Ruscillo 2000), birds, chicken and their eggs, dove eggs (Reese 2000a, Table 625, p. 563-4) and also a variety of sea food (Reese 2000b). Recovered remains of vegetal foods from the Kommos sanctuaries are scant and their publication does not elaborate on their exact spatial origin (Shay and Shay 2000).

Another case of fish consumption in the context of dining in a sanctuary is that from Corinth (Bookidis *et al.* 1999). The excavation of two Classical dining rooms at the sanctuary of Demeter and Kore has produced, among other finds, 49 fish bones. They all belong to very small sea-breams (11-15 cm in length). The assemblage comprises both cranial and postcranial elements (Bookidis *et al.* 1999, pp. 38-39). According to Snyder who analyzed the assemblage, these small fish may have been prepared by boiling or frying or they might have been part of a fish sauce (Bookidis *et al.* 1999, pp. 44). The fish were part of a menu which also included pig, sheep and/or goat (Bookides *et al.* 1999, pp. 32-38) and various plant foods, such as grains, pulses and fruits (Bookidis *et al.* 1999, pp. 19-32).

Fish bones have been found in various sanctuaries. Because in most cases the bone collection was done only by hand or by dry sieving with a relatively large mesh, few items are reported in each case. In the Hellenistic deposits of the Zeus Sanctuary in Pilarou Cave on Santorini, the remains of a gilthead sea bream (*Sparus auratus*) have been found along with those of a bluefin tuna (*Thunnus thynnus*) and some indeterminate fish bones. According to Becker these remains probably represent food refuse (Becker 1997). Other fish remains of broadly contemporary strata have been found in the Hero and Demeter Sanctuary at Messene, Peloponnese. More specifically a bluefin tuna vertebra (*Thunnus thynnus*) has been recovered from a 3rd -2nd century B.C. deposit near the temple of Artemis Orthia, and four remains of grouper (*Epinephelus* sp.) from a 2nd -1st century B.C. building fill deposit south of the Asklepieion (Nobis 1994, p. 303). Whether these fish remains represent dining refuse or something else is uncertain because no contextual information or discussion of their specific associations is provided.

Evidence for the inclusion of fish on sacred menus, in formal dining events taking place in a sanctuary as part of cultic activities (festivals, memorials etc.), is provided by eight, almost identical inscriptions, temple accounts from Delos, dating to the first decades of the 2nd c. B.C. (Appendix 2: Homolle 1892, 1890; Linders 1994.) When women were celebrating the festival of Eilithyaia, provision was made for the purchase of some type of preserved fish and probably some fresh fish also, among other food items, which included sheep meat, cheese, sesame, honey, vegetables, walnuts, and wine. Fish, however, were not among the food articles required for the men's feasting at Posideia, at festival in honour of Poseidon (Linders 1994). The fish on the Eileithyaia menu strongly brings to mind the tiny fish remains from the Sanctuary of Demeter and Kore at Corinth, despite the chronological and geographical gap between the two cases.

Two other Hellenistic inscriptions from the islands of Thera and Cos contribute to the construction of an even more elaborate picture. The first inscription also known as "the Foundation of Epiktete", is a sacred law dating to around 210-195 B.C. which defines in detail how the heroes, ancestors of Epiktete, are to be worshiped. It is defined that three fish are to be offered to them alongside pastries and the customary divine parts of the sacrificial victim (Appendix 2: IG 12[3] 330). The other inscription, almost a century earlier (325-300 B.C.) is also a sacred law, known as "the Foundation of Diomedon of Cos" (Appendix 2: SIG[3] 1106). It is a religious ordinance which includes directions to the priests for the cult of ancestors. It specifies that the priest should provide the sacrificial animals and the wood for the sacrifice and from those he will receive a thigh and the skin. Also he will perform an "*apopyrida*" according to the ancestral custom. *Apopyrizo* means to eat some food just plucked out of the charcoal (Appendix 1: Hesychios *Lexicon*; see chapter 10.1). "*Apopyris*", as a noun, denotes a sacrifice which involves the burning of the offering in the fire (for an archaeological example see Forstenpointner 2003), but is often taken by researchers to mean a sacrifice in which fish were totally burned (Liddell-Scott v. I, p. 341; Appendix 2: SIG[3] 1106). If this is so, then the directions in the above mentioned inscription, in essence, arrange for the consumption of fish by the hero. Whether this was followed by fish consumption by the worshipers, i.e. the members of the association, is not clear, but I consider it plausible.

The management of the fish kept or bred in "sacred lakes" or "sacred rivers" by some sanctuaries (see discussion in chapter 7.1.3) is not perhaps irrelevant to the role of fish consumption in cult. When fish were kept or bred in "sacred waters" they were systematically used in the processes that took place within the sanctuary, be they dinning, sacrifice, divination or other. Such an arrangement is not very different from the breeding and keeping, by sanctuaries, of sacred herds of animals, in order either to sell them or to use them in cult (Jameson 1988; Psaroudakis 2004).

The issue of the identity and social status of the fish consumers in cultic contexts is of interest, but difficult to tackle. Fish were definitely peripheral to the formal cult practices which involved a blood sacrifice. But communities which appear to have held fishing as a central feature in their lives (e.g. Halae Aexonidae and Boeotia, see chapter 5.3) might have thought differently. Tuna and eels, often the cornerstone fish in the economic life of such communities, were indeed sacrificed.

Were the diners men or women? The Delos inscriptions (Appendix 2; Homolle 19892, 1890, Linders 1994) and some literary sources (Appendix 1: Rufus of Ephesus *Medical Questions* 20.1-2, 13, 17; Semos *History of Delos*) indicate that gender might be an important factor in determining whether fish would be consumed or not. The Eileithyaia festival, a women's affair, was an occasion for preserved fish to be consumed, while, strangely enough, in Poseideia, a men's affair, fish was missing. The Kalaureia case does not give us any clues on the issue. Archaeological finds and inscriptions indicate that the sanctuary housed cults and activities which involved both sexes, separately or together. Here there is evidence for the cult of Zeus Soter, Aphrodite, Asklepeios (Wells *et al.* 2003) and perhaps Demeter in Builidng D (Wells *et al.* 2005) and of honoured ancestors (Hjohlman forthcoming). The only thing that can be conclusively stated is that the participants, whether men or women, were numerous.

Did the identity of the God connected to the event of dining determine whether fish would be consumed or not? Again this question is difficult to answer. Poseidon is often associated with fish offerings and apparently eating (e.g. Appendix 1: Antigonos of Carystos, *On Diction*, fr. 56A-B; Crates *Sacrifices at Athens*; Paus. iii.21.5; Plut. *Quest. conv.* 730e; Appenix 2: SIG[4] 1024) but there is evidence that this was not part of a standard pattern (e.g. Appendix 1: Agatharchides *European History*; Apollonides *AP* 6.105; Hierophytos, *Cronicles of Colophon*; Appendix 2: SIG[4] 1024; SIG[4] 1000). Rose, in his survey of fish, both real and representational, in sanctuaries, suggests that some gods, Poseidon, Hermes, Artemis and Priapos might have had a stronger association with fish (Rose 2000, pp. 520-536 and Appendix 1: Archippos, *Fishes*, fr. 18; Ath. vii 287a; Hesychios, *Lexicon*). This might or might not be true, depending on what picture will emerge once more data, especially archaeological, come to light. As a whole it seems that the consumption of fish is not tightly related to the identity of the God.

The case of exotic Atargatis and her cult is an exception to this rule. Fish was an essential element in her cult, being a taboo food for her worshipers under certain conditions (for details on the cult of Atargatis and the fish see Simoons, 1994, pp. 269-70 and Lightfoot 2003). The cult of Atargatis is relevant to fish-eating in a Greek context for two reasons. Firstly Atargatis became, at least in certain locations, a familiar religious figure from as early as the 3rd century B.C. as her cult spread throughout the Mediterranean and the broader Greek world, through Phoenician merchants (Simoons 1994, p. 271). The 2nd century B.C. ban on fish-eating on Delos, three days before participation in cult activities of a Syrian deity, makes us think that perhaps that deity was Atargatis (Sokolowski 1962, pp. 108-9, Parker, 1983, p. 359, Simoons 1994, p. 271; Lightfoot 2003, p. 44, 66). Furthermore, the sanctuary at Thuria in Messene, Peloponnese, testifies to her cult in the Hellenic area at least in Hellenistic times (Valmin 1928-29; Morin 1960, pp. 54-63). Therefore, religious and social rules related to Atargatis's cult might have influenced the usual practices of the Greeks.

Secondly, narratives about Atargatis and her strange relationship to fish were widely circulating among educated circles from as early at the 6th- 5th century B.C. and well into the first centuries of the Common Era (Appendix 1: Lucian *De Syria Dea*, fr. 14, 45-7; Mnaseas of Patara, *On Asia*; Xanthos of Lydia *Lydiaca*, fr. 17; Xen. *An.* 1.4.9). An interesting and perhaps non-coincidental phenomenon can be observed; by the 1st -2nd century A.D., when the Atargatis cult, along with other eastern cults, had spread in Roman territories, travellers and historians started paying attention and recording instances of fish breeding and fish taboos in sanctuaries throughout Greece with an intensity higher than in previous centuries (Appendix 1).

Eating fish in the context of Christianity is another relevant issue. In the early stages, customs related to fish-eating had apparently been formulated in Israel, in the communities of fishermen around Lake Galilee (Hanson 1997). Their diffusion in other areas of the Roman Empire is a fairly complicated issue, which deserves a separate investigation. Fish consumption among early Christians will not be examined here (but see Wright 1990; Dölger 1922, Simoons 1994; Hanson 1997).

8.2 Secular fish-eating in houses and taverns.

Fish-eating in the realm of the city, especially at lavish banquets, has been extensively discussed in recent years (see references at the beginning of this chapter), based on literary sources of every sort but mostly comedy. Here follows a short synopsis of the issues in which modern scholarship has placed particular emphasis.

Fish-eating had been seen in antiquity (in an Athenocentric discourse) as an activity strongly connected to the modern, urban lifestyle and the market (Davidson 1996, p, 62; Wilkins 2000, p. xxiv and 168), and as a domain where wealth and sophistication could be exhibited. Fish consumption in Classical and Hellenistic Greece appeared to be outside the realm of religious rules which promoted equality among citizens, expressed in the equal share of sacrificial red meat. Fish, being free from such ideological restrictions, offered the means of marking social differentiations (Davidson 1997, pp. 15-16). This had to do with affluence, because fish in the Athenian market was expensive (Davidson 1993, pp. 55-56). It also had to do with knowledge of distant places. The ability to buy fish was as important as the knowledge of the fish's origin, best quality and best way to prepare it. This preoccupation has been viewed by modern researchers as symbolic of the Athenean's power in trade (Wilkins 2000, pp. 158-160). The way fish-eating was articulated was treated as an important element in defining one's identity, whether ethnic, social or ideological (Wilkins 2000, pp. 97-98).

Fish-eating, incorporating all the above, emerges as an arena of ideological contestation and control. The fish-eating experience in an elite symposium (Davidson 1997) or its diffusion in the eating gatherings of ordinary people (Fisher 2000; Wilkins 2000, p. 206) has been seen as part of this broader picture. Excess in fish-eating is presented in literary sources as an expression of lack of control in general, an attitude that was heavily criticized (Wilkins 1993, p. 186, 193; Davidson 1997, pp. 20-35, 206-210). Finally, the connection of fish-eating to sex in Classical Athens has been explored (Davidson 1997, pp. 8-10, e.g. Appendix 1: Anaxandrides, *Odysseus*, fr. 34; Antiphanes *She Goes Fishing*, fr. 27; Ar. *Lys.* 1065ff; Hyper. *Agains Arastagora*) desire being the common element connecting the two. Loosing control over desire, being it for fish or for sex, was in this discourse an equally condemnable attitude.

The idea of fish-eating as an activity which could bring the consumers beyond the limits of proper behaviour appears as a common theme over several centuries. From the caustic political commentary of Aristophanes in the 5th c. B.C. (e.g. Appendix 1: Ar. *Vesp.* 490-5; *Ran.* 701-3; *Nub.* 559) to the satirised figures of the fish glutton and fishmonger of the Middle Comedy in the 3rd century (e.g. Appendix 1: Alexis, *Man with a Cataracht*, fr. 16; *Dimitrios or Philetairos,* fr. 47; *The Meeting at Pylae*, fr. 204; *Phaedon*, fr. 249; Antiphanes, *Rich Men*, fr. 188) and finally to the moralizers of the Roman era (e.g. Appendix 1: Plut. *Quest. conv.* 631d and 728-30), the theme repeats itself under various guises. Therefore, propriety and its transgression appear to be diachronically central. But what was proper in fish-eating? What were the proper circumstances or the proper table company? What was right in fish-eating within the household, or at public gatherings? What was the right fish for women, for children, for slaves or for foreigners? Were there secular occasions that required the consumption of fish or sea food? Were the same rules applied in Athens, Macedonia and the Aegean Islands? Where all these rules unchanged over the years?

Fish-eating within the relative privacy of one's house for example, which according to some historians became more prominent from the 4th century onwards (Hill and Wilkins 1996) is quite elusive as indeed is any kind of eating in this context. Andrew Dalby built a picture of where and when food consumption took place within the household on the basis of admittedly few comedy fragments (Dalby 1996, pp. 2-22; Wilkins 2000, p. 54ff). The materiality of these arrangements is quite difficult to discern. Part of the problem is that so little relevant published archaeological material is available.

An even more serious problem however, has to do with our definitions. What was the nature of the "domestic" in the periods under study and how the everyday rhythm of life and the materiality of people's relations were articulated within it are not self evident issues. The problem, which in recent years has been variably dealt with by several specialized studies (Jameson 1990; Goldberg 1999; Nevett 1999; Antonaccio 2000; Ault 2000; Cahill 2002) is perhaps best encapsulated by Lyn Foxhall (forthcoming). Foxhall underlines the fact that the material culture of the "household" is not the realm of undifferentiated domesticity. On the contrary, household activities and domestic social interactions in different Greek societies were characterized by a range of variation. Meals for example might take place at different rooms within the house, depending on whether only family members or visitors were participating (for further discussion on the issue of activity variation within households see papers in Allison 1999). By looking at the available archaeological evidence (mostly pottery and architecture) in this light she suggests that, at least in Hellenistic and Classical times, cooking done in the house was relatively simple and limited. It appears, according to her, that meals were not prepared for large numbers of people simultaneously, on a regular basis, at home. So cooking and eating was probably done many times a day for different people depending on their different needs and daily routines (Foxhall, forthcoming).

How does fish-eating fit in such a framework? Are we in a position to discern these multiple rhythms of food/fish consumption in the assorted remains usually incorporated in floors and refuse piles and collected during the excavation of a house? In theory, patterns of repeated behaviour would imprint themselves in space and could be revealed by detailed excavation and recording (Allison 1999, p. 12). No such case concerning fish bones has yet been available. At the moment, the picture we gain from the archaeological evidence for fish consumption in private houses is rather sketchy.

The issue of the social identity of consumers within the household is difficult to tackle. A late source (10th century A.D.) describes the centuries old custom of *Amphidromia*, which served to initiate the new born into the life of the household (Appendix 1: Suda 1772). The *Amphidromia* involved a family gathering with food gifts brought to the baby. The custom prescribed that cuttlefish and octopus should be included among these gifts. A comedy fragment by Ephippus, preserved by Athenaeus (Appendix 1: Ephippus *Geryones*, fr. 3), gives a detailed list of what was consumed in these meals (the same passage is also attributed by Athenaeus to another comedy writer, Eubulus, Ath. ii 65c). Sprats, octopus and cuttlefish were part of the menu (Wilkins 2000, p. 63). In that case the family was involved in the eating of the fish and sea-food, and the context was ceremonial. The Roman villa in Mytilene witnessed the consumption of many fish and sea food on the premises (Table 5.1g). No more can be said about the spatial configuration and the social dimension of their consumption until full analysis and publication of the materials from the Villa is available. The same goes for the fish consumption in the merchant's house in Late Roman Itanos (Table 5.1a). At the Late Roman farmhouse at inland Pyrgoudi (Table 5.1d), despite detailed publication, the preservation of in situ material was poor and the single fish vertebra found offers a rather poor ground for elaboration.

An interesting insight into the domain of secular fish-eating however comes from Hellenistic Krania on the west coast of the Thermaikos gulf at the southern reaches of Mt. Olympus (Appendix 4). Several thousand fish bones from individuals of various sizes have been recovered from the floor of two rooms within the so-called "Building A" and from a rubbish pit associated with the same building. These remains represent a wide range of marine, and euryaline taxa (Table 5.1a). Many of the remains are fish scales. About 10% of the fish bones are burned black, as are indeed several of the animal bones from the same deposits. Cranial and postcranial bones are all present, although the analysis is still preliminary and no exact figures are available. Interestingly, the whole range of aquatic habitats is represented by the fish identified in this assemblage, except perhaps the lacustrine. Not only the rich coastal waters of the Thermaikos gulf were exploited, but also brackish, lagoonal areas, which correspond perhaps to the estuaries of Pineios river. The Pineios River proper was also the source of some of the fish consumed on this site. One of the most interesting features of the Krania assemblage is the large number of remains from cartilaginous fish such as the sharks, the angel-sharks and the stings rays.

Various fish hooks, lead weights and netting needles (Appendix 3a) and an indeterminate number of seashells accentuate the fishing theme on this site. Mammal and bird bones are also present in moderate quantities. The analysis of the various aspects of this excavation is still in progress. Stratigraphic analysis is still preliminary and rather generalized and the results of the analysis of pottery, architecture, small finds and other materials have not yet been concluded. The preliminary report of the finds describes a rich variety of artefacts, which include, except for the fishing tools mentioned above, large amounts of domestic pottery, local and imported (for references see Appendix 4). So far only the charred plant remains have been analyzed in depth; again, by necessity, on the basis of very basic contextual information (Margaritis 2006).

Building A was located fairly close to the coastline. The excavator argues that it belonged to the harbour sector of the ancient city of Irakleon (Poulaki-Pantermali 2001, pp. 335-340; Heuzey 1860, p. 88ff). The ethnobotanic analysis of carbonized remains of cereals, pulses, fruits and nuts, suggest that Building A, or at least part of it, was probably a place of consumption rather than consumption combined with production and storage. This assessment is based on comparison with other houses/farmsteads in the broader area of ancient Irakleon (Margaritis 2006). Most of the ethnobotanic samples which derive either from the rooms' floors and destruction layers or from the rubbish pit are mixed. Margaritis interprets this phenomenon as a result of their gradual accumulation and as reflecting a variety of different consumption episodes which involved a variety of plant foods. Concentrated samples of sesame and pine cones, both very unusual finds, represent short term storage for immediate use and indicate the preparation of special dishes (Margaritis 2006, chapter 11.3). Combining this observation with the unusually high number of cooking and drinking vessels she suggests that Building A might have been an inn or tavern which also served as residence of the owners (Margaritis 2006, chapter 13.4). Buildings interpreted as taverns on the basis of the abundance and type of pottery, have been found at Halieis (Foxhall forthcoming) and perhaps Olynthus (Cahill 2002, pp. 85-97). In those cases, however, there is no record of the food remains associated with the assorted pottery.

If we accept the above interpretation, the very many fish remains (and seashells) found in this particular part of the Krania settlement (or ancient Irakleon), are not the remains of household eating. They are probably the abundant remnants of the menu of a harbour tavern, on the shore of one of the most eutrophic seas in the Eastern Mediterranean. But the theme of this menu was not narrowly sea fish. It was fish in general. Many varieties of them from different origins have been identified, marine and purely riverine fish as well as fish from the estuaries and those which move in and out of the river on a seasonal basis. Could this diversity reflect the harbour's reputation for certain types of fish and other delicacies? Such a reputation might not be unlike that reported by Archaestratos of harbour sites such as the nearby Pella or Olynthos (Appendix 1: Archaestratos in Ath. various lines and Table 5.3). It appears possible, that had Archaestratos's work been preserved more fully we might read about the harbour of Irakleon and its marine and delicacies.

Alternatively, this diversity of fish (and seashells) could be viewed as reflecting the identity of the clientele of such a public establishment. Irakleon was a port of trade, which probably attracted people from the broader area who were involved in some type of trade or travelling. The customers in the tavern could originate both from the marine world and from the world of the hinterland and the river, from the mountains and the northern part of the Thessalian plain through which the Pineios River runs. This would perhaps explain both the availability of all these types of fish and the demand for them.

To this Hellenistic harbour tavern, one could juxtapose a Late Roman tavern from Mytilene. That establishment, not far from the sea, again in one of the eutrophic areas of the northern Aegean, was housed in a building which, in earlier days, was an affluent Roman villa (see Appendix 4). In striking contrast, both to the Krania tavern and to its previous function, no fish or other marine remains whatsoever have been found there. Clearly, the menu in this tavern had no interest in fish. By inference, the customers of this place had no interest in the sea either, at least as far as their culinary experiences went.

A different case has been located in Classical Athens. Fish bones, a variety of sea shells, various mammal bones, cooking pots and serving dishes along with wine amphorae from both Athens and abroad, have been found in large numbers in the fill of a well in the Athenian Agora (underneath the Roman Stoa), which dates to the end of 5th and the first quarter of the 4th c. B.C. The excavator has interpreted the assemblage as the refuse of a nearby *kapeilion* (tavern) (Shear 1975, pp. 341-61). The establishment has been characterized as a seafood restaurant (Shear 1975, p. 357), although the mixed nature of the bone assemblage indicates more a tavern with a varied menu. It has been claimed that this establishment probably catered for a well-to-do clientele, based on the quality of pottery, the presence of the fish bones and the imported wine amphorae. A second example, of such an urban public eating establishment, has been found on Delos. Fish (and plant) remains have been collected from deposits associated with the Lake House on Delos, a high class brothel in the urban centre of the island (Zapheiropoulou 1991, p. 24). A final case is the butcher's shop which, at special occasions also catered its clientele with cooked food, meat and fish, but not wine in the Theatre area at Corinth in the 2nd century B.C. (Williams and Zervos 1986, pp. 136-7, Reese *et al. 1987*).

The Mytiline, Krania and Delos establishments were situated near the harbour with Krania being probably on it, while the *kapeilion* in the Athenian Agora and the Corinthian food outlet represent urban eating places in the heart of busy markets. Clearly a totally different type of clientele frequented these establishments. The published data at our disposal are rather inadequate for an extensive and detailed comparison of the four cases. Keeping the above discussion in mind however, it is tempting to trace in the Krania case the elusive mariners, those people who tied their living to the sea: sailors, fisher folk, traders, shipbuilders, manufactures of fishing equipment, etc. Again, as so often in Classical archaeology, as it stands at the moment, the Krania case just shows us exciting possibilities.

8.3 Conclusions.

This chapter has discussed those social contexts of fish-eating, for which archaeological documentation is available. Therefore, the consumption of fish in private banquets that appears so prominent in the ancient discourse on fish-eating (and in the modern historical treatment of the issue) has not been discussed here. On the contrary emphasis has been placed in two other contexts, little analyzed in the aforementioned research.

One is the domain of cult. Fish as a sacrificial animal for gods and heroes, fish as a prescribed menu item for formal feasting, and fish as animals kept or bred by sanctuaries are the main lines of investigation. Although these are perhaps only three of many possible connections of fish to cult some patterns emerge. First, fish was much more intensively involved in cultic processes than previously thought. They played a role not only in feasting but also in sacrifice. It appears that although their use in this context was not as widespread as that of domestic animals, fish sacrifices were, nevertheless, prescribed as necessary on specific occasions. We know very little of the rationale behind choosing one food item instead of another for the communal eating in sanctuaries. So, in most cases we can not tell whether the inclusion of fish in the menu was deliberate, or just a chance choice. The case of the Kalaureia dining episode suggests that the first possibility was probably the case. The systematic representation of marine elements in the Kalaureia menu, with fish of various kinds, seashells and seawater indicates that care had been taken to emphasize the marine theme, although we can not tell with certainty if this emphasis was related to the identity of the participants (mariners) or the cultic occasion which resulted in the dinning episode.

Secular fish-eating, especially in the semi-public context of a tavern is the second theme explored in this chapter. First, it has been established that fish-eating in such places was quite common, especially, in areas close to the sea. The existence of only one well-documented fish-bone assemblage from such a context and the lack of any detailed publications of relevant excavations do not facilitate the exploration of the whole menu of such establishments. We could however suggest that proximity to the sea did not automatically imply the serving of fish. Apparently, a fish or a meat menu in these taverns was an issue of specialization. Even less can be inferred from the data at our disposal on fish-eating in domestic contexts. Although the consumption of fish is clearly documented, nothing can be said about its intensity, regularity, etc. It is interesting however to note that eating fish at home could be ceremonial, depending upon circumstances.

The discussion in this chapter did not offer an overview of the social contexts in which fish-eating took place. Based on the realization that these contexts could be as numerous as the human affairs, it focused only on a few of them, those we are in a position to investigate at the moment. Some patterns observed however are particularly interesting. Even though the discussion in this chapter was organized, for the sake of convenience, around the division "cultic versus secular", it is obvious that the distinction between the two is far from clear-cut (for an extensive discussion of this dichotomy see also Dasen and Piérart 2005). *Amphidromia*, a cultic event, not as grand as a public blood sacrifice in a temple but very important in the life of a family and a community, took place at home, while on several occasions, eating, an everyday activity, was done in the sanctuary as part of a ceremonial process. Fish was part of both spheres, secular and cultic. The dichotomy between the public and private in fish-eating is another problematic distinction. The gathering of friends and the eating of fish in a family affair, the serving of fish in a tavern and the consumption of fish in sanctuaries by very large (Kalaureia) or smaller groups (Sanctuary of Demeter and Kore) are cases which indicate that these definitions were rather blurred.

Finally, other issues such as the way fish-eating was articulated around distinctions of gender, age, ethnicity, calendar of yearly activities and events etc., remain for the moment obscure.

CHAPTER 9

CHOOSING FISH FOR THE TABLE

The body of surviving texts that refer to fish-eating makes it abundantly clear that in Classical, Hellenistic and Roman Greece there were in place distinct hierarchical and non-hierarchical evaluations of fish on the basis of their taste, texture and other features. Such categories form the special-purpose classifications which are fashioned by cultural interest, experience and use (Ellen 1993; Atran 1998). Despite the fact that the world of the fish had been the object of "scientific" observation, at least as early as the 5th century B.C., with fish being classified in a manner which is different from, yet familiar to modern biological taxonomy, the classification of fish as food followed different lines (on the issue of Aristotelian and modern animal and fish taxonomies see Balme 1961; Solmsen 1978; Lee 1985; Pellegrin 1986; Green and Depew 2004, p. 11; on 5th century classifications see Mair 1963, pp. xxiii-xxv).

Often the criteria for placing fish in categories related to consumption were flexible, co-existing or transforming in different contexts of reference. Not all fish were equally desirable nor were the same type of fish from different locations considered of the same quality; not all people had the same preferences. Ethnicity, status, gender and educational background appear to be parameters relevant to fish preferences, as indeed was the case with all types of food and drink. Besides, it seems clear that different fish were suitable for consumption on different occasions and medical/dietetic sources, which cover a wide chronological span, make it clear that taste was only one of the qualities under consideration when a choice of fish dish was the issue. Choice of fish clearly depended upon context. Alternative fish classifications will be discussed in this chapter. The discussion will start from a somewhat unlikely point of departure, the Hellenistic fish price list recovered at Akraephia, an inland town on the coast of Lake Kopais in Boeotia (Appendix 2: Salviat et Vatin 1971).

9.1 The Akraephia fish price list.

This inscription, incised on two slabs, heavily damaged at places, preserves a fish price list (Appendix 2, no. 19). It has been claimed that the need for such a regulation of fish prices arose during the celebration of the Ptoia, a musical festival in honour of Apollo at the nearby oracle (Vatin 1971, p. 109), but this remains speculation. The Akraephia inscription is a standard point of reference when food prices and related regulations are the issue (Gallant 1985, pp. 39-42; Boeckh 1886; Schaps 1985-8, pp. 68-70; Davidson 1993). Several attempts have been made to calibrate the fish prices referred to in the inscription with others encountered in literary texts, mostly comedies (e.g. discussion in Davidson 1993). Such attempts invariably face serious problems. Many of the fish names are unidentified. Furthermore, there is a difference in currency and probably in goods' prices between Boeotia and Athens. These, along with the unreliability of the comedies as a source of fiscal data and the problem of the chronological incompatibility between the different sources, make such conversions highly problematic (Davidson 1993, pp. 55-56; 1997, pp. 186-90).

Here, the Akraephia inscription will be analyzed in terms of the internal features of the price list. The emphasis will be placed not on the prices themselves but on the categorization of fish and their relative value, as this is reflected by the prices (for a hierarchy of fish on the basis of taste, as revealed in comedies see Davidson 1993, p. 54, note 14).

The entry text of the inscription, albeit heavily damaged, permits its dating to the Hellenistic period. What survives best is a large part of the price list. The fish have been divided in two broad categories: the marine fish and the frsh-water ones (Table 9.1). In the first case, the fish are presented in alphabetical order. The fresh-water fish are mentioned without any order. Each entry refers to a different fish species, or different quality or size range of the same species. The fish name (and occasionally the size or quality definition) is followed by an acrophonic numeral which represents the price. In some cases the price refers to the individual fish and in others to a determined weight unit.

The marine and fresh-water fish are clearly distinguished in this list, with euryaline fish grouped with the first and the eel (catadromous) with the second. The quality and price of certain fish, such as the gray mullet (*kestreus*), the scorpion fish (*skorpios*) and the common sea bream (*fagros*) was judged by size, with the largest individuals being more expensive. The quality of others was judged by different criteria. Some fish are characterized as of the "original variety" (*kotharo*) which is differentiated from another (unknown to us but obvious to the buyers) variety, apparently less valued, if we judge by its lower price. In the case of large tuna (*thounnokeitos*) the price is determined by the body part, with belly parts being more expensive.

Although the price list is partly illegible, it becomes apparent that it refers to distinct classes of fish. Quite a few of those mentioned in the inscription bear a top price and cost 1 obol and 4 coppers or thereabout. This is the highest price in the list with the exception of the large tuna and probably the largest of the eel (*egxelious*) and the unidentified *varakos*. Of the cheaper fish, two groups may be observed. One group is the cartilaginous fish

Table 9.1 Fish and prices in the Akraephia Inscription (end of 3rd-beg. of 2nd c. B.C.).

Boetian name	**Common English name**	**Size-quantity-quality**	**Price**	
		SEA FISH		
Alfeistas	wrasse		XX	2 chalkoi
Amia	pelamid		XX	2 chalkoi
Agnathos	lamprey?			
Arkanos				
Allopias	fox shark?			
Alak[...] (ilakati)	a type of tuna	original variety		
An[.]ko[ ]	another type of tuna			
Vogglot[tos]	flatfish			
Vatis	skate or Ray	original variety		
Avoratos	skate or Ray	a less valued variety	ΠΧΧ	5 chalkoi
Vopaτos	skate or Ray	a less valued variety		
Vatrachos	angler fish	large size	[.] Π	incomplete
(Vatrachos)	angler fish	?different size?	] XX	incomplete
Vemvras (in plural)	sprat, smelt or anchovy			
Goggros	eel	original variety		
Goggros	eel	variety *omfaloi*		
Goggros	eel	other variety		
Gelavrias	a member of Gadidae family?		I Π X	1 obol and 4 chalkoi
Galeos	dog fish	large	[...] X	incomplete
Galeos	dog fish	1 mina	H XXX	9 chalkoi
Erouthros (erythrinos?)	pandora	small size		
Epseitos (in plural)	small fry			
Thratta (in plural)	Clupaeidae		IX	1 obol and 1 chalkous
Thounnokeit[os]	large tuna	belly part	I I XX	2 obols and 2 chalkoi
Thounnokeit[os]	large tuna	other parts	I [.]X	>1 obol?
Thounnides (in plural)	tunas		IXX	1 obol and two chalkoi
Ixla (kichla) (in plural)	wrasse		I Π XX	1 obol and 5 chalkoi
Kottoufos (in plural)	wrasse		I Π X	1 obol and 4 chalkoi
Ithoulis (ioulis) (in plural)	rainbow wrasse		I Π X	1 obol and 4 chalkoi
Ippouroi	Dolphin-fish?			
Fiops (Iops)	unidentified		I Π X	1 obol and 4 chalkoi
Koris	unidentified			
Kitharos	unidentified	> 1 mina	I Π X	1 obol and 4 chalkoi
Kitharos	unidentified	< 1 mina		
Kestreus	grey mullet	large	I [ ] (per ½ mina)	> 1 obol
Kestreus	grey mullet	other sizes	I Π X	1 obol and 4 chalkoi
Kounopreistis	type of shark		H	½ obol
Kouon karcharias	type of large shark		Π	3 chalkoi
Kantharos	black sea bream	(all types or sizes)	H	½ obol
Kokkoux	gurnard		I Π X	1 obol and 4 chalkoi
Korakinos	meagre			
Kallionoumos	star gazer			
Lavrax	sea-bass	> 1 mina		
Lavrax	sea-bass	< 1 mina		

Table 9.1 (Continued)

Mouros	moray			
Melanoros (in plural)	black sea bream			
Rinovatos	skate			
Rina	monk fish	original variety	Η Π	
Rafis	garfish (in plural)		Ι Π Χ	1 obol and 4 chalkoi
Skarina (in plural)	parrot fish		ΗΧΧ	
Skorpios	scorpion fish	large 1 mina		
Skorpios	scorpion fish	1 mina		
Skorpios	scorpion fish	smaller		
Fagros	common sea bream	> 1mina		
Fagros	common sea bream	1 mina		
Fagros	common sea bream	smaller		
Charakias	unidentified			
LAKE FISH				
Varakos	unidentified	1 mina		
Varakos	unidentified	½ mina	Ι Π Χ	1 obol and 4 chalkoi
Varakos	unidentified	smaller	Ι (per mina)	1 obol
Lavrichos	sea-bass	1 mina		
Lavrichos	sea-bass	smaller	Π Χ (per mina)	4 chalkoi
Poukris (in plural)	unidentified	small	Η (per mina)	½ obol
Valler[os]	unidentified			
Charax	unidentified		Π	3 chalkoi
Egxelious (in plural)	eel	large		
Egxelious (in plural)	eel	smaller	Ι Π Χ	1 obol and 4 chalkoi
Gastrimargas	unknown			

*The table is based on the transcription of the two fragments of Akraephia inscription by C. Vatin (1971). The fish species identification follows Thompson 1941 and the transcription of the acrophonic numerals into the Boetian currency of the time follows Schaps (1987). The mina is a weight unit and equals to 0,56 kilograms.

(e.g. *avoratos, voratos, galeos* of 1 mina, *kounopristis, kyon karcharias, rina*) and the smaller Scombridae (*amia*). Another group is the assorted small fish (*alfeistas, kantharos*). A variety of prices among the fish can not be evaluated, because most of the fish mentioned remain unidentified.

If we consider the price of each fish as a broad reflection of its desirability and "value" in the consumers' perception, then the Akraephia price list confirms the high reputation of tuna belly parts and eel which is repeatedly emphasized in other sources as well (Dalby 1996, p. 226, note 60). We can not evaluate the position of the small schooling fish, because of lack of data. It is interesting however that in Akraephia, the *thratta*, a type of Clupaeidae, usually caught in large quantities, was in terms of price, close to the top fish. Whether this is a reflection of its desirability or of the fact that Clupaeidae, a type of easily spoiling fish, had to be transported to the inland Akraephia, thus acquiring a higher price, is not possible to tell. The cartilaginous fish, which had some reputation in certain parts of the Aegean, if we accept Archaestratos' authority, appear to have been among the more affordable large fish (for the cartilaginous fish see Dalby 1996, pp. 69-70). The Akraephia price list also confirms another impression, derived from the literary sources. That is that the largest fish were more expensive than the smaller representatives of the same species.

The fresh-water fish price list also offers an interesting insight. Among these fish some are very expensive, falling into the same category as the belly parts of tuna. Most of the fresh-water fish however, are quite cheap. While this may be a reflection of the proximity of their source, with Akraephia being located on the shores of Lake Kopais, the price differentiation between species definitely indicates a hierarchical evaluation of their taste/quality.

The analysis of the Akraephia fish price list offers a range of interesting insights into the way fish preferences were articulated in the Hellenistic period. Unfortunately it is not at the moment possible to compare this with similar contemporary evaluation systems for other foodstuffs. The Roman agoranomic inscription found in Piraeus, which regulates the prices of several meat cuts sold there (Steinhauer 1994) could be viewed in the same light. There, several body parts, such as feet, belly parts, liver, lung, head and brain from different animals (pigs, goats and cattle) attain a different price. It should be noted here, however, that a direct comparison of the two inscriptions is not possible, not only because of the chronological and geographical distance between the two inscriptions, but also because the Piraeus inscription only includes certain parts of the animals and not the fleshy meaty ones. The possibility that it refers to meat sold after sacrifices (Isenberg 1975) is quite plausible. The point of relevance here is that, even among meats, there is a hierarchy based both on the type of animal and on the anatomical part sold, i.e. the texture (Wilkins 1993, p. 194, note 9) which is reflected on price.

9.2 Edible fish categories beyond Linnaean taxonomy.

The analysis of the Akraephia price list makes it obvious that, at least in Hellenistic Boeotia, there was a hierarchical evaluation of fish tastes, which echoes similar evaluations from Athens and perhaps other parts of Greece. Large versus small fish, seawater versus freshwater fish, local versus imported or non-local varieties and more generally fresh fish versus preserved fish are some such categories. The investigation of how these categories were articulated may facilitate the understanding of fish-eating in Greece from 5th century B.C. to 7th century A.D.

Contemporary anthropological research has made it clear that food consumption is a subject that touches not only upon economy but also upon issues of social and cultural values (see discussion and references in chapter 3). Here I will attempt to explore the nature of such social and cultural values, looking at the way ancient Greek societies formulated the categories of edible fish (or any food category), especially as these are reflected on the material and literary remains of these choices (Serjeantson 2000, p. 179). This will be done by looking both at the internal properties of fish and at the social conditions that have ascribed value to them (following Hamilakis 1999b, p. 57; Appadurai 1986a).

The choice of fish implies a number of explicit or implicit statements on the part of the consumers. Davidson's (1993; 1995; 1997) and Wilkins's (2000) works discuss extensively the nature of some such statements, mostly those connected to a discourse around proper social behaviour, status, ethnicity or identity more generally. Such works make it clear that a choice of fish often implies choices of self-presentation or reflect established knowledge and opinions about the world. Desiring small cheap fish for example, when you are in a position to buy larger more expensive ones, appears in comedies as a sign of simplicity (Appendix 1: Alexis *Principal Dancer*, fr. 200; Antiphanes, *Butalion*, fr.69). In this case, the size of fish becomes a social marker. Wealth here does not count. It is the proper desire that makes a difference.

The first problem in our quest is to explore how alternative fish categories beyond the ones imposed by modern analysts through Linnaean taxonomic identification can be discerned and documented. The small sea breams (Sparidae) for example, are usually viewed as distinct from gobies (Gobidae) or picarels (Centracanthidae), when a fish-bone assemblage is analyzed. For ancient Greeks, however, these three biologically distinct taxa, along with others, were grouped, when young, in a single taxon, a fish mix called *afyi*, which was specifically targeted by fishermen. It was cooked in a special, highly distinctive dish. For this dish, the exact composition of the fish mix in terms of fish species was irrelevant. The size of the fish and their fishing grounds were of importance (see Thompson 1947 for references to the relevant sources).

The structuring of our contemporary thinking about animals (and in this case fish) is a basic problem in our research. Limitations imposed by the dominant archaeological field techniques are equally important. Direct archaeological material testimonies of alternative fish categories are few. Most of the archaeological data to date, especially fish bones, originate from conventional excavations, where attention to bio-archaeological remains and excavation micro-contexts is minimal. Material originating from such excavations is incapable of illuminating issues such as the ones discussed here. Any preference for large rather than smaller fish in a particular context, for example, could only be detected if excavation and recording were detailed and oriented towards the identification of remains of the same temporal horizon and of contextual coherence. Grouping of archaeological deposits in broad chronological or functional groups, such as "Late Hellenistic floor deposit" for example, can not help.

As a consequence, the present discussion by necessity relies heavily on what certain people, from specific social, chronological or geographical contexts, chose to say/write about the issue. The following analysis is based on the understanding that this intellectual discourse, meaningful as it may be, does not necessarily reflect the whole range of past practices and opinions on the issue of fish taste evaluations.

9.2.1 *Edible and inedible fish.*

A discussion of these categories is probably best introduced with the most basic distinction of all; the differentiation between edible and inedible fish. Were all fish considered suitable for the table? Was there a set of rules which determined what fish could be consumed and what not, or under which circumstances did these rules take affect? Contemporary anthropological literature has

Fig. 9.1 Dolphin mosaic from the "House of Dolphins" on Delos, end of 2nd –beginning of 1st c. B.C. (from Asimakopoulou-Atzaka, 2003, fig. 18).

shown that the categories of "edible" and "inedible" are highly flexible and closely linked to social and ideological conditions within individual societies (de Boeck 1994; Hamilakis 1999b, p. 57). For a fish to be edible it has to occupy a certain position within the realm of human relations and human perceptions of the aquatic environment (to paraphrase Hamilakis 1999b, p. 57)

The dolphin offers a good example of an essentially inedible fish (Fig. 9.1). For ancient Greeks, the dolphin, which was considered one of the large sea fish, was not regarded as edible (Opp. *Hal.* v. 416), even though Greeks were aware of the fact that the Thracians ate it (Opp. *Hal.* v. 519-88; Ael. i.18). Thracians however, were considered by Greeks to be a people with many strange habits (Hrdt. Book 5) and the inclusion by Thracians of dolphin among the edible fish, was part of a broader discourse which emphasized their otherness (Hartog 1988, chapter 6). Dolphins were among the most popular sea-creatures in Greek mythology (compilation of these myths in Thompson 1947, pp. 52-56) and a common motif in representational arts (Burr Stebbins 1929; Vidali 1997), its presence being connected to Poseidon and Aphrodite, as well as to a number of lesser gods and heroes along several parts of the Greek coast (Burr Stebbins 1929; Thompson 1947, pp. 52-56; Rabinovitch 1947; Diez 1957). The dolphin was invested with anthropomorphic characteristics, such as love for its offspring and loving kindness in general, love for music and reigning powers over other fish. Its physiology was well studied (Ar *HN*; Opp. *Hal.*). The only context in which dolphins could perhaps be consumed was the medical. Only certain parts of its body, the fat, the liver and ashes, were used. The way these elements were used remains unclear and it is possible that they were applied in a way that precluded actual consumption, i.e. entering into the body through the mouth (for a summary of the above and references see Thompson 1947, p. 56). To date no dolphin bones have been discovered anywhere in Greece (Mylona 2003 a, table 19.1).

To judge by the geographical distribution of the dolphin-related myths, the taboo on dolphin flesh was fairly widespread on the Hellenic peninsula. The reason why the dolphin was regarded as inedible is not clear but, according to Oppian (*Hal.* v. 416), to hunt dolphins was sinful (*apotropos*) and displeasing to the Gods. Its association with Poseidon and other deities may have made it inaccessible, according to public perception. Its anthropomorphic features made the dolphin closely related to humans, again placing it on the side of the inedible creatures. Herbert Hoffman, following a wider anthropological tradition (e.g. Douglas 1966), attributes the taboo status of dolphins in Classical antiquity to their zoological anomaly and their unusual habits: the dolphin is anomalous since it disregards the boundaries between the elements (air and water) and in addition bears its young in a womb unlike other fish (Hoffman 1977, p. 528). Hoffman's position reflects an idea which has been popular among some anthropologists, that animals can be categorized in terms of their social distance from human beings as "tame", "game" and "remote". Animals that transgress the borders between these categories are anomalous, which results in their consumption becoming taboo (Leach 1967; for a critique see Halverson 1976; cf. also M. Douglas on the categories of animals and their taxonomies in the Leviticus – Douglas 1966).

The ban on dolphin eating is not a case unique to the Greek world, although it is perhaps the most widespread and rigid one. In several parts of Greece, the consumption of certain marine creatures is banned due to their association with a god. At the end of the 2nd century A.D. the fishermen from the island of Serifos threw back to the water every lobster (*tetyx enalios*) caught in their nets because they considered them the playmates of Perseus (Appendix 1: Ael. 13.26; Burkert 1983, p. 211), while, according to another contemporary source, the escort fish (*pompilos*) was regarded sacred to the Samothracian gods and its consumption was severely punished. The culprits, were eaten by a sea monster (Appendix 1: Pancrates *Occupations at Sea*; Burkert 1983, p. 212). Here the rules of eating are reversed adding to the sense of dread.

Occasionally, fish appear to become inedible only under certain circumstances. In some cases, fish avoidance was part of a more general abstinence from animal flesh (on vegetarianism in ancient Greece see Osborne, 1995). This was the case with mystic groups such as the Orphics (Guthier 1993; Detienne 1989a, pp. 5-8) or the special group of female devotees to Demeter called "the Bees" (Detienne 1989b, p. 145). Similar but less clear is the case of the Pythagoreans' food restrictions (Appendix 1: Plut. *Quest. conv.* 728-30; Porph. *Vita Pythagorii* 45). According to the extant information (e.g. Yonge, date unknown; Theodorson 1996), Pythagoreans, a philosophical sect which advocated among others that the soul was passing between humans and animals on death, appear to have adopted a vegetarian diet and avoided the consumption of animal flesh. We can not be certain about the exact nature of these prohibitions and it is not clear if the animal flesh avoidance was generalized or not (for a commentary and relevant references, see Kirk, *et al.* 1983, 214-238; Parker 1983, pp. 358-9 and 360-3; Guthier 1993 and Garnsey 1999, pp. 85-89). In the Pythagorean regime however two fish are specifically mentioned as forbidden; the pandora (*erythrinus*) and the saddled sea-bream (*melanourus*) (Yonge, date unknown). Other foods mentioned as unsuitable for consumption are beans and perhaps wine during the day time. Further Pythagorean prohibitions appear to concern behaviour rather than food.

The initiates to the Eleusinean Mysteries are another group of people who, according to Porphyry were forbidden to eat fish (Appendix 1: Porph. *On abst.* 4.16). Red mullet is the only species referred to by name. Other foods, the consumption of which was prohibited to the initiates, were house-birds, beans, pomegranates, apples, eggs, egg-laying animals, and the meat of animals that had died naturally (Arbesmann 1929, p. 76). According to the antiquarian Semos in his *History of Delos* around 200 B.C., the women from the island of Delos who prayed to Vrizo, the interpreter of dreams, offered to the goddess all sorts of goods except fish (Appendix 1: Semos *History of Delos*). Fish was not consumed either by them or by the god. Semos explains this exclusion by the fact that the women were praying for the safety of their ships at sea. In this case a "magic" strategy of protection against sea dangers seems to have been followed (for a similar phenomenon among present day communities of fishermen in the USA see Poggie and Gersuny 1972). In both the above examples, fish became inedible only temporarily, under certain circumstances.

Ancient Greek literature abounds with references to various restrictions on fish-eating which are related to the cult of Atargatis (Burkert 1983, pp. 204-212; see also discussion in chapter 8.1.2). Greeks chose to focus on the fish-related attributes of the oriental goddess (Appendix 1: Lucian *Syr.D.* 14, 45-7; Mnaseas f Patara *On Asia*, Xanthos of Lydia, *Lydiaca*, fr. 17) downplaying other aspects f her cult (Lightfoot 2003, pp. 65-6). These stories are usually presented as narratives of strange religious practices and most, but not all, date to Hellenistic and Roman times, when the Atargatis cult had spread from Syria to Egypt, Greece and the West (Martin 1987, p. 81; Lightfoot 2003, p. 44, 76). The implication in these narratives is that the rules relating to the cult of Atargatis differ fundamentally from Greek formal practice in which fish sacrifice is supposed to be an exceptional phenomenon (Detienne 1989, p. 221; Davidson 1997; Wilkins 1993; 2000, p. 303).

In summary, in Classical Greece, no fish appear to be completely inedible. Even in the unusual case of the dolphin, its flesh became suitable for some type of consumption under certain conditions. The inedibility of fish was in most cases temporary or geographically restricted or applied to a very small section of the population or specific fish. In any case it was justified on the basis of the fish's close association with a god or as part of a broader regime which involved a number of other food restrictions.

9.2.2 *Large fish versus small fish.*

Among the edible fish, several categories were formulated. One of the most consistent and pervasive was the distinction between large and small fish, which, as the Akraephia price lists indicates, was often only loosely connected to taxonomy. The largest individual within a species attained a higher price, apparently being more highly esteemed. This idea of the high desirability of the largest fish permeates the extant fish-related literature and its implications have been extensively analyzed by ancient historians (on the distinction on the basis of size see Purcell 1995, p. 142).

On a primary level the issue appears to be one of price. Large fish were more expensive. The ability to buy them on a regular basis was presented in comedies as a sure sign of affluence (Appendix 1: Alexis *Odysseus at the Loom* fr. 159; Wilkins 1993). Buying large fish in quantity, however, was considered rather vulgar, and the uncontrolled craving for fish (commonly associated with large fish) as a sign of loose morals (e.g. Davidson 1995, p. 209; Wilkins 2000, pp. 69-70, 293-296; 1993, p. 199).

Price, however, was not all. In comedies, the desire for large fish and the appreciation of its qualities (whatever these might be) appears to have been a sign of cultivation, of familiarization with the cosmopolitan, urban lifestyle of the elite (e.g. Appendix 1: *Odysseus at the Loom* fr. 159; *Principal Dancer*, fr. 200; Antiphanes *Butalion*, fr. 69). The widespread focus on large fish, as it emerges from the Classical and Hellenistic literature, has been discussed by Davidson and others who have emphasized the cultural and political insinuations of such a desire (Davidson 1993, p. 54; 1995; McDowell 1971, p. 495). These observations, which are basically related to the discourse around fish-eating in a large urban centre such as Athens, will not be further elaborated here. Certain issues however need to be brought forward.

Wilkins has observed that the discourse around fish-eating as a social marker (a notion presented in detail by Davidson 1997) refers to the elite and develops around the *symposium*. According to him, the symposium was not actually the preserve of the elite but it concerned larger parts of the polis citizenry (Wilkins 2000, pp. 65-66; Fisher 2000). Therefore, the distinction qualities of fish-eating could have existed in domains other than the elite symposium, such as the tables of ordinary people who could not perhaps afford the regular consumption of large fish but who included fish in their diet on a fairly regular basis. If the large size was considered a highly desirable fish trait in this discourse of distinction, then the popularity, but relatively moderate price of the cartilaginous fish (see chapter 9.1) might be seen as one feature of the modification of the elite lifestyle (consumption of large fish) by the middling citizens that Wilkins talks about.

Zooarchaeological data could perhaps illuminate the issue. In carefully collected and well documented fish-bone assemblages (e.g. Itanos, Kalaureia, Table 5.1a), remains of small and large fish are found together. This is true both for domestic assemblages (Itanos) and for assemblages representing more formal public events (Kalaureia). Having the above archaeological observation in mind, certain comic passages which enumerate a number of different fish on the table may be viewed in a different light to the one described above. In the Old Comedy "*Muses or The Marriage of Hebe*" by Epicharmus of Syracuse (Appendix 1) the gods, in a Sicilian context, eat dozens of species of fish, both expensive and cheap ones, large and small (Wilkins 2000, pp. xxi-iii). Such citations of different fish in the form of lists (also of other foods) have been viewed as a typical literary *topos* especially in Middle and New Comedy which is related to a desire to place emphasis on material culture (Wilkins 2000, p. xxi). For the purposes of the present study, what is of interest in the lists is that the writers find it appropriate to list large and small fish side by side in a manner which echoes the size variety of the fish consumed in a single dining event at the Sanctuary of Poseidon on Kalaureia (see chapter 8.1.1).

A further observation may be of interest. Small fish appear, in literary sources, as a very common feature of the Athenian fish-market. Renowned varieties of such fish, as the whitebait from Faliron occur with the frequency of a stereotype (Table 5.3). When fish from other parts of Greece are referred to, they are always the largest, most impressive ones. The issue here is whether such literary emphasis on small fish in Athens reflects the high demand for fish in the market, which triggers a more efficient exploitation of every available fish resource or whether it has a different explanation. Going back to the zoo-archaeological evidence, it becomes clear (Table 5.1a) that in all well-sampled and water-floated sites, admittedly none being from Athens, fish-bone assemblages are dominated by the remains of small to very small fish. Emphasis on the catching of this type of fish therefore is widespread and equally true for small and larger settlements (as far as our sample goes). The literary emphasis on the Athenian small fish, as opposed to the large fish from other parts of Greece, is probably more related to the peculiarities of the literary genres which are the source of this information. Most comments on the size of fish elsewhere originate from travel literature. The mechanisms of memory were probably at play when authors remarked on the large or peculiar fish from each visited place instead of the ordinary. After all, the largest, most impressive or unique fish, is more memorable and worth writing about.

9.2.3 Local versus non local and imported fish.

The culinary distinction between local and imported or non-local fish is probably related to the above mentioned discourse on the marvels found abroad. Taking Athens as the centre, a discourse of excellence (best taste or best size) appears to apply more to fish from elsewhere than to fish of local origin. It is interesting that when best taste is the issue, Attic excellence focuses mostly on the small fish discussed above (Table 5.3). Our main literary source for this discourse is of course Archaestratos. Scholars appear to agree that Archaestratos travelled himself to all those port towns referred to in his treaty as sources of excellent fish (Dalby 1995). His 4th century B.C. evaluations were apparently of interest to 2nd century intellectuals, such as Athenaeus and his readers. But we are in no position to ascertain either how widespread these evaluations were in the 4th century Greek world when Archaestratos lived, or how broad his readership was (Wilkins 1994). A preference for non-local fish is a very elusive subject to investigate. Given the restrictions in keeping fish fresh, fish from distant places could only be known either in a preserved form, having lost their initial taste or through literary descriptions and by reputation.

Fish from distant places are quite inconspicuous archaeologically. They are occasionally identified, as, for example, in the case of the catfish from Kommos which had been imported from either Egypt or Syria-Phoenicia (Rose 2000, pp. 512-514). In this case however, we can not be sure whether fish had been imported live (for the possibility of long distance transportation of live catfish see Greenwood and Howes 1973-4, p. 259) for its value as a fresh exotic fish, or preserved as food. The issue of imported preserved fish will be discussed below.

There is one parameter concerning the consumption, physical or literary, of fish of non-local origin which appears to have been quite important. That is the knowledge that is involved into its selection and consumption. Being able to appreciate the imports from a particular origin, or knowing for example which part of the Aegean produced the best dogfish (Appendix 1, nos. 31, 34, 37) implies a broad and comprehensive perception of the physical and cultural world. It may also imply travel experience, or at least association with people who had such experience (Helms 1988). In Classical Corinth, a trading establishment, the so-called "Punic Amphora Building", specialized in the importation and sale of preserved fish from the Atlantic (see section 7.2.3) and of wine brought in from places renowned for this product, such as Chios and Mende in the northern Aegean (see Appendix 4). This special combination of goods in the same shop probably emphasizes the interest of the consumers in the far away and, perhaps, the exotic. In the case of this trade in the Punic Amphora Building we can see just such an interest in travelling and the wider physical and cultural world.

9.2.4 *Fresh-water versus marine fish.*

The medium the fish live in provides another ground for fish distinction. In the Akraephia inscription, fresh-water fish are listed separately from the marine ones. How this category was articulated in practice and in various parts of Greece is not possible to define on archaeological grounds. Where remains of both kinds have been found (e.g. Krania, Table 5.1a), no obvious spatial or contextual distinction is observed in their deposition (but see chapter 8.2 for problems of inadequate publication). In literature fresh-water fish which are referred to by name are very scarce (see discussion in chapter 6.2). Eel is the only fish of any widespread reputation which, on the basis of the Akraephia inscription, was considered to be of the freshwater kind. Fresh-water fish are mentioned in a generic manner in medical texts (e.g. Galen, in Powell 2003).

In the literature there is an ambiguity concerning the euryaline fish. In the Akraephia inscription they are grouped with the marine fish, but an evaluation of the quality of their flesh by Galen in the 2nd century A.D. makes it clear that people were aware of these fish's ability to live in fresh waters. It is interesting that Galen links the quality of the grey mullet's flesh to the degree of salinity and cleanness of its water habitat with grey mullet living in closed ponds being of the inferior quality, while those living in the open sea being the best (Appendix 1: Galen *On the Powers of Food* iii.709-13). The same text exhibits an inability (or unwillingness) to distinguish between the various species of grey mullet on the basis of taxonomy, and chooses to do so, on the basis of habitat (Wilkins 2005, pp. 24-25).

The literary sources are laconic on the variety of consumed fresh-water fish but very eloquent on the marine ones. The Akraephia inscription makes it clear that there were fish, which at least in certain parts of Greece, were as highly esteemed as the best of the marine fish. The *varakos* that had a price equal to that of the renowned tuna belly parts is one such example. It could therefore be suggested that there was in place an evaluation system for fresh-water fish that worked in parallel to that of marine fish but never entered the written records. The reason behind this exclusion had probably to do with lack of familiarity of urban consumers, who were the main producers of the written record, with fish. The issue of the cultural distance between riverine or lakeside communities and the inhabitants of large cities, which has been discussed in chapter 6.2, is perhaps also relevant.

9.2.5 *Fresh versus preserved fish.*

An important fish distinction was that between fresh and preserved fish. The variety of possible preserved fish products and by-products and their different textures has already been discussed (see chapter 7.2.1). For ancient Greeks this was a subject of considerable interest. So much so, that special treatises were composed on preserved fish both in Classical and Hellenistic times (Appendix 1: Diocles of Carystos *Hygieina*; Euthydemos *On Salt Meats*). At this point it is perhaps of interest to note that preserved fish fill a different culinary "niche" than the fresh ones. Fresh fish were considered as ingredients for the main dish or as appetizers (in the case of small fish). Preserved fish on the other hand, were either consumed as appetizers (Dalby 1996, pp. 75-76; Curtis 1991, p. 29), or in the case of fish sauces, as condiments, that played a role in themselves and not as substitutes for fish. Fish sauces were added to both fish and meat dishes, to greens, cereals and pulses (Wilkins 1993, pp. 198-9). Fresh and preserved fish were treated differently and this is reflected in their pricing. In both cases a range of prices, from very high to very low indicate the existence of various quality categories.

These developed independently for the fresh and the preserved fish (see chapter 6).

An interesting point concerning the development of the taste for preserved fish concerns its chronological pattern. Preserved fish is by no means a phenomenon of the historical era in Greece. Examples such as the remains of dried (and salted?) fish and of fish paste found in the 17th century B.C. destruction layer at Akrotiri, Thera (Mylona, personal analysis; Trantalidou 1990) illustrate the long ancestry of fish-preserving practices in the area. However, both literary references to preserved fish and archaeological finds in the form of fish remains in vessels, specialized pottery or fish preservation establishments, have multiplied in the Roman period. Such evidence is relatively common in Greece, but particularly visible in other places at the periphery of Greek world (e.g. Bekker-Nielsen 2005, Curtis 1991). Some forms of preserved fish, garum more specifically, appear to become particularly fashionable during the Roman era. While only a few literary references can be found to it in the extant comedies of Classical date, it is very common in later periods (Dalby 1996, p. 76).

9.2.6 *Fish of dry flesh versus fish of moist flesh.*

Distinction between fish (but also meat) on the basis of their texture is often found in literary texts (Wilkins 2000, p. 18) but is more regularly encountered in medical literature. Fish evaluations on the basis of their flesh texture are found in the *Hippocratic Regimen* II (around 400 B.C.) or the works of Diokles of Karystus (late 4th century B.C.), Mnesitheus (4th century B.C.), Diphylus (3rd century B.C.) and Galen (2nd century A.D.) (Wilkins 2005, pp. 22-3; Wilkins 2003). The fact however, that excerpts of these works found their way into sympotic literature such as the *Deipnosophistae* of Athenaeus, indicates, perhaps, that the issue of texture was of wider relevance (Wilkins 2000, p. 18). According to the *Hippocratic Regimen II* fish are divided on the basis of characteristics such as dryness, lightness or heaviness. These categories are related to the nature of the waters from which the fish originate (Wilkins 2005, p. 22). Similar categories applied also to salt fish. In that case, salting was considered a process which could drastically alter, and often improve, the texture of fish (Appendix 1: Galen *On the Powers of Food*s iii.709.13).

The dry and the moist, the light and the heavy, or the digestibility of fish flesh were often used in the context of medical writings as criteria which rendered specific fish suitable or otherwise for a diet oriented towards healing, restoration of the body's balance or just for general health. Such an approach did not only concern fish. In ancient Greek and Roman medical thinking a person's good health was considered the result of a balance between his/her bodily fluids (Powell 2003; Craik 2000; Grant 1995). Food was elemental in the preservation of the body's balance and the person's good health. Fish and fish sauce appear as central elements in this restorative diet (Curtis 1991, pp. 29-35).

It is interesting that the choice of fish as food on the basis of the above mentioned qualities appears to be gender specific. According to Rufus of Ephesus (Nutton 2000), a leading physician, who flourished during the reign of Trajan (98-117 A.D.), women, due to their physiology should avoid the eel, the sturgeon, the turbot and the river fish in general (Appendix 1: Rufus of Ephesus *Medical Questions* 20.1-2, 13, 17). In other words, they should refrain from eating fish which elsewhere are described as oily, heavy and moist, apparently because their consumption would bring imbalance to their body which was by its nature moist and cold or according to other authorities moist and hot (Craik 2000). In this case gender is a crucial factor in the choice of fish as food.

The issue of texture, which is best exemplified through the categories "moist" and "dry" as these have been discussed above, was, however, much more intricate. Preference for certain anatomical parts of fish is aptly relevant (Davidson 1996, p. 64; Wilkins 1993, p.194). The heads of large fish, especially rocky inshore fish, ranked very high in culinary preferences (Wilkins 2000, p.18). The archaeological record verifies this idea. In the merchant's house of Late Roman Itanos, when they were getting large fish for the table, such as groupers, meagre or parrot-fish, they were exhibiting a distinct preference for the head (Mylona 2003b, Table 7.1). On the contrary, heads of the migratory fish, are rarely mentioned in literature and, interestingly enough, they are also absent from the Hellenistic dining deposit at the Sanctuary of Poseidon at Kalaureia, despite the prominence of bones from the rest of the body in the same assemblage (see chapter 8.1.1).

The head of the rocky fish tends to be fattier, softer and more gelatinous than the rest of the body, which as a rule has firm, lean white meat. Nowadays, heads are customarily boiled in Greece, to create fish soups or jellies (Mylona, pers. observ.). We could perhaps assume that the high esteem for expensive belly parts of tuna and eels, as well as the more affordable cartilaginous fish, was due to the soft, fatty, gelatinous flesh which characterizes them all. This conclusion leads us to another interesting issue, which has to do with the mechanisms by which gender relations may shape the popular taste.

In ancient Greece, or rather Athens, for most of the centuries covered by the present book, the discourse around fish taste evaluations was generated by men, who were writing for a mostly male audience, be it in the theatre or the formal/informal symposium. In this same society, over most of the same period, females were, at least in principle, excluded from the consumption of fatty, gelatinous fish (see above) for health reason. If we accept that these medicinal rules had some bearing upon widespread beliefs, then we might expect that the male-dominated discourse on culinary tastes would have incorporated these rules, giving them a culinary, taste-related dimension (for the function of taste and other senses in relation to fish see chapter 10.1). So the discourse on the excellence of eels, fish heads and tuna belly parts, might, in fact, have been the discourse of male consumers, defining their distinct gender position in a subtle but quite efficient way, ascribing taste value to the foods which were considered suitable for them for other reasons.

9.3. Conclusions.

The study of fish's biology was part of Greek culture, at least from the 5th century B.C. and classifications of fish, based on their biological characteristics were available. However, such classifications appear to have been an issue of academic concern, with no bearing upon the way Greeks viewed and classified fish that were to be consumed. In that context, different, often complementary classifications had been constructed, which were based on those characteristics of the fish that were relevant to consumption.

For ancient Greeks (especially Athenians) texture, size, provenance and processing, were all qualities of fish that had certain social and ideological undertones, which were often quite irrelevant to the fish themselves in the biological sense. In those cases, fish, like every category of food, served mostly as vehicles that transmitted particular messages to the people participating in or excluded from their consumption. The ability to afford a large or an imported fish, or the exclusion from the consumption of eel for example, not only spoke of personal preferences, but made clear statements about who the consumer was, and what was his/her position in society.

These categories however can be viewed in another light, which focuses on individuals in a strict sense. It focuses on the consumers' bodies and senses and on the ways in which the social and ideological undertones mentioned above become incorporated by the members of a society. The cooking and eating of fish is the domain where such issues can most fruitfully be explored.

CHAPTER 10

EATING THE FISH. FISH DISHES AND THE SENSES

At Iassos, according to Strabo, the arrival of the fish was announced by the ringing of a bell, which proved to be more exciting than the performance of a musician in the agora, for all but a deaf man (Appendix 1: Strabo 14.2.21 and also Plut. *Quest. conv.* 667ff). In Arisotphane's *Knights* the Sausage Seller curses the Paphlagonian to choke on a hastily swallowed squid, which has just been snatched out of the sizzling frying pan (Appendix 1: Ar. *Eq*.927ff)! And when a comic chef in Philemon's *Soldier* describes how, by use of a special recipe, he transforms a humble mud-eating grey mullet into a divine dish, he claims that the aroma of it could even raise the dead (Appendix 1: Philemon *Soldier* fr. 82). In a different field, the famous Classical fish plates are often decorated with representations of naturalistic fish, which almost appear to be swimming around the plate (see chapter 10.2). Sounds, smells, and verbal or real images of fish constitute an integral aspect of fish-eating (Messer 1984, pp.218-222). Eating involves pleasure or disgust, closely linked to sensual perceptions, which, to the consumers, are experiences of an immediate, direct, bodily impact (Falk 1994; Hamilakis *et al.* 2002a).

Such issues have been very little investigated by researchers of past societies (but see papers in Hamilakis *et al.* 2002a). Recent anthropological thinking can function again as the starting point in the exploration of this issue. The role of the sensory experiences involved in eating, that is smell, hearing, taste, touch and also memory, have, in recent years, become focal points in anthropological, but also some archaeological research (see extensive discussion in Chapter 3). In the theoretical framework of embodiment, the senses are perceived as the media by which individuals attend to the world and by which they incorporate social relations (Ingold 2000a). Through the senses, the world, physical and social, becomes comprehensible. Furthermore the senses are a domain for sharing the experience of living with fellow members of a group, be it family, community or other in the most personal manner. Some aspects of the impact of the sensory experiences around fish-eating and their implications in the material, social and ideological life of the consumers will be explored in this chapter.

10.1 A liking for specific fish tastes as shared culture.

Perhaps one of the most touching aspects of a review of the comic and sympotic literature is the recitation of recipes, especially in works set in Athens between 400-300 B.C. Vivid details of raw ingredients, preparation, ways of cooking, condiments and dish presentation enables the reader or listener (depending on the literary genre) to create an image of the specific food, to almost smell and taste it (e.g. Appendix 1: Sotades *Locked up Women*, fr.1; Davidson 1993). These verbal sensual experiences certainly permit the sharing of the dish among large numbers of people, not unlike the present day popular cooking magazines and TV shows. The large audience of the theatrical plays, which, at times, reached several thousands (Wilkins 1996; Pickard-Cambridge 1968, pp. 263-69) indicates that these images were relevant to a wide range of the (Athenian) population.

The basic categories of fish, i.e. large and small, local and non-local, fresh and preserved etc., which have been discussed in the previous chapter, certainly have considerable bearing on the issue of taste. Other parameters were, however, also at issue when a choice was to be made. A fish was not eaten by itself in its natural state (raw) as was done by the uncivilized *Ichthyophagoi* (Appendix 1:Arr. *Indike*; Diod. Sic. iii. 15-17; Strabo 15.2.2, see also discussion in chapter 6 and Levi-Strauss 1997). It was cooked using a variety of ways and ingredients. Boiling, stewing, grilling or frying were the basic cooking methods, and in each case different fish or fish parts were considered most suitable. The *afyi* for example was always fried, while the head of *glaukos* was boiled (Davidson 1996, p. 64). Comedy fragments and lexicographic works such as Hesichius's *Lexicon* preserve words which refer to special fish cooking methods or preparations. *Epanthrakides* are fish cooked on the coals; *darta* means flayed, and was a generic word for fish such as skate whose rough skins were removed before cooking (Mnesitheus, in Ath. 357c); *epsitos* was fish boiled, perhaps in a shallow open shaped pot called *lopas*, which, interestingly, means limpet (Sparkes and Talcott 1970, p. 3 and 227; Amyx 1958, p. 210, note 76). We may assume that these methods of preparing fish were so common and repeated that a single word came to denote the whole preparation or cooking process with no further elaboration. The issue of texture, which has been discussed in chapter 9.2.5-6, comes again into the foreground (Wilkins 1993, p. 194, note 9; 2000, p. 18; for detailed discussions on specific fish dishes see Dalby 1996, p. 66-75 and Dalby and Grainger 2000; for pottery shapes suitable for fish cooking see Sparkes 1962 and Wilkins 2000, p. 30-36).

From the above procedures what remains and can be seen archaeologically are the fish bones, which happened to be burned during the process of cooking (Nicholson 1995; Colley 1987) or the cooking vessels themselves with the traces of food cooked in them in the past (Evershed *et al.* 2001). Cooking vessels offer a promising, yet still unexplored field of research. The case of *lopas* is a good example of how certain issues related to cooking are interpreted on the basis of modern common sense, instead of vigorous analysis. The *lopas* (shallow cooking pot) of various sizes is considered to be a vessel used to fry fish or make fish stews (Amyx 1958, p. 210, note 76). Such an assumption is based on rather shaky grounds which mostly consist of literary references combined with illustrations, which associate *lopas* with fish. Lipid analysis on pottery could demonstrate the use of certain types of pots to cook fish, but no such studies are available as yet. So, for the moment, the association of certain pottery types with specific types of food remains crude and fairly unreliable.

Smell was another central issue in fish consumption (Wilkins 1996, p. 50, note 20 and more generally Thiercy 1993). Freshly caught fish has a delicate subtle aroma which intensifies as the fish looses its freshness. The highly distinctive smell of the fish-market is actually the smell of spoiled fish. In the literary sources, fish smell is presented as exciting or repulsive. Roman authors, for example, hold a highly ambiguous position towards fish sauces and their main characteristic, i.e. their production through fermentation. They find its odour both exalting and disgusting, often at the same time. They make derogatory and enthusiastic remarks about it, depending on the context, on the specific type of fish sauce or on the nuances of the argument (see Curtis 1991, pp. 155-56). Again, qualitative characterizations depend on context.

Most modern scholars assume that the pungent aroma produced by fermentation of fish during fish sauce production (but also during the production of purple dye) is repulsive, something to get rid of (Curtis 1991, p. 3; 2005, p. 36). In this spirit, the consecutive floors of crushed amphorae fragments, overridden by crushed lime-stone in the Punic amphora Building in Classical Corinth (see Appendix 4) has been interpreted as a means to hide "the potentially overpowering stench of fish clinging to some of the amphorae" (Zimmerman-Munn 2003, p. 198). Such a statement assumes a universally uniform qualification of smell. Yet, in other instances, such as the town of Tyritake in the Crimea, Black Sea, we find fish processing establishments within the settlement or even, in small scale, within houses (Curtis 1991, p. 122; Højte 2005, p. 142-148). Similarly, but on a smaller scale, in Roman Lusitania, on the Atlantic coast of modern Portugal, small-scale processing establishments have been found annexed to wealthy villas (Edmondson 1987, p. 129). We may assume that the strong odours of the trade were not bothersome for the inhabitants of those settlements or those people not directly involved in the process. This could be so because fish processing was an integral part of their life and the source of their wealth. The strong odour of the ongoing successful business might be the smell of well being rather than disgust. What the above archaeological examples show is just how culture specific the qualifications of odours may be.

Ancient Greeks, at least as far as our sources go, seldom ate fish with no accompaniment. The recipes at our disposal list a number of spices, herbs and sauce ingredients (see various recipes in Dalby 1996, p. 66-75). Olive oil, vinegar, fish-sauce, cheese, marjoram and the expensive imported *silphium* are some such additions. Certain of these accompaniments do not alter the taste/aroma of fish greatly (e.g. olive oil). Others are so strong, that we may assume that they covered the aromas of the raw material (e.g. marjoram, cumin, cheese). Smell in this case, and a palate for spicy tastes is central. Olfactory images were fairly strong in food-related literary genres.

Vision and the colour and shape of food, and fish in particular, appear to have been crucial as well. Classical sources, or even earlier ones (Hom. *Il.* xxi.203, 353) seem to hold eel as the supreme fish. Eels wrapped in beetroot leaves, was presented as a verbal image that aroused the appetite and we find it repeated in comedies as a stereotypical phrase (for eel in comedies Gilula 1995, p. 390; for eel in Aristophanes, Davidson 1993, note 3; Degani 1995). A particular colour or a distinct "packaging" for fish appears to elevate it, to make it more desirable or special. The exaltation of red mullet could be seen just in this light. This fish, which has an undistinguished light brown pattern on its body when alive, turns pink when caught and a bright coral-red when fried. The importance of bright red colour and its multiple social and ideological associations have been discussed in relation to purple dye (Jensen 1963; Reinhold 1970; Moatsos 1932), one marine resource which has only been discussed in passing in this book. The high esteem for red mullet could possibly be part of this discourse around red colour.

The importance of certain colours as enhancers of fish desirability is vividly illustrated by the case of preserved sardines, produced exclusively in the city of Volos in Thessaly at the beginning of the 20th century (Athanassopoulos 1926). During the salting process, red, iron rich soil, imported by boat from the Sporades, especially the island of Skopelos, used to be added to the salt, in order to produce salted sardines with a vivid red flesh, quite unlike the pale pink flesh attained by the normal salting process. This type of sardine was produced mostly for export to countries of the Middle East. Chemical analysis, performed by state officials, to establish the dietetic and taste related effects of the use of red soil, showed that apart from the flesh colouring no other chemical change occurred in the fish. Nevertheless the producers insisted that the use of the red soil made the salted sardines not only tastier but gave them a more appetizing aroma as well.

The place of each type of fish (or sea food) dish in the order of a meal was apparently also important, bringing to mind Douglas's work on the sequencing of meals and menus (Douglas 1997). Again, from the literary sources, it becomes evident that certain fish were suitable as entrees to the meal, spicy food to stimulate the appetite and prepare the palate for the more elaborate tastes of the main meal. Here an interesting distinction was made.

Sea food such as, squid, octopus and sea urchins are repeatedly referred to as appetizers (Appendix 1: Alexis *Crateia or The Drugist*, fr.115; Demetrius of Scepsis *The Trojan Battle-Order*). Large fish, but even smaller ones with fuller, rounder taste were considered suitable for the main course (see recipes in Dalby 1995, pp. 66-75 and in Dalby and Grainger 1996). An interesting distinction appears to involve the role of different processed fish products. While pickled fish (in solid form and with distinguishable flesh texture) were considered suitable for entrees, fish sauces (fish in a liquid form, with no flesh texture preserved) were but one ingredient in main dishes based on fish, meat or vegetables, grains etc. (Curtis 1992, pp. 29-30; Bekker-Nielsen 2002, p. 31).

The above information derives from comedies and gastronomic literature. If we accept Wilkins's conviction that comic food and eating reflects eating in the city, even if it is presented according to comedy's own codes and priorities (Wilkins 2000, p. 99), then we may assume that the patterns emerging from these sources are patterns of shared behaviours, shared ideas about what is proper, what is right around the table, not only in terms of behaviour (see discussion in chapter 8) but also in terms of the food itself, and its properties. The reference to pickled fish and sea food, along with other strongly tasted foods, in the literature recounting the beginning of the meal rather than the meal proper, may thus be seen as a reaffirmation of the proper order of things. When this order is disturbed, when, for example, a participant of the meal eats *opsa* (Davidson 1997, pp. 20-26), the spicy food, as main course, then criticism follows which involves the social and ideological conduct of the person (Davidson 1997).

While assumptions like the above appear quite well supported by the literary sources, a basic reservation must be brought forward. Most discussion about fish and sea-food recipes, about appetizers and sauces, is focused on luxurious meals, on festive events, where several people participate and where an elaboration at food and drink, manners and conversation can be amply developed (Dietler 1996; Hayden 1996). The ways and preferences of less affluent or poor consumers are indeed elusive. Similarly, aspects of the fish-eating of various groups within the population of a city or the countryside remain unknown. Women, slaves, children, foreigners could be expected to have distinct sets of proper tastes, smells, food forms. The case of the fatty gelatinous and soft fish, being, in theory, unsuitable for women (see chapter 9.2.6), offers just a glimpse to this largely invisible world of alternative sensual priorities on the food front.

Archaeologically, the consumption of certain foods can be ascertained, but not the order or particular combinations of consumption. David Reese has used the term "marine meals" to describe concentrations of shells and fish bones, which have been recovered in the same spot or restricted area/level in Iron Age Sanctuaries at Kommos (Reese 2000b). Such a term presupposes the consumption of the two classes of marine food together or in the same meal, a coincidence that can not always be documented. The detailed analysis of the bio-archaeological remains from the two dining rooms at the Classical Sanctuary of Demeter and Kore has successfully demonstrated how remains of edible plants and animals found together do not necessarily constitute ingredients in the same menu. The sea-shells found in the dinning rooms, along with mammal and fish bones, as well as plant remains, were collected for reasons other than eating (Bookidis *at al.* 1999, p. 39), therefore, they were not part of a "marine meal". However, the high frequency of the association of fish bones and sea shells in the same strata which is observed archaeologically (Table 5.1a-g), requires a more detailed examination of these remains and their context, in order to confirm a real association of them as food or not.

The above discussion has demonstrated that the senses, this aspect of being which is often considered so elusive and perhaps difficult to document archaeologically, are in fact crucial parameters in the formulation of specific choices around fish-eating. Consequently the senses may justifiably be considered as one of the parameters which should be examined when the importance and the role of fish as a food resource is explored along with the more traditional topics such as resource availability, nutritional needs, social parameters in fish consumption etc.

10.2 Fish-eating in 4th century as a multilevel sensual act.

In the 6th century B.C. a vessel shape was invented in Athens, and spread across the Greek world. It is known in archaeological literature as the Gallatin plate. It is essentially a black glazed dish, with a low foot, a rim and a sloping floor. At the very end of the 5th century this shape developed into what came to be known as the "fish plate" (Fig.10.1). At around 400 B.C., this type of pot was heavily decorated in the red-figure style. Decoration followed two distinct forms. One group of these dishes was decorated with one mythological scene, that of the abduction of Europe, with fish and Neireids giving an intensely marine character to the composition (McPhee and Trendall 1987, p. 19). The other group was decorated with fish. Certain fish and other marine creatures, rendered in two different techniques, covered the floor of the plate and gave to this pottery form its characteristic name (Fig. 10.2). At the same time the Gallatin plates continued to be produced and remained in demand until the 2nd century B.C. The size of the fish plates varies greatly from micrographic examples found in sanctuaries, apparently as dedications, to large plates with a diameter reaching 55 cm found in settlements and graves (McPhee and Trendall 1987, pp. 19-20). Most of the fish plates however have a diameter of about 20-25 cm.

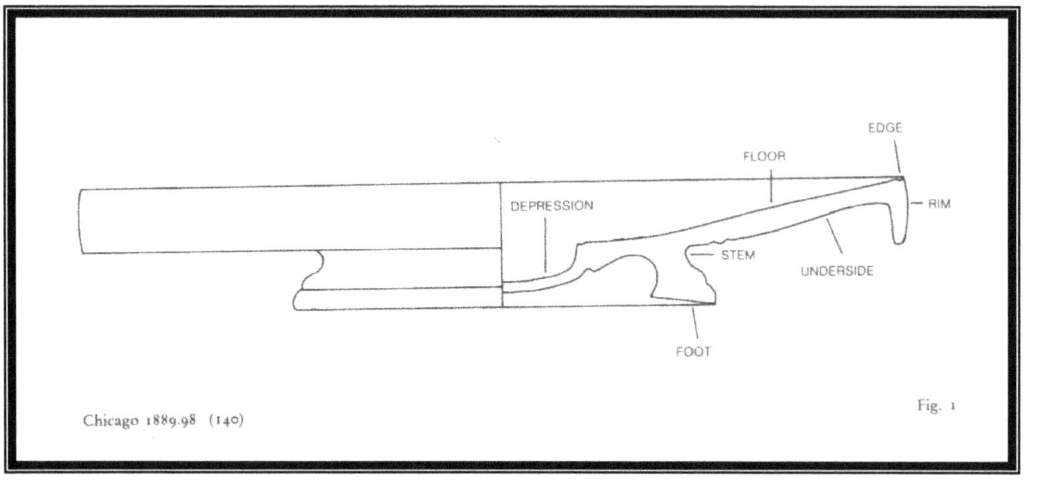

Fig. 10.1 Fish plate outline (from McPhee and Trendal 1987, p.174, fig.4).

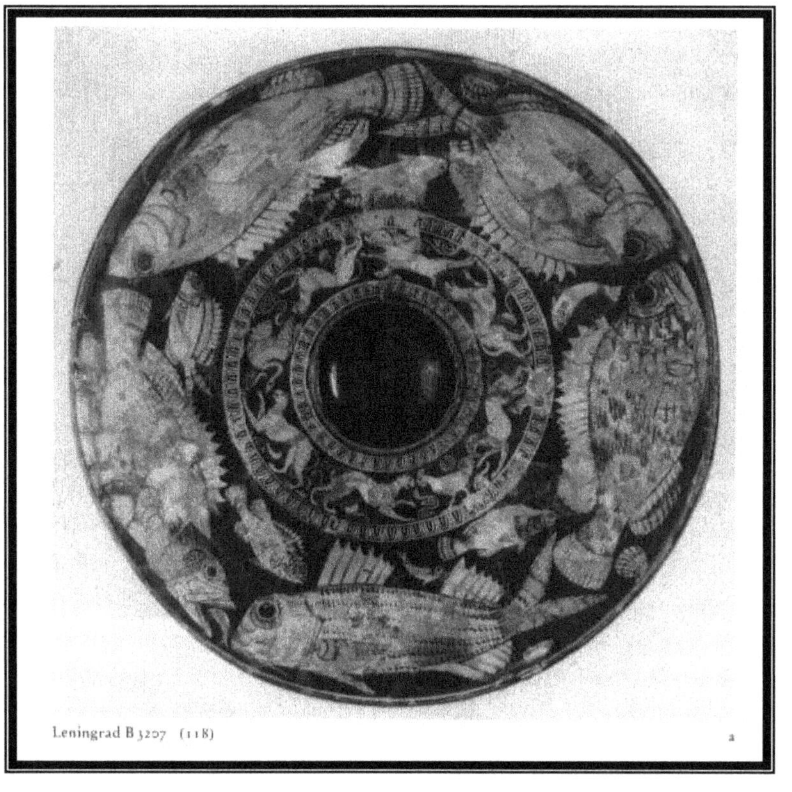

Fig. 10.2 Attic fish plate from Taman (from PcPhee and Trendal 1987, pl. 8a).

The most common fish depicted on Attic fish plates are various sea breams and the grey mullets (Fig. 10.2). The head morphology of some of the sea-breams, especially the curving of the frontal (Fig. 10.3), suggest that the fish represented are probably members of the *Dentex*, *Diplodus* or *Pagrus* genus (*Dentex gibbosus, Diplodus sargus, Diplodus vulgaris, Pagrus pagrus, Pagrus caeruleosticus).* Also, cuttlefish, squid, octopus and john-dory occur less often. In the later examples, scorpion fish, sea perch (*Serranus scriba*), red mullet, and angler fish are also introduced (Fig. 10.4). All these are marine creatures commonly found in the waters of Eastern Mediterranean but also in Athenian comedies and sympotic literature, and in food refuse recovered in archaeological excavations throughout Greece (see Table 5.2). Surprisingly, the migratory Scombridae, both small and large varieties, are absent.

The production of the fish-plates was initially an Attic affair and they were manufactured by different potters, some of whom are known from other, high quality works. Beazley's Meleager painter is one of them (McPhee and Trendall, 1987, p. 19). What makes the Athenian fish-plates particularly interesting is that they were produced for only half a century and they were almost certainly produced for export. The available evidence suggest that

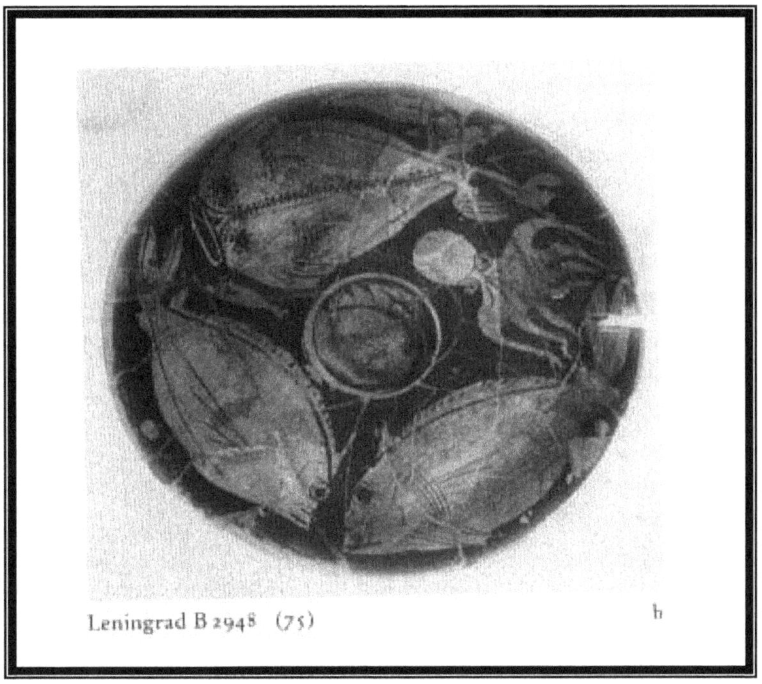

Fig. 10.3 Attic fish plate from north Black sea (from McPhee and Trendal 1987, pl. 6b).

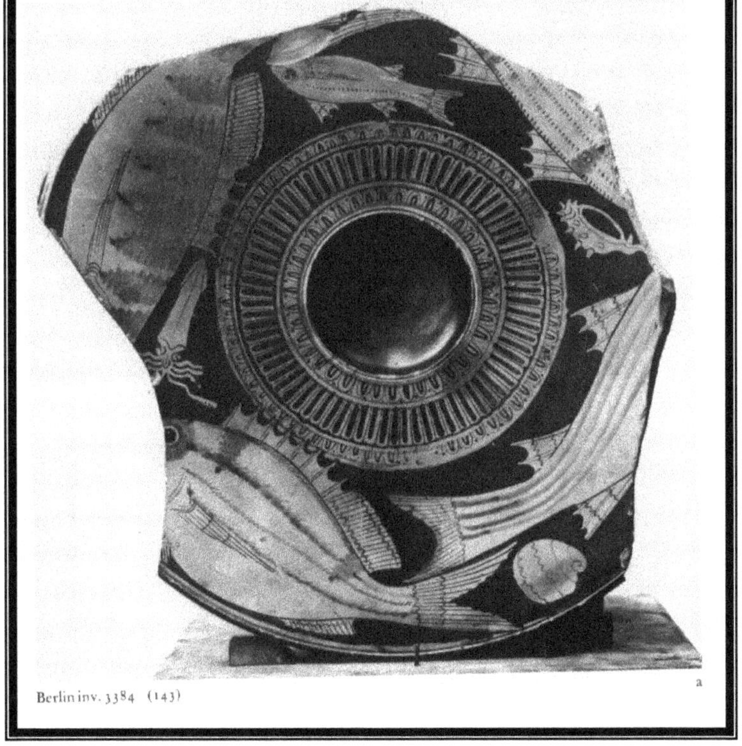

Fig. 10.4 Attic fish plate from Tanagra, now in Berlin (from McPhee and Trendal 1987, pl.11a).

the exportation of fish plates was, at least initially, fairly focused, with most of the surviving fish plate fragments found in settlements within a small geographical area on the north coast of the Black sea, in Crimea, at the Taman peninsula (Fig. 10.5). The Taman peninsula is located at the Eastern coast of the Kerch Straight, the only opening of the Sea of Azov to the Black sea. A few examples are also recorded from Avdera and Olynthos in Northern Greece, Athens, Tanagra, Asia Minor, the Iberian Peninsula and Cyrene on the North African coast (McPhee and Trendall 1987, p. 21).

It is not clear what prompted the modification of the Gallatin plate and the creation of this distinct decoration.

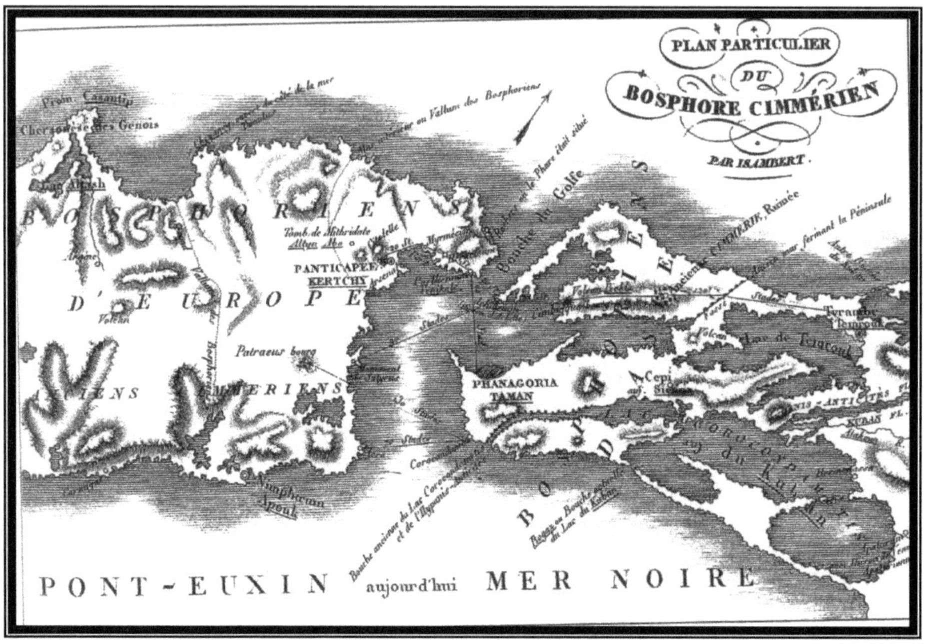

Fig. 10.5 Taman peninsula. Engraving, 1823 (from Koromila, 2001, p. 217).

Catering for a specific clientele in the Crimean area, who demanded the particular mythological scene of the abduction of Europe for its wedding and/or funeral related symbolism, is what is suggested (McPhee and Trendall 1987, pp. 30-1; Barringer 1991). The fish decoration remains enigmatic.

Although the production of the fish-plates began in Attica for exportation to the Black Sea, it eventually found very fertile ground to develop in Southern Italy. Analysis of the southern Italian production reveals the existence of several local traditions around Sicilia, Appulia, Paestum and Campania (Lacroix 1937a; McPhee and Trendall 1987, p. 54). In southern Italy and Sicily, the fish-plate production covered the third quarter of 4th century B.C. The Europa myth is almost absent here but the fish theme has developed and has even been used in decoration of other types of vessels such as amphorae. After the fish decorations ceased, before the end of the century, the similarly shaped black glazed stemmed dishes which were in use before the fish plates appeared, continued to be produced in large quantities (McPhee and Trendall 1987, p. 57).

Strangely enough, the southern Italian fish plates consistently depict the fish with their back towards the rim, while the Attic equivalents had taken the opposite position. More types of fish are depicted in the south Italian fish-plates than in the Attic ones, and the sea food creatures appear much more varied (McPhee and Trendall 1987, p. 57). Here we find gurnards, flat fish, rays, cephalopods, shrimps and lobsters (Figs. 10.6, 10.7). Interestingly, the scorpion fish which is one of the favourite fish on Attic fish plates, is very scarce in Southern Italian examples but the tuna is present (McPhee and Trendall 1987, p. 57).

The southern Italian fish plates, like their Athenian counterparts, have been found in both domestic and mortuary contexts. In southern Italy, fish plates have been found in far larger numbers than anywhere else, indicating an intense local demand for them. The nebulous issue of their use remains enigmatic. Most scholars agree that they were expensive objects, used in formal dining, as a token of luxury. It is usually assumed that they were especially used for serving fish with the central depression serving as a receptacle for the fish juices or the accompanying fish sauce, or as a base for a small handless bowl that served the same function (McPhee and Trendall 1987, pp. 21-2).

The written sources preserve some, admittedly obscure, references to serving dishes which could be fish plates. Aristophanes in *Pluto* (813-4) refers to "pinakiskous saprous tous ixthyrous" which can best be translated as "the malodorous fish platters". A stronger argument in favour of the fish plates' use for serving fish comes from Sicily. There, several fish plates have been recovered during the excavation of a Punic cemetery at Palermo (for the dating problems of these fish plates see McPhee and Trendall 1987, p. 66). Three of them contained the remains of an indeterminate number of unidentified fish, certainly more than two (Fig. 10.8; Mylona, personal observation). The published photograph of the remains in one of these plates (McPhee and Trendall 1987, p. 66, pl. 15c) indicates that these were some medium/large sea bream and probably a scorpion fish (personal identification), fish which were not only popular as

Fig. 10.6 Apulian fish plate from Tatanto, fourth quarter of the 4th c. B.C., from the antiques market (from Zindel 1998, p. 77).

Fig. 10.7 Apulian fish plate from Tatanto, fourth quarter of the 4th c. B.C., from the antiques market (from Zindel 1998, p. 57).

Fig. 10.8 Fish plate form a Punic cemetery at Palermo (Tomb 36), S. Italy, with fish remains found in it (from McPhee and Trendal 1987, pl.15c).

motifs in fish plates decoration but was also held in high esteem in both Greece and southern Italy. In reality the above evidence just suggests that fish had, in fact, been served in the fish plates, but other foodstuffs could equally well be served instead. It remains a fact however, that these flat stemmed vessels, even when empty of food are actually filled with live-looking fish, which almost look as if they are swimming in a pond.

The question remains. What prompted the production of the fish plates in the first place, and what made them so hugely popular in certain places at the periphery of the Greek world for such a short period of time? It appears that what was important about this type of vessels was not the shape, which in a similar but unadorned form was in circulation both before and after the fish plate period, and probably not the function, which could be served by other similar vessels. The important element, clearly highlighted, was the decoration; the fish and the sea food (the mythological scenes will not be discussed here). Keeping in mind that the fish plates were serving vessels of high quality, we can safely connect their use to the table, to formal dining, either for the living or for the dead. We need to re-examine the fish plates in such a context, but taking into account other parallel sympotic developments, always in the background of the material conditions of fishing, fish availability and fish-eating in the particular areas.

The first question to be asked is whether there is any element common to the northern Black Sea and southern Italy (and to a lesser extent all other locations where fish plates have been found), that prompted the desire for fish plates. Is the motivation behind this particular consumption desire the same in both places or was it differentiated?

The Crimea and the Taman peninsula in particular, where the only outlet for the Azof Sea waters to the Black Sea is located, have extremely eutrophic seas, which even nowadays sustain large fish populations. Due to the discharge of large rivers into the shallow seas of the area, salinity levels are very low, and the ichthyofauna of the area is quite restricted to fresh-water fish or euryaline species tolerant of this type of conditions (Slastenenko 1955; Kruglov 1994 and for the marine fauna of the Black sea see Svetovidov 1964 and Anonymus 2005). On the Taman peninsula, which in antiquity was either one big island or many smaller ones (Kuznetsov 2003, p. 897, 900 and references therein), Greek colonies had already been founded in the 6th century B.C. Those belonged to the Bosporan kingdom. The 4th century was a period of development and well being for the Greek cities of the area (Kuznetsov 2003). The fertile hinterland appears to be interspersed with a large number of settlements (Kuznetsov 2001). In antiquity fishing and fish processing in this area had reached the level of a distinct industry which produced preserved fish for export (Curtis 1991, p. 121ff; Højte 2005). This industry focused on a few species only, more particularly anchovy, herring and red mullet as well as sturgeon, carp and catfish (Højte 2005). Fish therefore was, in this specific area, a central element in life.

Apart from the grey mullet, and perhaps certain euryaline sea-breams, the other fish or sea food depicted on Attic fish plates exported to the Taman peninsula are either rare or totally absent from its waters. The ichthyofauna of those parts was/is dominated by fresh-water fish or fish that tolerate very low salinity. And yet, the consumers of the fish plates imported from Attica, obviously asked for fish and sea food of the rocky salty waters such as those found on Greek and other Eastern Mediterranean coasts. This phenomenon is relevant to an observation made by Lewins on his analysis of Athenian pictorial pottery found in southern Italian contexts. Lewins emphasizes the fact that these vases' decoration should be viewed as reflecting the interests of the consumers who demanded specific decorations and not as representations of life in Attica (Lewins 2003). Similarly in Taman, it was a need of the consumers of the fish plates that dictated the choice of the particular fish.

With the above thoughts in mind, we could probably suggest that Greeks (or people of Greek origin) living in the Taman peninsula, brought a small piece of their mother country's sea to the rich but relatively uninteresting, in terms of variety, coasts of the area. In this context the use of the fish plates in symposia, or other gatherings involving commensality, could be viewed as mnemonic devices, which revived, even temporarily, the memory of the Greek seas and its delicacies. These people were perhaps unable to taste and smell the fresh and succulent flesh of the rocky fish and sea food of the Aegean coasts on their tables, but at least their eyes could feast on them! Here Bakhtin's dictum, that by taking food into the body we take in the world (Bakhtin 1968, p. 281), takes another dimension which involves not the actual eating of food but its incorporation through sight. The occurrence of fish plates in other coastal settlements, many of which are also situated on lagoonal eutrophic coasts, might be seen as an application of the same principal on a smaller scale. If we accept the above interpretation, then the fish plates in the Taman peninsula testify not only to the immense importance of food in defining or preserving identity (Garnsey 1999, pp. 62-81; Wilkins 2000, pp. 282-288; Messer 1984, pp. 226-229), but also the central role played by the senses in this process (Sutton 2001).

The Southern Italian market never introduced any fish plates from Attica. Although in the second quarter of the 4th century B.C. many other Attic red figure vessels reached areas such as Campania (e.g. Trendall 1967; 1989), no fish plates did. On the contrary, fish plates were produced locally, the earlier being found in Sicily (McPhee and Trendall, 1987, p. 21). Why did the Southern Italians developed a desire for fish plates in the middle of the 4th century? Here the Taman model can not be applied. In the first place, the marine environments of Southern Italy are very similar to those of the Greek seas, with inshore rocky fish being very common and as far as our sources go, very much appreciated in the 4th century B.C. (Appendix 1: Clearchos of Soli *Lives*). Secondly, the exclusively local production of fish plates, just after the Attic production ceased, despite the existence of mechanisms for the introduction of the original Attic version in earlier years, probably implies some sort of a statement, of a desire to deal with the fish plates in an independent, locally determined manner. Perhaps, the consistency in the different orientation of the fish depicted on the plates is an expression of this desire.

Southern Italy in the 4th century B.C. was the theatre of some unique developments. At that place and time a distinctive local gastronomy developed. Andrew Dalby (1996, pp. 108-111 and chapter 5 and 6) and more comprehensively John Wilkins (1996) have discussed the economic, social and ideological parameters which led to this unique phenomenon. Abundance and variety of products, both local and imported, the existence of a variety of recipes and developed agriculture and commerce are some of them; also the demands from critical eaters and the association of eating with pleasure were important (Wilkins 1996; criteria based on Goody 1982). Southern Italy is the place where famous cooks lived and developed their art and where cookery books were first written. Mithaikos, Glaukos of Locri, the two Heraklides from Syracuse and Hegesippos of Tarentum are the most famous of such cooks/cookbook writers (Appendix 1: Philarchos *Histories* fr. 45; Wilkins 1996b). Archaestratos's *Life of Pleasure*, although of a different literary genre, should be understood within this discourse (Wilkins and Hill 1994; Dalby 1996, p. 109 ff; Dalby 1995).

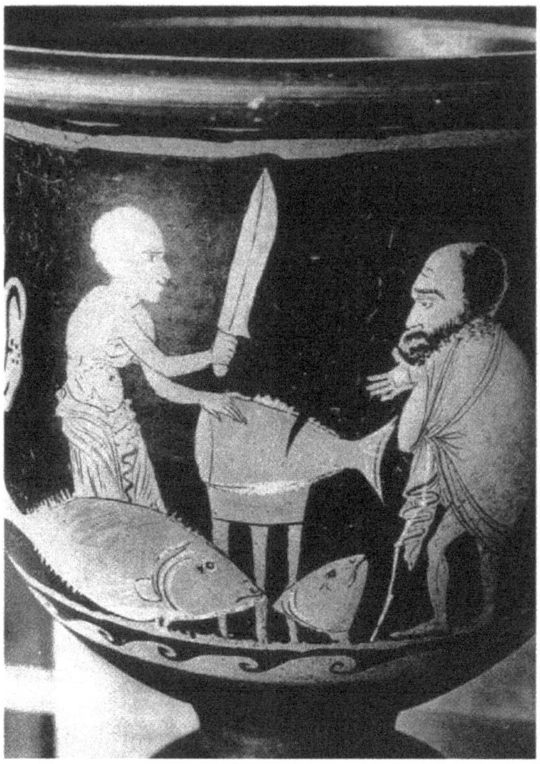

Fig. 10.9 Tuna in the market. The fishmonger slices the fish. Sicilian bell crater from Lipari, 4th c. B.C., now in Museo Mandralisca at Cefalù (Trendall 1967, p. 208, no 54; photograph from Sparkes 1999, fig. 11.7).

To return to the fish plates, they seem to have been in vogue in an area and time where people, or at least the more affluent, paid much attention to fish, to its provenance, its preparation and presentation in a dish. Not only did they pay attention, but they were relied on a developed detailed discourse, which set the agenda for the proper conduct of culinary fish lore. The southern Italians were tapping rich fishing resources, eating fish and sea food and discussing it in an elaborate manner at table and elsewhere. Cookery books with emphasis on fish had been compiled from at least as early as 400 B.C., which, following the model of other scientific treatises (medical, botanical etc.), gave clear instructions for the cooking of fish (Hill and Wilkins 1996). Representations on pottery of fish in the market are known from southern Italy of that period (Fig. 10.9). It appears that in the fish plates the southern Italians found a way to accentuate this fish experience.

In southern Italy the fish plates full of sea treasures would probably be best understood as objects which offered a double sensual stimulus. Both the mouth and the eyes were taking fish in and probably at the same time, the ear was occupied with verbal images of fish and fish dishes through poetry such as Archaestratus's. Fish was embodied in the fullest possible way.

The fish plates, one of the most recognizable yet enigmatic pottery forms of the Greek world, offer a good starting point in the investigation of the role of the senses in food consumption, especially fish. What has become apparent in the above analysis is that although the same type of vessel, the fish plates, are involved in the same act, that is fish-eating in a formal setting, the meanings attributed to them change in each different case depending on a variety of culturally specific parameters.

10.3 Conclusions.

This chapter discussed how the senses shaped the experience of fish-eating in the context of Greece during the centuries under study. It is claimed that experiencing fish-eating through the senses, far from being subjective and person-specific, is culturally determined. The way people define something as sweet or foul smelling for example or the definition of what is appetizing in terms of taste, smell and appearance are based on the experience of the senses but is tightly linked to the culturally approved standards of every society. In this framework, the exploration of smell, hearing, taste and vision in relation to fish-eating may be an insightful way to gain access to cultural, social or other aspects of the societies under study. The examples used in this chapter illustrate the potential of such an approach in the study of food-related issues.

The perception of fish through the senses and its incorporation through actual eating (but also through vision or hearing) could take an extra dimension when memory and imagination were also at play. The case of the fish plates, both Attic and south Italian, illustrates just how the varied sensual stimuli related to fish-eating function as elements in a discourse of memory and self definition. Eating fish, looking at fish even when the dish has been emptied of its food contents and hearing about fish are ways, not only to fill one's stomach, but to fill one's mind. By incorporating fish in all these manners (often combined), the consumers actually incorporate other, non edible spheres. For the inhabitants of the Taman peninsula, eating off the imported Attic fish plates perhaps meant incorporating the Aegean Sea, the distant homes of their memories.

The archaeological documentation of these issues is still problematic. In certain cases, such as the ones discussed in this and the previous chapter, the senses emerge as not only relevant, but often predominant in the interpretation of archaeological finds. Again, as so often in this book, it become clear that such issues concerning the past can become visible only if the appropriate questions are asked and if a detailed, context-specific excavation and recording takes place. In such cases, the combined use of archaeological and literary data could open up new ways of understanding past societies.

CHAPTER 11

CONCLUSIONS

This book set out to investigate the nature of fish consumption in Greece from the Classical period to Late Antiquity and the way in which fish and its consumption was incorporated in the economic, social and ideological life of Greeks. To do so, certain basic questions have been examined: what was the contribution of fish to the economy and how was its production and distribution/marketing organised, what was thesocial context of fish-eating and finally, how was the act of consumption linked to the formation of cultural categories and fish symbolism?

To deal with these research questions a combined approach to the issue has been devised. Fish bones, written evidence, fishing related artefacts and structures have all been used against a background of anthropological ideas on fishing and on food consumption. Contemporary fisheries research in Greek waters has also been brought into the research. This combined approach has proved to be particularly fruitful, offering insights into previously unknown or little explored issues. Additionally it highlighted the need for a certain methodological re-orientation of the excavation and finds analysis within Classical archaeology. Such a re-orientation would give more emphasis to the collection of bio-archaeological remains and to the contextual analysis of food-related finds. A change along these lines has already begun, and some such cases have been discussed in this book.

The analysis of the data at hand have shown that what characterised fishing in ancient Greece was variability, both in the nature and abundance of the exploited fish and in the manner of their exploitation. Although the scale of fishing operations can only be indirectly known in certain cases, our research shows that fishing was widespread across Greece, and it was focusing both on marine and inland waters, thus having an impact, albeit a variable one, to the local economies. Evidence for fishing, often intensive, originates both from areas of eutrophic and of poor waters. It appears that what prompted the practise of fishing was not the abundance of fish in an area but rather the desire for it for various reasons. At those locations were fishing was important, fishermen communities had formed, with distinct cultural traits and world views. Although these communities appear to have been quite real, they nevertheless, were often marginalised socially but also economically.

Where fishing was productive enough to allow the consumption of fish by wide sectors of an area's population, the full economic potential of the fish was exploited not by the fishermen but by others, after fish was incorporated into the mechanisms of the urban market. Through these, fish became available as a foodstuff to many people who lived close or even at a distance from the source of fish. One basic distinction in this sphere was made between fresh and processed fish. It appears that these two categories were following different trajectories as consumable goods and were eventually considered as different types of food. Therefore, fresh and preserved fish have to be examined separately. In the context of the market, fish size, price, provenance, preservation and texture became central distinctive traits, which were further elaborated in the various contexts of preparation and consumption.

At certain areas of Greece, fish were regularly consumed in private and in public, in secular and in cultic contexts. It appears that fish was part of cultic processes more often and in more ways than previously thought. In some occasions fish-eating was a prerequisite and it was regulated by specific rules, while in others fish was eaten on a regular, less formal basis. Our evidence suggest that there was a proper manner of fish consumption both in the cultic and secular spheres, which were dictated by issues of gender, status, fish-eating occasion etc.

Those characteristics of fish that had an impact when fish were sold in the market were further elaborated as fish moved through the different stages of acquisition, preparation and consumption. Complex schemes of classification appear to have developed. Size, provenance, preservation and texture were variables that grouped fish according to their culinary and cultural relevance rather than the biological one. In this process, the senses involved in fish consumption, smell, hearing, taste, touch and sight, as well as the memory generated in the process of eating, were basic tools in the construction of alternative fish taxonomies.

The quest in this book for the consumers' fish choices, preparation and consumption lead to an exploration of the minutiae of fish-eating. The way fish were grouped together as a distinctive dish according to their size rather than species, the importance of fatty texture in fish preferences or the reasons for the artistic representation of live fish in dishes wherein fish were probably served, are some of them. The way such choices were organised and negotiated, however, followed certain patterns, which were geographically and temporally specific. These reveal broader patterns of cultural classifications, which focus on issues of wealth, status, identity, gender, etc.

This research has demonstrated that the ideas developed around fish acquisition and consumption were used not only to define social and cultural categories within the consuming community, but also to define this community's identity in relation to strangers and also in relation to the producers of the fish themselves and the sea (or other water bodies). The issues of edibility, of size, of provenance, all became material for the construction of myths and anecdotal stories which not only divided different groups of people on cultural grounds but also functioned as causal explanations for these divisions. The cultural and social distance between fishermen and fish consumers that has been observed in several occasions, was a fact that fuelled the formulation of a distinct set of conditions and ideas around fish consumption and at the same time was explained by these conditions and ideas.

Concluding, it appears that fish production, distribution/marketing and consumption are related in a cyclical manner, each affecting the others in crucial ways. In order to be able to appreciate the impact of fish-eating on any ancient society we need to be able to explore all these interconnected aspects of it.

An implication of the conclusions reached in this book, and summarily presented here, is that the archaeological research on food may prove a very effective way to explore various aspects of past societies that touch upon economic, social and cultural issues. The ideas and prospects offered by anthropology and discussed at the beginning of this book proved to be within the grasp of archaeological research and indeed quite rewarding. Zoo-archaeology has much to offer in such a quest. Fish remains in this case provided the means to explore the materiality of fish consumption and the zoo-archaeology's methodologies and ideas helped to relate it to broader issues of social archaeology.

This research has also important implications for Classical archaeology and the study of Classical world in general. The combined use of different types of evidence, documentary and archaeological, and specially those related to food eating, has provided the opportunity to overcome the shortcomings of each separate class of data. This book has demonstrated that through such a combined approach, elaborate and case sensitive interpretations can be reached, which moreover are supported by a robust base of empirical data. For Greek archaeology, an approach such as the one adopted in this book has the extra benefit of illuminating aspects of the past that have been little explored to date. Issues of everyday practise, such as the function of the food market, the consumption of food and the practical and ideological organisation of these activities, are often taken for granted. This book has shown that once such topics are scrutinised in a certain manner, then our knowledge about life in ancient Greece is much enriched.

This book attempted to tackle a difficult issue in Classical archaeology, by reconsidering a well-known topic, for which established, but divergent, ideas already existed. A methodologically and theoretically novel approach has been adopted. In this process this book has offered new data on Classical fish-eating as well as many new or better documented insights into the issue, and has demonstrated that fish-eating was an activity deeply embedded in various aspects of life in ancient Greece. In doing so, it has also opened up new, mostly un-trodden paths for future research. Fish was just one prominent food stuff in Classical Greece. Red meat, game, bread, wine, spices and fruits are others, equally challenging cases. Their analysis in a manner similar to that adopted for fish here, would not only enrich our knowledge about life in Classical Greece, but would also enhance our understanding of fish-eating by providing a context of relevant values. Also the exploration of these foodstuffs in the biographical manner adopted in this book would also permit the uncovering and better understanding of various spheres of activities, such as hunting, herding, gardening and agriculture and of the worldviews connected to each. In this way, the position of the fishermen and their distance from the fish consumers would be better evaluated.

Another major area for future research is the archaeology of the Classical fishing communities and the mariners more generally. The lack of such research is quite evident. This book has demonstrated that such communities existed, and often had distinct cultural characteristics. Where they could be found, what was their way of life, what were their worldviews and how were they interacting with other groups of people, are all issues of importance for the understanding of Classical society. Maritime archaeology and anthropology have already developed methodological and theoretical frameworks for such research and they could be used as a base for the exploration of the Greek maritime communities. The combined contribution of material remains and written sources would provide depth in such an approach. The still surviving, but swiftly disappearing, fishing traditions in modern Greece is another domain that urgently requires documentation and analysis. This would further contribute to the formation of an insightful archaeology of fishing in Greece.

Appendix 1. Literary sources related to fish, fishing and fish-eating related.

AUTHOR	TITLE	TYPE	DATE	DESCRIPTION
Aelian	*On the Characteristics of Animals* 6.1	Philosophical/ Natural history	2nd-3rd c. A.D.	On the invention of artificial fly at river Astreus (Axios) in order to catch a specific spotted fish which feeds on a particular fly, that lives on the river's waters. The artificial fly consisted in a fishing hook, wrapped with scarlet wool, with two wax coloured cock feathers attached to it.
Aelian	*On the Characteristics of Animals* 13.20	Philosophical/ Natural history	2nd-3rd c. A.D.	Many sea-monsters on the eastern coast of Athos, Chalkidike peninsula.
Aelian	*On the Characteristics of Animals* 13.21	Philosophical/ Natural history	2nd-3rd c. A.D.	On the Triton myth at Tanagra. Triton, a child of Poseidon, is sacred and any profane act against him is severely punished. Demostratus had witnessed the dead body of a Triton.
Aelian	*On the Characteristics of Animals* 13.26	Philosophical/ Natural history	2nd-3rd c. A.D.	*Tetix enalios* (a sea crustacean). It was not caught by the fishermen at Serifos because it was considered Perseus's playmate.
Aelian	*On the Characteristics of Animals* 14.20	Philosophical/ Natural history	2nd-3rd c. A.D.	Old fisherman from Rethymno, Crete, used sea-horses to cure his sons who were also fishermen and suffered from mad dog bite.
Aelian	*On the Characteristics of Animals* 15.3-6	Philosophical/ Natural history	2nd-3rd c. A.D.	The tuna fishing begins with a prayer, in the area of Hipponicum, on the west coast of Brutium, Southern Italy.
Aeschylus	*Net-drawers* (Wessels and Krumeich 1999, pp. 107-124).	Satyric play	5th c. B.C.	The title of the play, *Dyctyoulkoi* (net-drawers), indicates a special function within the fishing enterprise.
Aeschylus	*Glaukos Pontion* (Wessels and Krumeich 1999, pp.125-130).	Satyric play	5th c. B.C.	About Glaukos in Anthedon.
Agatharchides	*European History* FHG III, p.192 (in Ath. vii 297d)	Historical treatise	2nd c. B.C.	Boeotians sacrifice large eels, according to a specific ceremony, following an old custom.
Alexis	*The Epidaurian* 77 K-A (in Ath. iii 119f)	Middle comedy	3rd c. B.C.	Two brothers, the sons of Chaerephilos, became Athenian citizens because their father imported *tarichos* (preserved fish) in Athens.
Alexis	*Man with Cataracht* 16 K-A (in Ath. vi 224f)	Middle comedy	3rd c. B.C.	Fishmongers' attitudes.
Alexis	*Phaedon* 249 K-A (in Ath. viii 340b)	Middle comedy	3rd c. B.C.	One character imagines a second one as an *agoranomos* (market controler) trusting him to stop Callimedon's incessant assaults on the fish-stall. The second speaker balks at this, considering such an intervention as a tyranic act.
Alexis	*Cauldron* 130 and 131 K-A (in Ath. vi 226a-b)	Middle comedy	3rd c. B.C.	A rich citizen introduced a law for set, reasonable prices of fish.

Appendix 1 (Continued)

Alexis	*Odysseus at the Loom* 159 K-A (in Ath. vii 302f)	Middle comedy	3rd c. B.C.	Certain fish, such as hairies, squilding and small fry are fit only for the freedmen and not for wealthy men.
Alexis	*Crateia or the Drugist* 115 K-A (in Ath. iii 107a)	Middle comedy	3rd c. B.C.	On the preparation of a special dinner with a lot of fish and sea- food. Detailed description of the various dishes. Distinction between appetisers and main dishes.
Alexis	*Principal Dancer* 200 K-A (in Ath. vii 287f)	Middle comedy	3rd c. B.C.	A *parasite* character associates the consumption of large fish with sophistication and wealth and the eating of small fish with simplicity.
Alexis	*Heiress* 78 K-A (in Ath. vi 227d)	Middle comedy	3rd c. B.C.	If a poor looking man is often seen to buy fish, he should be arrested as a criminal, unless he can prove that he has the means for his extravagance.
Alexis	*The Greek Woman* 76 K-A (in Ath. vi 226f-227a)	Middle comedy	3rd c. B.C.	On the perils brought by fish: they can eat a man after a ship-wreck, but, equally, dead in the market, they can destroy a man financially.
Alexis	*The Meeting at Pylae* 204 K-A (in Ath. vi 225f)	Middle comedy	4th c. B.C.	On fishmongers' attitudes and high prices of fish. They are said to be like kings levying heavy tribute on customers.
Alexis	*The Woman who Smacks* 57 K-A (in Ath. iii 104c-d)	Middle comedy	4th c. B.C.	On fishmongers raising a bronze statue to honour a good client "who has saved their trade".
Alexis	*Dimitrios or Philetairos* 47 K-A (in Ath. viii 338d)	Middle comedy	4th c. B.C.	Satirises a fish-glutton who buys all the fish in market without consideration of other buyers.
Anonymus	*Anthologia Palatina* A.P. 6.23	Epigram	Hellenistic	A fisherman dedicates his tools to Hermes.
Anticlides	*Inversions* Müller 1979, fr.7, p. 148 (in Ath. xi 466c)	Historical treatise	early 3rd c. B.C.	About the myth of Enalos on Lesvos.
Antigonus Carystius	*On Diction* Dorandi 1999, fr. 56A-B, p. 42 (in Ath. vii, 297e)	Anecdotological work	3rd c. B.C.	Fishermen of Halieis (Argolid) used to celebrate a festival dedicated to Poseidon, during the tuna season and sacrifice the first tuna of the catch to him. This offering was called "*thynnaeon*".
Antiphanes	*The Lady's Maid* 127 K-A (in Ath. vii 303f)	Middle comedy	4th c. B.C.	A farmer says that the countrymen eat nothing from the sea but shore fish like conger eel, sting rays and the tail of the tuna. Exploitation of lacustrine resources in a farmstead by Lake Kopais.
Antiphanes	*She Goes Fishing* 27 K-A (Ath. viii 338e)	Middle comedy	4th c. B.C. (336)	Heteras (courtesans) nick-named after fish in an ambiguous description where it is not clear whether reference is made to *heteras* or fish.
Antiphanes	*Butalion* 69 K-A (Ath. viii 358a)	Middle comedy	4th c. B.C.	Eating of large fish is an urban trait while peasants in their ignorance prefer small fish. A fishmonger travels in the countryside selling small fish.
Antiphanes	*Adulterer* 159 K-A (Ath. vi 225d)	Middle comedy	4th c. B.C.	On spoiled fish sold in the market.

Appendix 1 (Continued)

Antiphanes	*Rich Men* 188 K-A (in Ath. viii 342f-343a)	Middle comedy	4th c. B.C.	Reaction to a rich man, Euthynnus (tuna), who buys many fish in the market causing *anopsia* (luck of fish).
Antiphanes	*The Sheep-owner* 191 K-A (in Ath. vii 295c)	Middle comedy	4th c. B.C.	A praise of eels from Boeotia, *glaukos* from Megara, sprats from Karystos and breams from Eretria.
Antiphanes	*Young Men* 164 K-A (in Ath vi 224c)	Middle comedy	4th c. B.C.	On the extraordinary high prices of fish.
Apollonides	*Anthologia Palatina* A.P. 6.105	Epigram	Hellenistic	A fisherman called Menis offers a grilled red mullet and a hake to Artemis along with some wine and bread in return for a rich catch.
Archaestratos	*Hedupatheia* (in Ath. at various lines)	Mock epic poetry	mid. 4th c. B.C.	He recommends eels from Kopais lake and Strymon river, grey mullets from Avdera, dog-fish from Rhodes, boar-fish from Arrathus river in Amvrakia, *glaucos* from Olynthus and Megara, liver-fish from Delos and Tenos, dolphin-fish from Carystos, dog-shark from Toroni, sea-bass from Amvrakia, Kalydon and Lake Volvi, salema from Mytelene, red mullet from Thassos, sea bream from Delos or Eretreia, *chromis* from Amvrakia and Pella.
Archaestratos	*Hedupatheia* (in Ath. vii 304d)	Mock epic poetry	mid. 4th c. B.C.	Comments on the richness of waters around Karystos.
Archaestratos	*Hedupatheia* (in Ath. vii 310e)	Mock epic poetry	mid. 4th c. B.C.	On the dog fish as a man-eater.
Archaestratos	*Hedupatheia* (in Ath. vii 327e)	Mock epic poetry	mid. 4th c. B.C.	The common sea-bream (*fagros*) should be eaten "at the rising of Sirius" in Delos and Eretria, but only the head of it and the tail portion. The rest is not worth eating.
Archippos	*Ichthyes* 14-34 K-A (in Ath. at various lines)	Old comedy	5th c. B.C.	Fish declare war to Athens, in order to put to the sword the fishmongers and sea-food epicures.
Archippos	*Ichthyes* 18 K-A (in Ath. vii 328)	Old comedy	5th c. B.C.	Gild-heads sacred to Aphrodite of Kythera.
Aristophanes	*Wasps* 490-5	Old comedy	5th c. B.C.	Salted fish are the cheapest. A seller of sprats accuses for elitism those who show disdain for her wares.
Aristophanes	*Knights* 927 ff.	Old comedy	5th c. B.C.	Curse involving choking with fried squid just taken out of the sizzling pan.
Aristophanes	*Acharnians* 671	Old comedy	5th c. B.C.	Reference to the production of fish-sauce on the island of Thassos.
Aristophanes	*Acharnians* 884cf, 894	Old comedy	5th c. B.C.	About eels from Kopais lake.
Aristophanes	*Acharnians* 1156-61	Old comedy	5th c. B.C.	A curse according which the longed-for squid is snatched away by a dog before the cursed man can taste it.

Appendix 1 (Continued)

Aristophanes	*Frogs* 701-3	Old comedy	5th c. B.C.	Connection between wealth and fish-eating.
Aristophanes	*Peace* 1013-14	Old comedy	5th c. B.C.	Eels from Kopais lake brought in Athens wrapped in beet leaves.
Aristophanes	*Lysistrata* 1065ff	Old comedy	5th c. B.C.	Connection between fish and sexuality. A prospective lover is referred to as an eel from Boeotia.
Aristophanes	*Clouds* 559	Old comedy	5th c. B.C.	Connection between fish and politics: Cleon is compared to an eel-fisher who can only capture his pray by stirring up the mud.
Aristotle of Stagera	*Rhetoric* 6.55	Philosophical treatise	4th c. B.C.	Refers on an epigram on the base of an Olympic Games winner's statue. He used to transport fish from coastal Argos, to Tegea, on the mountains of Peloponnese.
Aristotle of Stagera	*History of Animals* 602a.5	Zoological treatise	4th c. B.C.	Detailed study of the physiology and ethology of numerous sea- and fresh-water creatures.
Aristotle of Stagera	*History of Animals* 603a.20-25	Zoological treatise	4th c. B.C.	In Pyrrha lagoon on Lesvos, scallops had been exterminated by a period of draught and by the continual workings of the fishermen dredges.
Aristotle of Stagera	*Politics* Book 4, 1291b lines 22-24	Philosophical treatise	4th c. B.C.	He singles out Byzantium and Tarentum as ports with important fishing population.
Aristotle of Stagera	*The Constitution of Naxos* (in Ath. viii 348b)	Philosophical treatise	4th c. B.C.	Describes how the inhabitants of the inland city of Naxos go to the coast to buy fish. Animosity arises when they fail to get their supplies.
Arrian	*Indice* 29-30 (Müller, 1965, v. I, pp. 348-51)	Geographical treatise	2nd c. B.C.	*Ichthyophagoi* (fish-eaters) was a people of the north coast of Red Sea. Their customs and way of life are described in detail.
Artemidorus	*Geographies* (in Ath. viii 337f)	Geographical treatise	ca 100 B.C.	Divination taking place in a fresh-water spring at a place called Dinus, in Lycia. When the water bubbles, many fish approach and the sacrificers put in the water wooden rods with first fruits, boiled and roasted meat, barley cakes and pieces of bread.
Athenaeus	*Deipnosophistae*	Sympotic literature	ca 200 B.C.	Several topics related to fish consumption are discussed extensively throughout the 14 books of *Deipnosophistae*.
Athenaeus	*Deipnosophistae* (vi 224d)	Sympotic literature	ca 200 B.C.	Fish-mongers' attitudes in Rome are as bad as those in Athens.
Athenaeus	*Deipnosophistae* (vii 287a)	Sympotic literature	ca 200 B.C.	Bogue was a fish sacred to Hermes because of the grunting noises it produced.
Athenaeus	*Deipnosophistae* (viii 329b)	Sympotic literature	ca 200 B.C.	"Thassian pickle" was a special dish consisting in broiled fish plunged into a pickle.
Athenaeus	*Deipnosophistae* (viii 331 e-f)	Sympotic literature	ca 200 B.C.	Reference to Arethusa spring near Chalkis, where tame fish live. Also the eels there wear silver and gold earrings and receive food from the worshipers and bits of the entrails from sacrificial victims and bits of fresh fish.

Appendix 1 (Continued)

Cleitarchos of Alexandria	*History of Alexander* Müller 1977, fr.1a, p. 76 (in Ath. iv 148 d)	Historical treatise	end of 4th c. B.C.	Anchovies in Theban banquets as a sign of the city's poverty and destitution after its demolition by Alexander.
Chionides	*Beggars* 4 K-A (in Ath. iii 119e)	Old comedy	early 5th c. B.C.	Negative attitude towards salt-fish.
Chrysippus of Soli	*Inscriptiones Insanos Esse et Impios* (SVF III 167 fr. 667)	Philosophical treatise	3rd c. B.C.	Introduces the term *opsomanes* (fish-maniac) instead of *opsophagos* (fish-eater), suggesting an obsession for fish.
Chrysippus of Soli	*On Things to be Chosen on their Own Sake* (in Ath. vii 285d)	Philosophical treatise	3rd c. B.C.	Athenians' negative attitude towards small fish. It is explained as a result of the abundance of such fish.
Clearchos of Soli	*Lives* FHG II, p.307 (in Ath. 12.518c)	Historical treatise	4th – 3rd c. B.C.	The tables of Sicilians were famous for their luxury. Sicilians call their seas "sweet" because of the pleasure they take from fish and sea food.
Crates	*Sacrifices at Athens* FHG IV, p. 362 (in Ath. vii 297e)	Historical treatise	4th c. B.C.	On sacrifice to Poseidon of the first tuna catch of the year at the deme of Halae Aixonidae in Attica. The sacrifice was p art of a festival called "*thynnaion*"
Demetrius of Scepsis	*The Trojan Battle-order* (in Ath. iii 91d)	Antiquarian / historical treatise	mid 2nd c. B.C.	About a Spartan, not-knowing how to eat a sea-urchin in an Athenian symposium.
Demetrius of Scepsis	*The Trojan Battle-order* (in Ath. iii 91d)	Antiquarian / historical treatise	mid 2nd c. B.C.	About a Spartan, not-knowing how to eat a sea-urchin in an Athenian symposium.
Demosthenes	*On False Embassy* 19. 229	Dicanic speech	343 B.C.	*Opson*-eating is connected to depravity. Philokrates buys prostitutes and fish with gold he received for treason.
Diocles of Carystus	*Hygieina* (in Ath iii 116e)	Medical treatise	4th c. B.C.	On the best preserved fish.
Diodorus Sicilus	*Universal History* Book III, 15-17	Historical treatise	1st c. B.C.	*Ichthyophagoi* (fish-eaters) on the Red sea and their custom of throwing their dead to the sea to be eaten by fish.
Diphylos	*The Merchant* 31 K-A (in Ath. vi 227d-e)	New comedy	4rd c. B.C.	If a poor looking man is often seen to buy fish, he should be arrested as a criminal, unless he can prove that he has the means for his extravagance.
Diphylos	*The Busybody* 67 K-A (in Ath. vi 225a)	New comedy	3rd c. B.C.	On the fishmongers bad attitudes.
Empedokles	*Purifications* (Diels 1974, fr. 117, v. vi, p. 358-9)	Philosophical treatise	5th c. B.C.	About evolutionary relation of humans to fish. Fish are the progenitors of humans.
Ephippus	Geryones (3 K-A) (in Ath. ix 370 c-d)	Middle comedy	4th c. B.C.	A list of foods consumed during the *Amphidromia,* a family feast.
Ephoros	*Histories* FGrHist 70 F225	Historical treatise	4th c. B.C.	On the Triton myth at Tanagra.

Appendix 1 (Continued)

Epicharmos	*The Marriage of Hebe* 39-64 K-A (in Ath. at various lines)	Old comedy	early 5th c. B.C.	Description of a variety of fish-dishes for a feast in a Sicilian context.
Eupolis	*Flaterers* 160 and 174 K-A (in Ath vii 328a)	Old comedy	5th c. B.C.	Reference to gross expenditure on fish.
Euthydemos	*On Salt Meats* (in Ath. ii.116 a-d)	Sophistic work	2nd c. B.C.	Encomium on salt fish.
Galen	*On the Powers of Foods* III.709.13	Medical treatise	2nd c. A.D	The grey mullets living in inland, close waters are of inferior quality than those caught in the open sea. Grey mullets' flesh is improved by salting.
Hegesander	*Commentaries* FHG IV p. 420 (in Ath. vii 337f)	Historical/ anecdotological treatise	2nd c. B.C.	On the circumstances of seasonal fishing and fish- preservation in Lake Volvi. The event is related to the honouring of the dead
Heraclides Kritikos	*Description of Greece* Pfister 1951, p. 82 (in Ath. i. 23d)	Geographical treatise	late 3rd c. B.C.	About the town Anthedon which turned to the exploitation of the sea because its agricultural resources were inadequate.
Herodotus	*Histories* Book III, 41-43	Historical treatise	end 6th –beg. 5th c. B.C.	The story of Polycrates (tyrant of Samos in the 6th c. B.C.) and a huge fish delivered to him, as a present, by a fisherman.
Herodotus	*Histories* Book V, 16	Historical treatise	end 6th –beg. 5th c. B.C.	About the lake-dwellers of lake Prasias, probably modern-day lake Vistonis in Thrace. The lake was very rich in fish, which were even used as animal fodder. Mention is made of *tilones* and *peprakes*, two unidentified fish, which were caught by mere immersion of a basket in the water.
Herodotus	*Histories* Book VI, 151ff	Historical treatise	end 6th –beg. 5th c. B.C.	Korovios, the purple-fisherman from Itanos (Crete) led the Therans to Cyrene, their new colony on the northern coast of Africa.
Hesychios	*Lexicon* (Schmidt 1965, v.2, p. 382)	Lexicography	5th or 6th c A.D.	Kaveiroi: deification of crabs on Lemnos.
Hesychios	*Lexicon* (Schmidt 1965, v.1, p.255)	Lexicography	5th or 6th c A.D.	"Apopyrizo": a verb meaning "I eat anything plucked off the coals", often used in relation to fish.
Hierophytos	*Cronicles of Colophon* FGrHist. IV p.428 (in Ath. 297e-f)	Historical work		About sacrifices of smoked fish as a commemoration of the foundation of the city of Phaselis.
Hyperides	*Against Arastagora* Burt 1954, v.2, fr. 28 and 29 (in Ath. xiii 589a-b)	Dicanic speach	4th c. B.C.	Two *heteras* (courtesans) called Anchovies. Fish takes sexual connotations.
Lucian	*De Syria Dea* 14, 45-47	Sophistic work	2nd c. A.D.	Fish in Atargatis's cult. Real and gold fish are dedicated to her.
Lynceus of Samos	*How to Shop in the Market* (in Ath. vii 313f)	Sophistic work	4rth – 3rd c. B.C.	On the difficulties of buying fish in Hellenistic Athens. Need for a special strategy to avoid competition and fishmongers' tricks.
Lynceus of Samos	*Centaur* 1 K-A (in Ath. iv 131f)	Middle comedy	4rth – 3rd c. B.C.	Athenian simplicity and style in fish and sea-food presentation. Sea-food and preserved fish as appetisers.

Appendix 1 (Continued)

Menander	*Fishermen* Kock iii.13 (in Ath. iv 132e)	Old comedy	4th c. B.C.	On the long absence of fishermen from home.
Mnaseas of Patara	*On Asia* FGH III, p. 155 (in Ath viii 346d-e)	Historical treatise	3rd c. A.D.	About the Goddess Atargatis and the fish-eating taboos and fish sacrifices related to her cult.
Myrsilos	*History of Lesvos* FGrHist. IV, p. 459-460 (in Ath. xi 466e-d)	Historical treatise	3rd c. B.C.	The myth of Enalos.
Oppian of Cilicia	*Halieutika*	Didactic poem	late 2nd c. A.D.	Poem on the sea-creatures and how to catch them.
Oppian of Cilicia	*Halieutika* (III 620-648)	Didactic poem	late 2nd c. A.D.	Description of tuna fishing by communal effort.
Pancrates	*Occupations at Sea* (in Ath. vii 283a)	Didactic poem	A.D. 2nd c.	A legend about the sacred escort fish, which was especially honoured among the Samothracian gods.
Pausanias	*Geography* i. 38.1	Periegetic treary	ca. A.D. 150	Fishing rights to priests of the sanctuary of Demeter at Eleusis in the sacred river Rheittoi (Attica).
Pausanias	*Geography* ii.35.1 and cf II.34.9	Periegetic treatise	ca. A.D. 150	About a diving competition at Hermione in the cult of "Dionysos of the Black Skin" and about the cult of Poseidon and "Amphitrite Potnia and Limenia".
Pausanias	*Geography* iii.21.5	Periegetic treatise	ca. A.D. 150	In the town of Aegiae, the sacred fish of a lake belonging to Poseidon, were not eaten.
Pausanias	*Geography* iv.24.1	Periegetic treatise	ca. A.D. 150	Kopais lake produces excellent eels and other fish.
Pausanias	*Geography* vii.22.4	Periegetic treatise	ca. A.D. 150	In Pharae (Achaia) sacred fish lived in a stream, which belonged to Hermes.
Pausanias	*Geography* ix.20.4	Periegetic treatise	ca. A.D. 150	The Triton myth at Tanagra.
Pausanias	*Geography* ix.22.5	Periegetic treatise	ca. A.D. 150	Anthedon was the home town of Glaukos, who turned into a minor sea-god.
Pausanias	*Geography* viii.21.2	Periegetic treatise	ca. A.D. 150	In Arcadia, crying-fish could be found in Cleitor River.
Pausanias	*Geography* x.9.3-4	Periegetic treary	ca A.D. 150 (event dated at the beg. of 5th c. B.C.)	Fishermen from Corcyra dedicated a bull statue to Apollo at Delphi after a good tuna catch.
Pausanias	*Geography* x.37.3	Periegetic treatise	ca. A.D. 150	At Bulis, at the borders between Boeotia and Phokis, more than half its inhabitants are fishers of the shellfish that gives the purple-dye.
Phanios	*Anthologia Palatina* (A.P. 6.304)	Epigram	Hellenistic	The speaker on the shore hails a fisherman standing on a rock with a view to buying fish from him if his catch is suitable.
Pherecrates	*The Kitchen or Midnight Festival* 70 K-A (in Ath. xiii 612a)	Old comedy	5th c. B.C	Reference to a female fish-seller "ichtyopolaena", as an unsuitable occupation for a woman.

Pherecrates	*Deserters* 26 K-A (in Ath. 119d-e)	Old comedy	5th c. B.C.	Negative attitude towards small-fish.
Philarchos	*Histories* FGH I p.348 fr.45 (in Ath. xii 521b)	Historical treatise	3rd c. B.C.	Smyndirides from Sybaris visited Corinth, bringing along his fishermen and cooks, for fear of not finding in Corinth services to mach his standards and tastes.
Philarchos	*Histories* FGH I p.334 (in Ath. xii 521b)	Historical treatise	3rd c. B.C.	Divination with green figs and large fish.
Philemon	*Soldier* 82 K-A (in Ath. vii 288c)	New comedy	begin. of 3rd c. B.C.	Description of a feast on sea- and fresh-water fish. The best boar–fish are from Argolid, the *glaukos* from Attica and the best conger eel from Sycion.
Philostephanos of Cyrene	*On the Cities of Asia* FHG III p.29 (in Ath. vii 297f- 298a)	Antiquarian treatise	3rd c. B.C.	About smoked fish sacrifice to Kylabras at Phaselis, Asia Minor, as a commemoration of the city's foundation. The pastoralist Kylabras parted with part of his coastal lands, where the colony of Phaselis was established, in exchange for preserved fish.
Plato	*Charmides* 163b	Philosophical treatise	4th c. B.C.	On fishmongers' foul character.
Plato	*The Lows* vii. 823	Philosophical treatise	4th c. B.C.	Negative comments on activities related to the sea, i.e. fishing, piracy etc, juxtaposed to hunting, which is considered a noble activity.
Plato Pliny the Elder	*Republic* 372e-373 *Natural History* vi.30.5.32	Philosophical treatise Natural history	4th c. B.C. (375 B.C.) 1st c. A.D.	Eating of *opson* (any strong-flavoured food but very often fish and fish products) is associated with a life of luxury, injustice and immorality. Reference to *Ichthyophagoi* (fish-eaters).
Pliny the Elder Pliny the Elder	*Natural History* xxxii, xv, 42-103 *Natural History* xxxii, vii. 16	Natural history Natural history	1st c. A.D. 1st c. A.D.	On the medical uses of various fish. At Chios, in a natural pool by the Shrine of Old Men, eels wear earrings (a phenomenon known in other places also).
Plutarch	*Table-talk* 631d	Philosophical treatise	1st-2nd c. A.D.	On fishmonger's foul character.
Plutarch	*Table-talk* 668a	Philosophical treatise	1st-2nd c. A.D.	Connection between *opson* eating and political and moral depravity.
Plutarch	*Table-talk* 728-30	Philosophical treatise	1st-2nd c. A.D.	On the Pythagoreans' abstention from fish. Explanation of the taboo.
Plutarch	*Table-talk* 730e	Philosophical treatise	1st-2nd c. A.D.	The priests of Neptune (Poseidon) never ate fish because humans and fish are supposed to have a common origin.
Plutarch	*Table-talk* 667-669	Philosophical treatise	1st-2nd c. A.D.	Sea-food versus land-food. To be eaten by fish was the most terrible fate for Greeks. The superiority of sea-food is defended by Symmachus, who is judged suitable for this role due to his origins from Nikopolis, near Amvrakicos Gulf.
Plutarch	*Table-talk* 667	Philosophical treatise	1st-2nd c. A.D	Aedypsos rich in fish.
Plutarch	*Table-talk* 667ff	Philosophical treatise	1st-2nd c. A.D.	Definition of an *opsophagos,* and bell-ringing when a fish-load enters the city.

Appendix 1 (Continued)

Polemon	*Letter to Attalus* (in Ath. viii 346b)	Geographical comments	2nd c. B.C.	Apollo "the fish-eater" as the protector god of Elis
Polycharmos	*History of Lycia* FHG IV, p. 479 (in Ath. viii 333d)	Historical work	2nd c. B.C.	Fish diviners in Lycia.
Porphyry	*On Abstinence from Animal Food* 4.16	Medical treatise	3rd c. B.C.	Fish were taboo for the initiates at the Eleusinean mysteries.
Porphyry	*The Life of Pythagoras* 45	Medical treatise	3rd c. B.C.	Pythagorean's abstinence from all marine creatures.
Rufus of Ephesus	*Medical Questions* 20.1-2, 13, 17	Medical treatise	1st-2nd c. A.D.	Women, due to their physiology should avoid, among fish, the eels, the sturgeon, the turbots and river-fish in general.
Semos	*History of Delos,* FHG IV p.493 (in Ath. viii 335a)	Antiquarian/ geographical treatise	ca 200 B.C.	When the women of Delos sacrifice to Vrizo (the interpreter of dreams), they offer her all shorts of goods, except fish, as they pray for the safety of their ships.
Semos	*History of Delos* FHG IV p.494 (in Ath. viii 331f)	Antiquarian/geographical treatise	ca 200 B.C.	Fish appear as omens in Delian Temples.
Sotades	*Locked up Women* 1K-A (in Ath. vii 293 a-e)	Middle comedy	3rd c. B.C.	Recipes of various dishes, many of which are based on fish.
Strabo	*Geography* 6.1.5	Geographical treatise	1st c. B.C.-1st c. A.D.	In Gela (S. Italy) people turn to the sea for lack of terrestrial resources.
Strabo	*Geography* 7.7.8	Geographical treatise	1st c. B.C.-1st c. A.D.	On *taricheia*, salt-fish establishments, for lake fish near Lycnydon (Orchida).
Strabo	*Geography* 10.2.21	Geographical treatise	1st c. B.C.-1st c. A.D.	On the fish from Kalydon.
Strabo	*Geography* 14.2.21	Geographical treatise	1st c. B.C.-1st c. A.D.	In Iassos, the arrival of the fish catch is announced by a bell. The whole population of the city attends the event
Strabo	*Geography* 15.2.2	Geographical treatise	1st c. B.C.-1st c. A.D.	About the *Ichthyophagoi* (fish-eaters) of the Red sea.
Strabo	*Geography* 43.3.5	Geographical treatise	1st c. B.C.-1st c. A.D.	In Phokaia people were forced to make a living from the sea (e.g. fish pickling) due to the poor quality of their land.
Suda (Suidas)	*Lexicon* 1772	Encyclopaedia	End of 10th c. A.D.	*Amphidromia*: a ceremony similar to christening. It was customary to bring a cuttlefish as a gift to the baby and attend a feast where fish were served.
Theopompos	*History of Philip* (in Ath. xii 526e)	Historical treatise	4th c. B.C.	Fishing is described as an activity connected to a luxurious lifestyle. By contrast to agriculture it entailed gain without work.
Timocles	*Epichairekakos* 11 K-A (in Ath vi 241a-b)	Middle comedy	second half of 4th c. B.C.	On the art of shopping for fish. Connection between fish and wealth.
Xanthos of Lydia	*Lydiaca* FGrHist 765 F17 (in Ath. viii 346e)	Historical treatise	6th-5th c. B.C.	Queen Atargatis and her son consumed by fish.

Xenarchos	*Purple-shell* 7 K-A (in Ath. vi 225c)	Middle comedy	4th c. B.C.	Reference to a law forbidding the sprinkling of fish with water to make them look fresher.
Xenophon	*Kyrou Anavasis* 1.4.9	Historical treatise	5th-4th c. B.C.	Fish taboo and sanctity of fish in Syria.
Xenophon	*Hellenica* 5.1.23	Historical treatise	5th-4th c. B.C.	Returning from a dawn raid on the Piraeus in 388 B.C. during the Peloponnesian war, the Spartan Teleutias captured a large number of inward-bound fishing boats, which were apparently returning from the night-fishing in the Saronic Gulf.

Appendix 2. Epigraphical sources related to fish, fishing and fish-eating.

	LOCATION	DATE	TOPIC	REFERENCES
1	Korykus (Asia Minor)	Late Roman	Gravestone inscription refers to the dead person's occupation (*alieus*= fisherman)	Mentzou 1975, pp. 166-167.
2	Korykus (Asia Minor)	Late Roman	Gravestone inscription refers to the occupation of a relative of the deceased (*sagineus*=net-fisherman)	Mentzou 1975, pp. 166-167.
3	Ephesus (Asia Minor)	Roman (1st c. AD)	A fishermen's guild offers a fisheries toll-house, built on its own expense, to Artemis Ephesia.	OGI 496; Curtius 1870, pp. 186-9; Horsley 1989.
4	Ephesus (Asia Minor)	Roman (2nd C AD)	A dedication of a statue of Isis by the fishermen association and others.	IEphesos, V 1503.
5	Kyzicus (Propontis)	Roman	Inscription, probably from the tomb of a fisherman, who designates the fishermen's guild as the recipients of the fines to be levied, should his tomb be violated.	IKyzicos, I 211; Hasluck 1904, p. 32, inscr. 43; van Nijf 1997, pp. 55-56
6	Smyrna (Asia Minor)	Romam	Guidelines for the care of sacred fish kept in a sacred lake and for their sacrifice to the Goddess Atargatis	SIG[4] 584; Lupu 2005, p. 29.
7	Smyrna (Asia Minor)	Romam	A penalty is set to those who damage the sacred fish of Atargatis's temple or damage the temple's property. The culprits are to be eaten by fish.	SIG[3] 997.
8	Parion (Sea of Marmara)	Roman	A dedication to Priapos in the Neilaion at Parion. The inscription reveals an association of tuna-fishermen. Each member's name, social status and function within the association is given	Frisch 1983, no 5; Purcell 1995, pp. 146-7.
9	Pergamon (Asia Minor)	Roman?	An inscription, which describes modifications of the way prices are set and the taxation upon the price of goods. Detailed arrangements for fish prices	SIG[4] 484.
10	Thera	210-195 B.C.	"Foundation of Epiktete". The inscription defines how the heroes/ancestors are to be worshiped. It is prescribed that three fish are to be offered to them alongside pastries, and the customary divine parts of the victim.	IG 12[3] 330; Lupu 2005, p. 86; Wittenberg 1990.
11	Myconos (Cyclades)	ca 200 B.C.	Inscription describing the sacrifices offered by Myconians to Poseidon *Phycios* (Seaweed) from which women are excluded. The cost of the sacrificial lamb is defrayed from the tax on fishing.	SIG[4] 1024.

Appendix 2 (Continued)

12	Delos (Cyclades)	4th c. B.C.	*Tarichopoleion- salted fish shop.* The term is found in a catalogue of properties of the Temple of Apollo.	Henning 1983; SEG 33(1983), p. 181, no 624.
13	Delos (Cyclades)	2nd c. B.C.	The sacred menu for the Eilithyaia festival. It included among other foodstuffs preserved fish and probably some fresh fish as well.	ID 440 (a. 173) lines 60-71; Linders 1994.
14	Delos (Cyclades)	279 B.C.	"The Catalogue of Hepsocles". Income generated from the fish caught in the Sacred Lake on Delos. Similar references in other inscriptions.	Homolle 1882, p. 67. Homolle 1890, p. 392.
15	Cos (Dodecanese)	325-300 B.C.	"Foundation of Diomedon". A religious ordinance which includes directions to the priests for the cult of ancestors. It specifies among others, that the priest should perform an "*apopyrida*", a burned fish sacrifice, according to the ancestral custom.	SIG[3] 1106. lines 177.42, 62; Paton and Hicks 1990, no 36, p. 75; Lupu 2005, pp. 86-87.
16	Cos	200-170 BC	A list of compulsory sacrificial obligations of several individuals or groups to Poseidon. It includes the obligations of the sellers of purple-dye, of traders who specialise in fish, those leasing watch-towers associated with fishing of migratory fish, also naval personnel, technicians connected to shipbuilding, and people working on the docks.	SIG[4] 1000; Sherwin-White, 1978, pp. 229-235.
17	Cos	? check	Funerary inscription on a tombstone referring to the profession of the deceased and his wife: *Porfyropolis* and *porphyropolidos* (purple-dye sellers).	Paton and Hicks 1990, pp. 203.
18	Piscokephalo (Crete)	2nd c. A.D.	The city-state of Praisos granted to the inhabitant of Stalis, the right to exploit a certain area and keep a fraction of the incomes from the port, the purple-dye and the fish	SIG[4] 427; Chaniotes 1996, pp. 388-393; Perlman 1996
19	Akraephia (Boeotia)	Hellenistic	Prize list for sea and fresh-water fish	Salviat et Vatin 1971, p. 95 109; Feyel 1936.
20	Kopais Lake (Boeotia, near Akraephia)	3rd c. B.C.	Boundary stone at Phtelio, between Akraephia (Karditsa) and Kopais (Topolia), on a promontory of the lake.	IG VII, 2792; Jamot 1889; Lauffer 1976: 40; Fossey 1988, p. 288.
21	Kopais Lake (Boeotia,near Akraephia)	6th-5th c. B.C.	Boundary stone from the shores of Kopais Lake.	SEG XXX. 440; Ager 1996, p. 71, note 1.
22	Troizen and Arsinoe (Argolid)	Late Hellenistic (163-146 B.C.)	Agreement between Arsinoe (Methana) and Troizen, which refers, to sharing of various common resources. These include the tuna catch from the commonly owned waters and the production of the salt-pans.	IG 4[2] 1. 76-77; Nikitski 1964; Peek 1969; Ager 1996, p. 381.
23	Epidauros (Argolid)	Classical	A merchant, called Amphimnastos, who transported and sold fish to Arkadia (Tagea), seeks help from Asklepios, promising to pay with a fraction of his earnings from the fish trade.	IG IV2 [1], 123; Herzog 1931, C xlvii, pp. 26-28.
24	Corinth (Corinthia)	1st c. A.D.	Two Latin Inscriptions from Agora in Corinth. They describe the owners and the identity of the *tholos* building where meat and fish were sold	Kent 1966, p. 127-8, no 321; West 1931, no 125

Appendix 3a. Fishing gear.

Area	Site	Date	Context	Finds	Comments	References
Attica	Vouliagmeni	$6^{/5th}$ c. B-C. to Roman	"The Priest's House" near the sanctuary of Apollo Zoster. The building is situated on the coast.. The fishing tools were found in Room Z	6 ring shaped lead net-weights, 26 cylindrical lead net-weights, l: 2-4 cm, diam.: 1-2 cm. 5 fish-hooks (3 joined) 4 bronze net-needles. 1 obsidian knife 1 small bronze arrow point, 22 clay loom-weights (in this context they could be interpreted as net-weights).	From the same building: one roof tile, inscribed after firing, with the picture of a boat. The fish-hooks are of various sizes.	Stavropoulos 1938.
Corinthia	Isthmia, Temple of Poseidon	Archaic	Debris deposit from the *pronaos*.	1 iron fishing-spear. 1 set of 12 lead net-weights.	In the same deposit there have been found a variety of dedicatory objects, some of which are tools. Among them, there have also been found two flaked stone blades. The fishing tools are probably also dedicatory objects	http://humanities.uchicago.edu/orgs/isthmia/publications/sm-ded/sm-dedications.html Gebhard 1998, p. 108
Corinthia	Isthmia. Sanctuary of Poseidon	Probably end of 5th-beginning of 4th c.?	Large circular pit in the sanctuary area, filled with objects cleared out of the sanctuary after the Temple fire ca 470-450, B.C.	A concentration of 19 fish-hooks	Dedicatory objects	Raubitchek 1998, p. 121, pl. 71, no 453
Corinthia	Corinth	Uncertain	Unspecified	Several fish-hooks		Davidson 1952, p. 193, pl. 88, nos 1447, 1448.
Argolid	Halieis. Akropolis	6th and 4th c. B.C.	Unspecified	4 fish-hooks, unspecified number of lead pieces, probably net-weights.		Jameson *et al.* 1994, p. 315
Argolid	Halieis. Lower town	Unspecified historical date	Unspecified	A hoard consisting in several dozens of fish hooks		Jameson *et al.* 1994, p. 315
Delos	Various spots	Unspecified historical date	Various deposits		The dedication of several fishing tools are known from inscriptions which list the donations of supplicants.	Deonna 1938, pp. 200-202.

Appendix 3a (Continued)

Kommos	Temples C and related deposits	Classical, Hellenistic and Roman	Various deposits	1 bronze netting-needle 2 lead line-weights		Schwab 2000, pp. 391-395, pl. 5.48, 5.49
Olynthos	Domestic district	Classical	House floor and street deposits	An undetermined number of fishing- hooks, lead net-weights, netting-needles. 19 net needles found together in a street deposit.		Robinson 1936; Cahill 2002 accessed at http://www.stoa.org/hopper/text.jsp?doc=Stoa:text:2003.01.0003:chapter=6:sect=3:subsection=4
Torone		Classical-Roman-Post-Roman	Classical fill deposits with later contaminations and mixed deposit	2 fish-hooks, at least 4 lead net-weights (various lengths, weights: 11-2.5 grs), 1 netting-needle.		Cambitoglou *et al.* 2001, pp. 729-732.
Thessalonike - Foinikas	Macedonian tomb	Hellenistic	Mixed deposits from the interior of the tomb.	1 silver fish-hook 1 bronze netting-needle		Tsimbidou-Avloniti 2005, pp. 66.
Itanos	House complex	Late antiquity	Destruction levels	2 fish-hooks		Mylona 2003b; excavation archive.
Leuki island	House 2	Unspecified historical date	Floor deposits in various rooms	Netting-needles, lead net-weights; bronze fish-hook, pumice		Papadakis 1983
Nemea	Nemea	Mixed Hellenistic?-Late Antique-Bytzantine. (predominately Hellenistic)	Sector N16, and area at the periphery of the Sanctuary of Zeus. Deposits disturbed by Late Antique and Byzantine farming activities.	2 bronze fish- hooks		Miller 1976, p.184, pl. 33. Miller 1980, p. 194, No BR 728, pl. 44d
Elis	Makrysia	Hellenistic	Zeus Sanctuary, on the banks of Alpheios River.	1 fish-hook		Papathanasopoulos 1970, p. 192, pl. 168.
Thassos	Thassos	Late Antique and Byzantine	Agora Unspecified context	About 40 fish-hooks, 1 multiple fish-hook, 3 net-needles, 1 fishing-weight.		Dunant 1956, p. 456, pl. 544b, e.
Thrace	Avdera	Hellenistic/Roman	Mixed deposits in relation to the city wall.	1 net-needle, 3 fish-hooks of various sizes		Lazarides 1965, p. 456, pl. 544b, e.
Piereia	Krania	Hellenistic	Deposits associated with a tavern	Indeterminate number of fishing- hooks netting-needles, net-weights.		Poulaki-Pantermali 2001, p. 337

The data in this table have been collected from various publications. Inconsistencies in the recording detail reflect inconsistencies in the original publications.

Appendix 3b: Fish related structures.

Area	Site	Date	Context	Finds	Comments	References
Corinthia	Kenchreae, Eastern port of Corinth	Construction: 1st c. B.C. – A.D. 1st c. Use: end of AD 2nd –early 3rd c.	SE of the warehouses, on the wave line	Fish-tanks (*piscinae*). Six basins cut in the rock and linked to each other and the sea with channels	Construction typical of the *piscinae* as these are described by Varro and Collumella	Shaw 1978, p. 25-35.
Corinth	Agora	Roman (before 77 A.D.)	In the Forum area near Peirene fountain.	Fish and meat market building. Foundations of a circular, colonnaded building with a running water channel passing through it. Two Latin inscriptions found nearby identify the building as the meat fish and meat market.	The association of the building to the inscriptions which refer to a fish and meat market is based on architectural similarities between this and a definitely identified market of this type in Pompeii.	Williams II 1993, pp. 39-40; West 1931, pp. 103-4; Kent 1966, pp.127-8.
Crete Ierapetra	Ferma	Roman?	Dug on the rocky shore, nowadays only partly submerged.	1 large tank divided in two by a low wall. Dimentions: 5.45 -4.80 – 5.55 – 5.50m., depth: 4.80 m. Two openings on the tank's wall permit sea water to circulate in the tank.	Construction typical of the *piscinae* as these are described by Varro and Collumella	Davaras 1975
Crete Siteia	Siteia	Roman?	Dug on rocky shore, nowadays partly exposed.	A cluster of about 10 fish-tanks, connected to the sea and with each other with a system of channels	Construction typical of the *piscinae* as these are described by Varro and Collumella	Davaras 1974
Crete Siteia	Siteia	Roman?	Dug on rocky shore, nowadays partly exposed.	Fish-salting tanks? A pair of a trapezoid and an apsidal shallow cut in the rock, close to the fish tanks. Part of the construction made by masonry, internally plastered with hydraulic cement. Narrow channels connect the two and the apsidal compartment with the sea. No filling of the structure directly from the sea was possible.	The excavator interprets this feature as a bath pool. It could however be a fish salting vat.	Davaras 1974.
Crete Heraklion	Chersonessos	Roman?	Dug on rocky shore	3 fish-tanks of various sizes (4x3m, 2x1.25m, ≈ 3x2m) with channels connecting them to the sea. Traces of a super-structure.	The tanks are situated close to the Roman harbor.	Hood and Leatham 1958-9
Crete Siteia	Mochlos. On the mainland shore across form the island.	Roman?	Dug on rocky shore	One fish-tank divided in two by a wall. Channels connected the tank to the sea and the two compartments with each other.		Hood and Leatham 1958-9
Crete East coast	Zakros on the northern shore of the bay	Roman?	Dug on rocky coast	One fish-tank 3.3x3 m.		Davaras 1974

Appendix 4. Sites of special interest to the study of ancient fishing and fish consumption.

NORTHERN GREECE AND THRACE

Torone
Chalkidiki peninsula.

Torone was a colony of Chalcideans from Euboea, on the south-west coast of the Sithonia peninsula. Built on a hill, it had a harbour called Kophos (deaf). The excavations within the city produced a number of fishing tools. Two fish-hooks and a netting needle have been recovered from Classical fills, and from mixed fills which contained Classical to Roman and later material. They also produced a number of lead net-weights. These vary in length but are more uniform in folded width. They range in weight from 2.5 to 11 gr. None of the net weights have been found together with the fish-hooks, although they all belong to either adjacent trenches or to similar levels (Cambitoglou *et al.* 2001, pp.730-32).

Olynthos
Chalkidiki peninsula.

Olynthos had initially been a Boeotian colony in the 7th c. B.C., but it was given to Chalkideans in the Classical period. Olynthos is situated at the head of a deep bay between the central and westernmost finger of the Chalkidiki peninsula, surrounded by very fertile land. It occupies two hills. A different city planning is evident from each of them, in accordance with the dominant trend of the era in which it was built (Stillwell 1971, pp. 651-2; Cahill 2002).

The excavation of the houses, which are quite uniform in design despite their individual features, produced a large number of fishing tools, mostly fish-hooks and netting needles. 19 of them have been found together in a deposit of one of the streets. The largest of them measures 22 cms in length (Robinson 1936; Robinson 1941; Cahill 2002)

Krania
Pieria.

Krania is situated at the northern foothills of Platamon Hill, 200m from the sea and at 10 masl. The excavated site of Krania is possibly part of the harbour sector of the ancient city of Irakleon. In the vicinity of this town have been recovered several farmhouses and two major ancient cities, Livithra and Phila (Margaritis 2006, ch. 6.1 and 6.4). According to a limited archaeological survey of the immediate area, the main city of Irakleon may have extended beyond Platamon Hill (Poulaki–Pantermali 2001). The area makes a suitable harbour for larger ships as the coast has deep waters and is well protected from severe winds.

The earliest traces of occupation at the site date back to the 3rd millennium B.C. and the latest to the 4th century A.D. Hellenistic activity at the site started with the construction of various buildings at the end of the 4th century B.C. Three Hellenistic phases were identified and extensive destruction layers of various buildings were excavated (Poulaki–Pantermali, personal communication). The main Hellenistic period dates from the last quarter of the 4th century to the first quarter of the 3rd century B.C. The frequency of finds throughout the destruction layer of this phase is impressive. Different types of domestic pottery, both coarse and fine, pithoi, amphorae, and various female and animal figurines represent the vast majority of finds. Fishing hooks, net weights, needles, loom weights, arrowheads and personal jewellery have also been found throughout the excavated area (Poulaki–Pantermali 2001, p. 337). The analysis of the stratigraphy and the archaeological finds from the excavation is still in progress and little information has so far been published (Poulaki–Pantermali 2001).

The fish remains have been collected by water flotation of soil samples from the floor and destruction layer of two large rooms, possibly belonging to the same building, Building A and a rubbish pit of the main Hellenistic period. The pit was used as a well during the Classical period and later transformed into a rubbish pit (Poulaki–Pantermali, personal communication). Apart from the fish bones, the fill, especially the ashy layers, contained large numbers of carbonised material, among which cereals, pulses and fruits are the commonest. Olives, grapes, danewort, figs, walnuts hazelnuts and the occasional almond are represented throughout the fill deposits as is a variety of weed seeds and also nidentified flower heads. Concentrations of sesame seeds and pine cones in certain spots add to the picture (Margaritis 2006, ch. 8.3.4). The finds from the destruction layers also comprise large amounts of animal bones, seashells and fine, domestic pottery.

The nature of the finds from the two excavated rooms of Building A and their contemporary rubbish pit indicate that this building served as food preparation and consumption space, with minimal storage. According to E. Margaritis (2006), the special features of this particular building point towards its function as a "guesthouse"/inn, perhaps combined with habitation by its owners. Its function might be viewed in connection to its location by the harbour of the ancient city (Irakleon?).

The fish-bone assemblage is especially rich, comprising in several thousand remains. Many of these are fish scales. About 10% of the bones are burned black as are, indeed, several of the animal bones from the same deposits. Cranial and postcranial bones are all present. Because the analysis is still preliminary, no exact figures are available. The fish-bone assemblage features a variety of taxa. Interestingly, they represent almost the whole range of aquatic habitats, except perhaps the lacustrine. Not only were the rich coastal waters of the Thermaic gulf exploited but also the brackish lagoonal areas perhaps at the estuaries of Pinios river, which according to the excavator (Poulaki-Pantermali, personal communication) was nearer to the site than today and the river proper with fish such as the Cyprinidae. The cartilaginous fish such as the sharks are particularly common at this site.

WESTERN GREECE

Kassope

Preveza.

Ancient Kassope, the capital of Kassopaea, was founded before the middle of the 4th century B.C. on the edge of a fertile valley to the south. The city flourished in the 3rd century B.C., when the large public buildings were erected and the private houses rebuilt, but its prosperity came to an end in 168-167 B.C., when it was destroyed by the Romans (Schwandener 1999).

Excavations at the site have produced 61 hand collected fish remains, which originate from house and street fills. They are mostly fish of the euryaline waters (sea bass, gilthead sea bream, grey mullet) and fewer purely marine creatures such as tuna or grouper (Boessneck 1994). The excavation also produced a very large number of domestic and wild mammals, birds and sea-shells indicating a wide variety of exploited resources (Friedl 1984).

ATTICA

The Priest's House

Attica, Vouliagmeni.

A free-standing building situated on the narrow neck of the Kavouri peninsula, near Vouliagmeni, at a distance of about 150 m from the Temple of Apollo Zoster. The building was constructed at the end of 6th or early in the 5th B.C. and remained in use until Roman times. Because of its location, architectural features and cult-related finds, the building has been interpreted as a "Priest's house", i.e. the lodgings of the Temple's personnel and, perhaps, of worshipers (Brown 2002). The excavation of its numerous rooms and courtyard produced a large number of animal bones (cattle and sheep bones are mentioned), sea-shells, quantities of ash and several cooking pots. These, along with the existence of a room with reclining benches, support the idea that one of the building's functions was preparation and consumption of food (Stavropoulos 1938).

The excavation of the house produced, among other finds, a large number of fishing related tools (Appendix 3a) found together at one spot within Room Z. Furthermore, a roof-tile, dating to the earlier phases of the building, bears the image of a moored ship, incised with a pointed instrument (nail?) at an unknown date (Stavropoulos 1938).

The "Priest's House" and the Temple of Apollo Zoster (Kourouniotis 1927-28) belonged to the Attic coastal *deme* of Halae Aixonidae (Eliot 1962, pp. 25-26; otherwise known for their custom of dedicating the first tuna of the year's catch to Poseidon (Appendix 1: Crates *Sacrifices at Athens*). This particular *deme* was also known for its extensive salt pans along its coasts (Andreou 1994).

Fish-market at the *deme* of Aexone.
Attica.

Early excavations in modern day Glyphada (a suburb of Athens), at the area of the ancient *deme* Aexonae, on the western coast of Attica, revealed a substantial circular building (diameter: 30 m) with a rectangular protrusion on its perimeter. By the time of excavation it was situated about 10 meters from the sea (Keramopoulos 1919). The building has been interpreted by its excavator as the fish-market-place on the basis of its shape and coastal location (Keramopoulos 1919; Konsolaki-Giannopoulou 1990, pp. 28-29). Others disagree with this interpretation finding the location rather unsuitable for a market (Eliot 1962). No details of the excavation or finds from this building are available and the building itself is now buried underneath the modern city. Its identity as a fish-market building, though probable, can not be verified.

Altar of Aphrodite Ourania
Agora, Athens.

North of the Panathenaic Way, in the Athenia Agora, a rectangular monumental altar with a limestone base has been excavated and attributed to Aphrodite Ourania. It dates to around 500 B.C., with some repairs later in the 5th c. (Shear 1984, pp. 24-33). The single fish bone collected from there is part of a large assemblage of animal bones recovered from the fill inside the altar. This is considered to be a secondary deposit from dumps in the area, which was included in the altar during its construction. Because the majority of the bones are burned, the whole of the bone assemblage has been considered to be the remains of burned offerings (Reese 1989)

Tavern in the Agora
Athens.

The excavation of a cluster of Classical shops and small business establishments which has been recovered beneath the Roman stoa in the Athenian Agora, provided evidence for the existence of a tavern which served fish and sea food (Shear 1975, pp. 346-365).

This complex consisted of a series of two-roomed spaces, most of which had access to the street. Although their architectural history can be deduced by the excavation of the spaces themselves, their use is best illuminated by finds recovered from a well located within one of the rooms (Room 6). After the well went out of use (late 5th c. - early 4th century) it was used as a dumping place by the nearby businesses, until about 380 B.C. (Shear 1975, p. 356). The deposits in the well were extremely rich in pottery which consist of shapes and types associated with the kitchen, serving vessels and a large number of wine amphorae, many of which were imported from well known centers of wine production, such as Mende, Chios, Lesvos, Samos, Corinth, but also Attica. The shear quantity of pots represented in this deposit makes it unlikely that they represent domestic use (Shear 1975, pp. 356-357).

Among the materials found in the well there was a large quantity of mammal bones, some of which were connected to bone working. Most of them however, belonging to cattle, pigs, sheep and goats, had evident traces of butchering. The excavator interprets the observed patterns of the anatomical part representation for each taxon as a sign of the functioning, in this complex, of a meat selling business (Shear 1975, p. 359). However, the features of the assemblage that are described in the report do not preclude the possibility that these bones are the remains of consumption.

The same strata have produced a large number of fish remains, belonging to unidentified large varieties. Along with these, the remains of large quantities of sea-shells (oysters, mussels and murex shells) have been recovered. These remains, along with the good quality pottery found in the same strata, have been interpreted by the excavator as indicative of the food served in a tavern which catered for a well-to-do clientele (Shear 1975, p. 357).

The food remains in the fill of this particular well have been seen as representing two different activities: food preparation and consumption (fish and sea-shells) and meat selling (mammal bones). The only published observation which supports this distinction is the fact that the mammal remains have been found in "pockets" within the deposit (Shear 1975, p.359). No such observation is provided for the fish bones and sea-shells. On the basis of this, the identification of two food related businesses does not seem to be very well supported. An alternative interpretation of all these remains as tavern refuse could also be plausible. The observed concentration of the mammal remains could be the result of a particular dumping habit rather than a different origin altogether.

PELOPONNESE

Sanctuary of Poseidon
Isthmia, Corinthia.

The sanctuary is located east of Corinth, near the Isthmus. Established around 1050 B.C. it had a long history, lasting to the 3rd c. A.D. The Sanctuary of Poseidon had a pan-Hellenic character and it hosted the Isthmian Games every four years (Gebhard 1992; 1993). Excavations in the sanctuary precincts produced a large and varied body of dedicatory objects. Among these, fishing tools of various types are common. Such objects have been found, for example, in the debris deposit from within the Archaic *pronaos*, dating to the end of 5th c. B.C., or in a large circular pit at the lower level around the Sanctuary (Tr. H, level VII). This was filled at the end of the 5th c. with objects cleaned out of the sanctuary, after the temple fire, ca 470-450 B.C. Such finds are probably votive (Raubitschek 1998, p. 121; Gebhard 1998, p. 108).

Port of Kenchreai
Corinthia.

Kenchreai is the eastern port of Corinth, facing the Saronic gulf. Its function was mostly commercial. The pre-Roman port is very little known, as it is now silted up. The Roman port has been intensively investigated. Jetties, quays, a light house and a number of warehouses formed part of the harbour facilities (Scranton *et al.* 1978).

Near the SE warehouses there is a complex of fish-tanks (*piscinae*). It consists of six basins linked by channels to each other and the sea (Fig. 7.2). They had been built in the 1st c. B.C. to the early 1st c. A.D. and were in use at the end of 2nd – early 3rd c. A.D. Their sizes vary from 5.90x5.55 the largest, to 4.80x1.90 the smallest (Shaw 1978).

Sanctuary of Demeter and Kore
Corinth.

The Sanctuary of Demeter and Kore is situated on Akrokorinth and its remains date from the Late Helladic IIIC to the 4th c. A.D. The sanctuary developed on three terraces. On the Upper terrace there were two theatral areas, in the Middle terrace the sacrificial activities took place, and the Lower terrace accommodated the dining rooms. These dining rooms were numerous. In the late 5th c. B.C. there were 25-30 such rooms, accommodating up to about 200 people. The dining rooms were equipped with built couches and several of them had adjoining cooking facilities. Fragments of cooking pots and drinking vessels were very abundant (Bookidis and Stroud 1997).

The fish remains discussed here were found in water-floated soil samples from two such rooms, with the majority of bones originating from deposits from within Room N:21. These remains, all bones from small sea-bream (Sparidae), are considered to be dining leftovers, representing either a dish of fried/boiled fish or some kind of preserved fish (Bookidis *et al.* 1999).

Amphora Building
Corinth.

The Punic Amphora Building, so named after the impressive concentration of the remains of such amphorae, is situated west of the South Stoa. It is dated to the 5th c. B.C. and was relatively short lived. The building comprised a number of rooms around a central court, which opened to the street. It is considered a commercial building, dealing with imported wine and preserved fish (Williams 1978; 1979; 1980; Williams and Russell 1981).

The fish remains reported from here have been found in various spots in the building. The largest concentration has been found in the court, among Punic amphora fragments that had been deposited in thick layers alternating with crushed lime. These are the remains of tuna and sea-breams. In some cases, segments of fish scales were adhering to the pottery sherds' inner surface, still retaining the original shape of the fish fillet (Fig. 7.7). These fish remains, apparently represent preserved fish (Williams 1979). On contextual evidence this fish is considered to have been imported from the Atlantic, through the Western Mediterranean trade routes (Zimmerman-Munn 1983; 2003).

Fish and meat market
Corinth.

Excavations in the market area, South of Lechaion Road, North of the Peirene Spring have revealed a colonnaded court with a circular foundation at its centre, which was apparently supporting a *tholos* (a hemispherical roof) of a Roman date. The overflowing waters of the Peirene spring passed through the building (Williams 1993, pp. 39-40). Fragment of two similar inscriptions, in Latin, have been found in the area. These refer both to the function of the building and the names of its owners (West 1931, no 125; Kent 1966, pp. 127-8, no 321). On the basis of the architectural design of the building (Mau 1904, pp. 94-101) and the references in the inscription to a *macellum* and *macellum piscatorum*, the building has been identified as the meat- and fish-market. According to the inscriptions, the owners of the building were probably a certain Quintus Cornelius Secundus and his family (West 1931, no 125; Kent 1966, pp. 127-8, no 321). The running water, flowing through the building, probably served for cleaning the fish and removing the discards (Williams 1993, p. 40).

Butchers' and food preparation establishment
Corinth – East of theatre.

Excavations in the area to the east of the Theatre in Corinth revealed two adjoining structures dating to the 2nd c. A.D., the ground floor of which had been devoted to the preparation of food (Williams and Zervos 1986, p.129). One of the two structures, Building 3, especially one room (the Southwest Room) produced a very large number of animal bones, all concentrated in the corner of the room, which also contained the remains of three tile-floored domed cooking ovens. The bones were collected by dry-sieving of sediments with a 10mm mesh (Reese *et al.* 1987, p. 255).

The bone assemblage consisted of the remains of several hundred sheep and goats, as well as several dozens of cattle and fewer pigs and dogs. All these animals were apparently eaten as several of their bones bear cut or burning marks. This deposit also produced remains of birds, fish and sea shell.

The fish remains comprise 8 bones of gilthead sea bream (*Sparus aurata*), originating from at least five individuals ranging in weight from 250 gr to 2 kilograms. One of the fish bones belongs to a grouper (Serranidae) of about 500 gr. and another to a large tuny weighting about 50 kgs. This is a vertebra that bears a cut mark. Finally, 6 fish remains are unidentifiable (Reese *et al.* 1987, pp. 266-267). The marine remains include a number of different sea shells (Reese *et al.* 1987, p. 258).

These remains have been interpreted as relating to a special civil or religious feast taking place in the theatre (Williams and Zervos 1986, pp. 147-148). According to Reese, who studied the mammal remains from this deposit, the room where the animal remains have been found was perhaps the major dumping area for preliminary food butchery and the remains were accumulated over an indeterminate but relatively short period of time (Reese *et al.* 1987, p. 264). Such a scenario however would not explain the presence of the assorted remains of fish and birds. Most of these are from animals which are not large enough to require dismemberment and removal of the discarded bony parts. It should, perhaps, be considered that the pile of animal bones deposited in one corner of the southwest room in Building 3 represent the remains of consumption rather than food preparation.

Halieis
Argolid.

Halieis was a small town situated at a harbour site near the southern tip of the Argolid peninsula. It was occupied from the 7th to the 4th c. B.C. The settlement consisted of two parts: the acropolis and the harbour area, both fortified (Stillwell 1976, pp. 375-76). Excavations have revealed the now-submerged remains of a Sanctuary of Apollo (Jameson 1969). The few fish remains available from this site originate from the submerged area of the sanctuary, one from inside a room and the second, which dated to the 4th c. B.C., originates from a refuse pit near the "dining area" of the sanctuary complex (Rose 2000, p. 530). Besides the fish bones, several fish-hooks have been recovered from the Akropolis (6th and 4th c. B.C.) and a hoard of several dozens of them from the lower town, in the harbour area. Also some pieces of lead found in a votive deposit on the Akropolis might be net weights (Jameson *et al.* 1994, pp. 314-316).

Hermione
Argolid.

A coastal town between Troizen and Halieis, situated on a promontory, with access to two harbours. Remains of a temple, which still survive on the promontory, probably identifies with the Temple of Poseidon mentioned by Pausanias (Stillwell 1976, pp. 388-8).

Hermione was famous for its purple dye as early as the 6th c. B.C. The export of the dye to distant places (e.g. Persia) indicates a well organised production. The practicalities of such a production is a matter of scholarly debate, as the only material evidence for the process are concentrations of crushed shells in various spots in the countryside near Hermione (Jameson *et al.* 1994, pp. 316-319). Hermione was also known in antiquity for a diving competition, which formed part of the cult of *Dionysos Melanaigis* (Dionysos of the black goat skin) (Burkert 1983, p. 211).

Pyrgouthi farmstead
Berbati valley, Argolid.

The Pyrgouthi farmstead is situated in an inland valley of the Argolid. A tower, constructed in the Early Hellenistic period, served, initially, for the defence of both people and produce, taking advantage of its position on a knoll at the centre of Berbati valley. At the end of the 1st c. B.C. it became part of a farmstead and later, at the end of 6th and beginning of the 7th c. A.D., it was converted into a press house (Hjohlman *et al.* 2005).

Despite the systematic sampling and water flotation of soil samples from this site, the animal remains are few, mostly due to poor preservation in the soil. Among the surviving fish remains there have been recovered two unidentifiable bones of medium size fish (20-30 cm) and the vertebra of a large fish, a member of the Serranidae family, perhaps a wreck-fish or a grouper. This bone, which belonged to a fish about 1.5 m. in length originated from the later occupational phase from within the tower at Pyrgouthi (Mylona 2005). The distance of the site from the sea makes it possible that the vertebra originates from a slice of preserved, probably salted, fish.

Messene
Messenia, Peloponnese.

Messene is an inland city, the construction of which was begun in 379 B.C. It expanded onto the slopes of Mt. Ithome and the recent extensive excavations have revealed a number of impressive cultic and public buildings (Themelis 1999).

The few available fish bones have been found in different deposits. One shark vertebra centrum is from an altar of a Classical date (post-369 B.C.), the four pharyngeal bones of grouper are from a 2nd/1st c. B.C. bath deposit south of the Asklepieion, that probably represents building fill, and the single amberjack vertebra is from a mixed deposit of 2nd/1st c B.C. to A.D. 200, from the monumental fountain house at the NW corner of the Agora, north of the Asklepieion (Nobis 1994, p. 303; Nobis 1997, p. 108.). No further context information is available and we are not in a position to speculate with any certainty, whether the fish remains are offerings or dining leftovers.

THESSALY

New Halos
Pagasitikos gulf.

New Halos is a Hellenistic town near the coast of the Pagasitikos gulf, south of modern Volos. It led a very short life, being inhabited from 302 to 265 B.C., a flourishing period for Thessaly. New Hallos was founded by Macedonians, either Demetrios Poliorketes or the Macedonian king Kassandras. It was the army of one of these that was probably responsible for the foundation of the city. The destruction and abandonment of the city was probably caused by an earthquake. Part of the town was fortified. The domestic quarters were extensive. It is estimated that it comprised 1500 houses with a population of 9000 people. The excavators believe that the inhabitants of New Halos, mainly soldiers and their families, formed an urban society with little luxury. Some of them were probably farmers (Reinders 1988; Reinders and Prummel 2003). Six of the New Halos houses have been excavated.

The town was situated 1.5 km from the shore of Sourpi bay which is part of the Pagasitic gulf. Between the town and Sourpi bay was a salt marsh which was protected from the sea by a beach ridge. Animal remains from the site indicate that the inhabitants of New Halos mostly consumed the basic domestic animals (ovicaprids, cattle, pig) and a moderate quantity of seashells. Hunted animals were also eaten, but in small numbers. Among the seashells, which make up 57% of the total animal remains, 28 taxa have been identified. Analysis of the size and robustness of one of them, the cockle (*Cerastoderma glaucum*), and a comparison with Bronze Age specimens from the same area indicate an overexploitation of this resource during the period of Hellenistic occupation of the town (Prummel 2003a, 2003b). Interestingly however, only one, indeterminate fish bone has been collected, despite the application of some water sieving (Prummel 2003b, p. 175).

Although the very low presence of fish remains in New Halos may, to a certain degree, be artificial and accidental (due to the restricted water flotation) its relative unimportance, especially in relation to other sea-food, could be taken as real. This lack of interest in fish, on the part of the New Halos consumers, might be related to the identity of inhabitants. The majority of the population appears to have arrived from elsewhere as retired soldiers, with no local roots. It is reasonable to assume that they originated from agricultural or pastoral backgrounds, and had no familiarisation with the sea. If this was true, then the fact that the rich fish resources of the area were not exploited more intensively, might be due to the lack of any strong fishing tradition which would permit the accumulation and transmission of fishing knowledge necessary to perform the art of fishing effectively (this interpretation is based on the discussion in chapter 6).

CYCLADES

Delos

Situated at the centre of the Cyclades, Delos is one of the smallest islands of the group. In very close proximity are the even smaller islet of Rheinia and the substantially larger Myconos. In antiquity Delos was considered the birthplace of Apollo, and was the seat of a Panhellenic sanctuary. The island was under the juradiction of the Athenians during the Classical Period. Later it was declared by the Romans an international, tax free harbour and it became the centre of intense maritime trading activity attracting an international population (Bruneau and Ducat 2005; Rauth 1993). This cosmopolitanism led to the installation of the cults of foreign deities for whom sanctuaries of a non-Greek type were built. A large number of inscriptions of various types illuminate the daily life as well as the economic and political affairs of the island (Bruneau and Ducat 2005).

An indeterminate number and type of fish bones have been found in three locations on Delos. One is a farmhouse courtyard of one of the largest farmsteads on the island. Animal remains from this context are relatively few (NISP: 173). The other is a rubbish deposit, at an abandoned street, in the area of the “Terrace of the Lions”, which linked the Temples to the harbour on Skardana Bay. The excavation of the dump also produced over 9000 animal bones and numerous table ceramics (Leguilloux 2003). The third location is a cistern beneath the *atrium* of the “Lake House” which probably functioned as a high class brothel. In this case the fish remains were part of a deposit which also included walnuts, hazelnuts, almonds, pine nuts, chestnuts, egg shells, and olive pits (Zapheiropoulou 1991, 24).

Inscriptions reveal other aspects of life on the island related to fish consumption. Rent records of the Temple refer to a *tarichopoleio*, a salted fish selling establishment (Henning 1983). Also inscriptions refer to taxation of incomes from fishing and purple dye production and to the Temple’s incomes from the fish caught in the Sacred Lake (Deonna 1938, Homolle 1882; 1890). Inscriptions also refer to the dedication by fishermen of fishing gear and objects shaped like a sea-shell to various Temples by fishermen (Deonna 1983, pp. 200-201).

Regulations on fish consumption appear in several contexts. The consumption of preserved and, probably, fresh fish was prescribed for the sacred menus of certain festivals (Linders 1994), while in other cases, such as the celebrations in honour of a Syrian deity (Atargatis?) the consumption of fish was forbidden for a number of days before the festival (Sokolowsky 1962, pp. 108-9, Parker, 1983, p. 359). Prohibitions of fish as offerings are referred to in relation to Vrizo, the interpreter of dreams (Semos, Appendix 1, no. 136). Finally, fish appeared as omens on Delos (Semos, Appendix 1, no. 137)

Sanctuary at Pilarou Cave
Ancient Thera, Santorine.

The Pilarou Cave and the open space in front of it was one of the cult places of Doric Thera in the Hellenistic period. An inscription incised in rock, near the entrance of the cave, refers to Zeus Damartios, who appears to be the honoured god of this place. Dedicatory objects, an altar, as well as a quantity of dining pottery, animal remains and charcoal, bear evidence of both sacrificial and dining activities at the site. The excavation produced over 557 identifiable animal bones. Among these have been identified the whole range of large and medium standard domestic animals (equid, cattle, pig, ovicaprids) and also hare, chicken and a variety of birds and fish. Most of the fish remains have been collected from the terrace in front of the cave and probably represent dining refuse. These are the remains (6) of two gilthead sea bream (*Sparus aurata*) ranging in size between 45-50 cm. and one vertebra of tuna (*Thunnus thynnus*) of about 80 cm long (Becker 1997).

ISLANDS OF NORTH AEGEAN

Epano Skala
Mytilene, Lesvos.

The excavation at the Akropolis site on Mytilene revealed a large building, constructed in the Roman period which initially functioned as a villa. Later, it was transformed into a tavern/brothel. The Roman villa deposits were very rich in faunal remains. Over 7500 bones were recovered, belonging to a variety of animals. Apart from the domestic large and medium mammals, wild boar, deer, several large and small birds, sea food and fish have been identified. These include the red mullet, the eel, the tuna the sea bream, the scorpion fish and the sea bass. The sea food includes lobsters, crabs, sea urchin, scallops, clams, the mussels and the sea snail. In the later deposits, related to the tavern/brothel, almost no birds or marine creatures have been found (Ruscillo 2001; 1993).

The preliminary reports on the animal remains from the Akropolis at Mytilene do not provide information on the exact location or dating of the various fish bones recovered. Only the largest of the fish bones have been identified but the regular occurrence of the remains of smaller fish is also reported (Ruscillo 1993).

CRETE

Kommos
Crete, Southern Coast.

Kommos is situated on the shores of the Libyan Sea, on the coastal fringes of Mesara, the largest plain on Crete. The settlement is situated on a small hill next to the beach. It was a flourishing harbour town in the Minoan period, but occupation continued on the site through the Proto-Geometric and Geometric periods, when two temples were built. In later times a third temple was built over the second, altars were added in the court, which can still be seen today, and other buildings were erected (Shaw and Shaw 2000).

The excavation of the Hellenistic Temple C produced a sizable assemblage of fish remains. The majority of these were recovered from a slab floor in the main room of the Temple. The assemblage's special features, such as the taxonomic variety (Table 5.1a) and the lack of burning are similar to those observed in the Minoan fish assemblage, which has been collected from clearly domestic, secular contexts, and probably represent dining leftovers. In contrast, fish bones from the Geometric Temple B are considered to be the remains of burned offerings. Among the fish remains from Temple C have been found a bone of a catfish, a fresh-water fish, apparently introduced from either Egypt or Syria-Phoenicia (Rose 2000). Associated deposits from Temple C also produced several seashells (Reese 2000b), a bronze netting needle and an object which is probably a fish line sinker. These objects are considered to be dedications, having been found in the context of a temple (Shaw 2000, p. 391).

Itanos
Crete, Eastern coast.

Itanos was a harbour city on the Eastern coast of Crete (Greco *et al.* 1997; 1998; 2001; 2003). It was occupied from the Geometric period and, certainly, up to the Byzantine era. In the 3rd and 2nd centuries B.C. Itanos played an active role in local politics and, due to its position and harbour, was involved in international trade. During this period Itanos had interests on the Leuke Island, which appears to have been a territory exploited for purple shells (Papadakis 1983). Tradition also emphasises the Itanians's involvement in this trade. Korovios, a purple shell fisherman is said to have led Theran colonists to Cyrene, on the northern coast of Africa, apparently exploiting his knowledge of marine routes (Appendix 1, no 85). The maritime nature of Itanos is further accentuated by the fact that its 5th - 4th c. coins depicted the protector god of the city, Glaukos-Melikertes, in the form of a half man-half fish creature, striking downwards with his trident (Fig. 6.2; Spyridakis 1979, p. 11). On other coins sea monsters are depicted (Head 1887, p. 469).

Fish remains recovered from Itanos originate from a Late Roman/Early Christian (5th-7th c. A.D.) house complex, south west of the Akropolis (Greco *et al.* 1998). These remains form part of a very rich assemblage, which also includes mammal, bird, sea-turtle and seashell remains (Mylona 1997; 2003b). The excavation of this particular house also produced two fish-hooks. The analysis of the material recovered from this house is still in progress. However, a preliminary presentation of the pottery from the house complex and a nearby Basilica of a similar date (both assemblages are presented together with no differentiation of provenance), indicated the use of a variety of shapes, many of which are cooking and serving vessels, which are not only of local production but also imported from North Africa, Asia Minor, Cyprus and other parts of Crete. It is interesting that the pottery assemblage comprises both imported, high quality, decorated pottery and misfired local pots. The cookware is of particular interest as it includes cooking pots of varying shapes, frying pans, and a variety of lids (Xanthopoulou 2005).

The majority of the fish remains from this particular context in Itanos (494 identifiable specimens and 525 non identifiable ones) belong to members of four families: sea breams (Sparidae), parrot-fish (Scaridae), groupers/combers (Serranidae) and picarels (Centracanthidae). Other species are represented by one or two bones each. Most of the fish are coastal varieties of both small and large size. The fish remains exhibit a distinctive pattern of anatomical part representation. The larger fish are represented mostly by head bones, while the smaller ones by all parts of the skeleton. This is probably an indication of the deliberate purchasing and consumption of the fish heads on this site (Mylona 2003b).

Kouphonisi (Leuke)
Lybian Sea, SE Crete.

Kouphonisi or Leuke is a small island off the SE coast of Crete. Nowadays deserted, it was occupied from the Early Minoan period to the 4th c. A.D. In Hellenistic and Roman times it was a trading post with a flourishing settlement. It was known, for its purple dye production. This resource and its trading importance resulted in its involvement in various disputes among some of the most powerful city-states of Eastern Crete (Leonard 1972).

Crushed purple shells are visible among the houses of the settlement on the North coast of the island. Some of these have been interpreted, by the excavator, as typical houses of the purple fishermen. Fish-hooks and lead weights are common finds in various excavated houses of the settlement (Papadakis 1983). The publication of the excavation results is still preliminary and no precise data are available.

Eleutherna
Rethymno, Crete.

Eleutherna is a sizable city in the northern foothills of Mt Psiloritis, in the Rethymnon district. It is an inland site, some distance from the sea. The city's layout is quite complicated. It developed around a central hill where the Akropolis was situated. Eleutherna was occupied since at least the Late Neolithic, but most of the remains visible today are from the Geometric period to Late Antiquity (van Effentererre 1991, Stabolidis 1993, Themelis 2002; Kalpaxis 2000/1; 1989/90,).

The various excavation projects on site have produced a small number of fish remains. Undefined Late Roman strata from the Akropolis (Pyrgi) have produced fish bones, all of which are from large fish. These are shark, sting ray, grouper and grey mullet (Mylona 2003b). The excavation of a Roman villa in the western sector of the city, in the Katsivelos valley, has also produced several fish remains, again from large fish. In this case grouper, gilthead sea bream and tuna have been identified (Nobis 1993, p. 115). Although the total number of remains is small, they are particularly interesting because of the site's considerable distance from the sea.

Appendix 5. Fish bone recording.
For each identifiable bone the following variables are recorded. Here follows a list of these variables and the values ascribed to them. The range of taxa may differ among different case studies.

TAXON

1. Sharks
2. Dogfish
3. *Triakidae*
4. *Squalus sqcualus*
5. *Squatina squatina*
6. *Rajia* sp.
7. Myliobatidae
8. Dasyatidae
9. Clupeidae
10. *Alosa* sp.
11. *Clupea harengusd*
12. *Sprattus sprattus*
13. *Sardina pilchardis*
14. *Alosa* sp.
15. *Engraulis engrasicholus*
16. *Salmo salar*
17. *Salmo truta*
18. *Synodus saurus*
19. *Anguilla anguilla*
20. *Muraena helena*
21. *Conger conger*
22. *Muraenidae/Congridae*
23. *Belone belone*
24. *Cepola rubescens*
25. *Echeneis naucrates*
26. Exocoetidae
27. *Dactyliopterus volitans*
28. Gadidae
29. *Micromesistius poutassou*
30. *Phycis phycis*
31. *Merlangius merlangus*
32. *Brosme boisme*
33. *Pollachius* sp.
34. *Zeus faber*
35. Serranidae
36. *Seranus cabrilla*
37. *Serranus scriba*
38. *Epinephelus aenus*
39. *Epinephelus guaza*
40. *Epinephelus alexandrinus*
41. *Epinephelus* sp.
42. *Polyprion americanus*
43. *Dicentrarchus sp.*
44. *Pomatomus saltator*
45. Carangidae
46. *Trachurus trachurus*
47. *Seriola dumerili*
48. Litses
49. *Coryphaena* sp.
50. Sciaenidae
51. *Argyrosomus regius*
52. *Sciaena umbra*
53. *Umbrina cirrosa*
54. *Mullus barbatus*
55. *Mullus surmuletus*
56. Sparidae
57. *Dentex dentex*
58. *Pagrus pagrus*
59. *Diplodus annularis*
60. *Diplodus sargus*
61. *Diplodus puntazzo*
62. *Diplodus vulgaris*
63. *Lithognathus mormyris*
64. *Obladus melanura*
65. *Boop boops*
66. *Boops salpa*
67. *Pagellus erythrinus*
68. *Pagellus acarne*
69. *Pagellus* sp.
70. *Sparus aurata*
71. *Spondyliosoma cantharus*
72. *Centracanthidae*
73. *Spicara* sp.
74. *Centracanthus cirrus*
75. *Chromis chromis*
76. Labridae
77. *Labrus* sp.
78. *Coris julis*
79. *Crenilabrus* sp.
80. *Symphodus* sp.
81. *Xyrichthys novacula*
82. *Sparisoma cretense*
83. *Trachinidae*
84. *Uranoscopus scaber*
85. Scombridae
86. *Scomber japonicus*
87. *Scomber scombrus*
88. *Thynnus* sp.
89. *Euthynnus alletteratus*
90. *Auxis rochei*
91. *Sarda sarda*
92. *Xiphias gladius*
93. *Gobius* sp.
94. *Blennius* sp.
95. Sphyraena sp.
96. Mugilidae
97. *Mugil cephalus*
98. *Lisa* sp.
99. *Atherina* sp.
100. *Scorpaena* sp.
101. Triglidae
102. *Aspitriglia cuculus*
103. *Eutryglia gurnandus*
104. *Triglia lucerna*
105. *Lophius piscatorius*

106. Pleuronectiformes
107. Indeterminate small
108. Indeterminate medium
109. Indeterminate large
110. Unidentified
111. Centracanthidae/Sparidae (Boops)
112. Fresh water fish
113. Sparidae/Mullidae

ANAT.PART
1. Articular
2. Angulare
3. Dentary
4. Premaxilla
5. Maxilla
6. Palatine
7. Quadrate
8. Hyomandibular
9. Postemporal
10. Preopercular
11. Frontal
12. Parasphenoid
13. Vomer
14. Pharyngeal bone
15. Tooth
16. Basioccipital
17. Hypohyal
18. Ceratohyal
19. Epihyal
20. Cleithrum
21. Supracleithrum
22. Palatinum
23. Pharyngeal bone
24. Opercular
25. Preopercular
26. Urohyal
27. Dermal plates
28. Rays diagn.
29. Otolith
30. Scapula
31. Vertebra I
32. Vertebra II
33. Vertebra III
34. Vertebra IV
35. Vertebra V
36. Anterior abdominal
37. Posterior abdominal
38. Caudal vert.
39. Penultima vert.
40. Ultima vert.
41. Vertebra indeterminate
42. Special features

SIZE of fish
1. Small: 0-10 cm
2. Medium: 10-30 cm
3. Large: 30-50 cm
4. V. Large: >50
5. **Small: 0-50 cm**
6. **Medium: 50-80 cm**
7. **Large: >80 cm.**
8. Irrelevant

(bold: scale for species of large size)

SIDE
1. Right
2. Left
3. Central
4. Indeterminate

PRESERVATION
1. Whole
2. 50-99%
3. 25-49%
4. 0-24%

BURNING
0. Unburned
1. Burned brown uniform (l-d)
2. Burned brown on spots
3. Burned black uniform
4. Burned black on spots
5. Calcified
6. Unburned in ash

CUT
1. Uncut
2. Cut
3. Pierced

CHEWING/CRUSHING
0. Unaltered
1. Chewed/crushed
2. Rodent teeth marks
3. Other

EROSION
0. No alteration
1. Digested
2. Soil erosion

Unit: this is the excavation unit and represents the minimum group of finds of common origin. It is shared by a number of different objects. For example, imaginary Unit 3125 could designate the contents of a small pit (pit iv) in the Late Hellenistic house 1, Room 1. This unit could include various finds, such as fish bones, mammal bones, shells, bronze tools, pottery, loom weights and a quantity of ash. This number is what connects the fish remains to their context and the rest of the archaeological finds.

Anatomical part: The anatomical parts which are considered identifiable and recordable are those defined by Wheeler and Jones (1989, pp. 87-125). Fin rays, branchiostegal rays, spines and vertebral spines, as well as several bones of the cranium are considered non-identifiable. They are however recorded separately as their presence might provide indications on issues such as fish processing.

Taxon: Fish bones are identified and recorded on several levels of precision. For a typical sample of different taxonomic categories recorded for this variable, see above. The list includes the most common taxa in Greek waters and the ones most likely to be found in excavations, on the basis of a survey of fish bone reports from the Aegean and Eastern Mediterranean. It also includes some more general categories. Bones belonging to members of the Sparidae family, for example, may be either identified to species or to family level, depending on the identifiability of the bone and its preservation condition. In some cases, groups broader than the family have been created, e.g. "Sparidae/Centracanthidae", or "sharks" to cover those cases, which can not be more closely identified. Such a flexible system of taxonomic recording has the merit of extracting the maximum possible information from surviving fish bones. Subsequent groupings on the basis of family or habitat, for example, can spring from here.

Relative size of live fish: Because the taxonomic recording is so variable, and because fish within a family or even genus can vary considerably in size, a relative size scale has been devised (see above), which complements the information from the previous variable. Estimation of the life fish size from each bone is approximate and based on comparison with reference specimens. In certain occasions, published data, which resulted from a similar process of comparison are also used. Such information is useful, when issues of food quantity, food selectivity on the basis of size and fishing methods are discussed.

Side: This variable refers to the position of the paired elements on the left or right side of the skeleton. Such data could be useful in the calculation of the minimum number of fish (MNI) represented in the assemblage.

Fragmentation: Recording the degree of completeness of each bone gives a measure of the fragmentation of the assemblage. The relative frequency of each fragmentation category (see above) in the assemblage can be an indication of its preservation status and its taphonomic history.

Burning: This variable records not only the presence of burning but also the type (see above). Such information may be useful both in the exploration of cooking methods involved in the preparation of fish dishes but also in the investigation of the assemblage's taphonomic history.

Cut marks: The recording of the presence and type of cut marks (see above) are crucial in the investigation of preparation processes for both cooking and storing.

Chewing/Crushing: A deformation of the bone due to intense pressure is recorded. Because of preservation problems, often only the result is obvious, but the identification of the causal factor (e.g. tooth marks) remains obscure. So, the two features are recorded together.

Digestion traces: The recording of the presence or absence of gut erosion traces follow the criteria set by Jones (1984; 1986). This feature is important for the investigation of the taphonomic history of the assemblage.

Measurements are not included in the data used in this study, although they are usually part of the routine of fish bone analysis (Morales and Rosenlund 1979; Wheeler and Jones 1989, pp.139-148). This is done, mostly because in the preservation conditions observed in the area of Greece, only few bones have well preserved and clearly defined margins that permit an accurate measuring. It has been judged that the relatively inaccurate but more generally applicable calculation of the size range of the live fish is more suitable in this case.

For the non-identifiable fragments, the variables recorded are: total number of remains, number of remains in each anatomical group (cranial, fins and ribs, other), predominant relative size of fish from which the bones originate, presence of burning and cut marks and finally comments, on features of the non-identifiable bone assemblage which can not be recorded in a standardized manner.

REFERENCES

Acheson, J. M., 1981. Anthropology of fishing. *Annual Review of Anthropology*, 10, 275-316.

Ager, L.S., 1996. *Interstate Arbitrations in the Greek World, 337-90 B.C.* Hellenistic Culture and Society vol. 18, Berkeley: University of California Press.

Alavanou, A., 1972. *Brauron and Halai Araphanides*. Athens: Lycabettus Press.

Alcock, S., 1993. *Graecia Capta: the landscape of Roman Greece*. Cambridge, New York: Cambridge University Press.

Alcock, S., 2002. *Archaeologies of the Greek Past: Landscape, Monuments and Memories*. Cambridge and New York: Cambridge University Press.

Alder, A., ed. 1967-1971. *Suidae Lexicon*. Stuttgart: Teubner.

Allison, M.P., 1999. Introduction. In M.P. Allison, ed. *The Archaeology of Household Activities.* London and New York: Routledge, 1-18.

Amyx, D.A., 1958. Attic Stellae, part iii. Vases and other containers. *Hesperia*, 27, 163-254.

Anderson, G., 1993. *The Second Sophistic: a cultural phenomenon in the Roman Empire*. London, New York: Routledge.

Andren, A., 1998. *Between Artefacts and Texts: historical archaeology in global perspective*. New York: Plenum Press.

Andreou, I., 1994. O Dimos ton Aixonidon Alon. In W.D.E. Coulson and O., Palagia, eds. *The Archaeology of Athens and Attica Under the Democracy. Proceedings of an international conference celebrating 2500 years since the birth of democracy in Greece, December 4-6 1992.* Oxbow Monographs 37. Oxford: Oxbow Books, 191-209.

Andrews, C.A, 1949. Review: Sir D'Arcy Wentworth Thompson: A Glossary of Greek Fishes. *American Journal of Philology*, 70 (3), 335-336.

Anonymous, 2003. *Anaptyxi Typologias ton Ellinon Epaggelmation Alieon Mikris Paraktias Alieias me Vasi Alieytika, Oikonomika kai Koinonika Charaktiristika*. Mesologgi: IXTHYKA A.E. and Patras University.

Anonymous, 2005. *Black Sea*, Living Black Sea Marine Environmental Education Program. Available from http://blacksea.orlyonok.ru/blacksea.shtml (Accessed 4 October 2006).

Antonaccio, C.M., 2000. Architecture and behavior: building gender into Greek houses. *Classical World*, 93, 517-534.

Apostolides, N., 1883. *La Pêche en Grèce. Ichtyologie, migrations, engins et manières de pêche, produits, statistique et législation.* Athènes: Ministère de L'Intérieur, Typographie de Perris Fréres.

Appadurai, A., 1986a. Introduction: commodities and the politics of value. In A. Appadurai, ed. *The Social Life of Things. Commodities in cultural perspective.* Cambridge: Cambridge University Press, 3-36.

Appadurai, A., ed. 1986b. *The social life of things. Commodities in cultural perspective*. Cambridge: Cambridge University Press.

Arbesmann, P.R., 1929. *Das Fasten bei den Griechen und Römern*. Religiosgeschichtliche Versuche und Vorarbeiten, 21.1, Giessen: Alfred Topelmann.

Arnold, C. 1986. Archaeology and history: the shades of confrontation and cooperation. In J. Bintliff and C. Gaffney, eds. *Archaeology at the Interface: studies in archaeology's relationships with history, geography, biology and physical science*. Oxford: BAR International Series 300, Oxford: Oxbow Books, 32-39.

Arnott, W.G., 1996. *Alexis. The Fragments. A Commentary*. Cambridge: Cambridge University Press.

Asimakopoulou-Atzaka, P., 2003. *Psiphidota Dapeda. Proseggisi stin techni tou archaeou psiphidotou.* Thessalonike: Unversity Studio Press.

Astaras, T.A. and Sotiriadis, L., 1988. The evolution of the Thessaloniki-Giannitsa plain in northern Greece during the last 2500 years - from Alexander the Great era until today. In G. Lang and Ch. Schlüchter, eds. *Lake, Mire and River Environments during the Last 15000 Years*. Rotterdam, Brookfield: A. A Balkema, 105-114.

Athanassopoulos, M., 1923. Sur les Thonnides en Grèce. *Comptes Rendues de l' Academie des Sciences*, 117, 501-502.

Athanassopoulos, M., 1924. Note complementaire sur les Thonnides en Grèce. *Bulletin de l'Institut Océanographique, Monaco,* 440, 1-3.

Athanassopoulos, M., 1925. Sur la pêche en Grèce. *Bulletin de l'Institut Océanographique, Monaco,* 454, 1-11.

Athanassopoulos, M., 1926. Notes sur la faune en Grèce. *Bulletin de l'Institut Océanographique, Monaco,* 480, 1-7.

Atran, S., 1998. Folk biology and the anthropology of science: cognitive universals and cultural particulars. *Behavioural and Brain Sciences,* 21 (4), 547-569.

Ault, B.A., 2000. Living in the Classical polis: the Greek house as a microcosm. *Classical World,* 93, 483-96.

Austin, M. and Vidal-Naquet, P., 1998. *Economic and Social History of Ancient Greece.* Berkeley: University of California Press.

Badham, R.C.D., 1854. *Prose Halieutics or Ancient and Modern Fish Tattle.* London: Parker.

Bakhtin, M., 1984. *Rabelais and his World.* Cambridge Mass.: MIT Press.

Balme, D.M., 1961. Aristotle's use of differentiae in zoology. In S. Mansion, ed. *Aristotle e les Problèmes de Méthode.* Louvain: Publications Universitaires de Louvain, 195-212.

Barrett, J., 1993a. *Fragments from Antiquity. An archaeology of social life in Britain, 2900-1200 B.C.* Oxford: Blackwell.

Barrett, J., 1993b. Bone weight, meat yield estimates and Cod (*Gadus morhua*) a preliminary study of the weight method. *International Journal of Osteoarchaeology*, 3, 1-18.

Barrett, H.J., Beukens, R.P. and Nicholson, A.R., 2001. Diet and ethnicity during the Viking colonization of northern Scotland: evidence from fish bones and stable carbon isotopes, *Antiquity*, 75 (287), 145-154.

Barringer, M.J., 1991. Europa and the Nereids: wedding or funeral? *American Journal of Archaeology*, 95 (4), 657-667.

Barthes, R., 1997. Toward a psychosociology of contemporary food consumption. In C. Counihan, and P. van Esterik, eds. *Food and Culture. A reader*. New York, London: Routledge, 20-27.

Bates, D., 1936. Animal remains, fish, etc. *Excavations at Thermi, Lesvos*. In W. Lamb *Excavations at Thermi in Lesvos*. Cambridge: Cambridge University Press.

Baumann, H., 1993. *Greek Wild Flowers and Plant Lore in Ancient Greece*. Translated from the German by W.T. Stearn and E.R. Stearn (original ed. 1982), London: The Herbert Press.

Beazley, H.D., 1963. *Attic Red-Figure Vase-Painters*. Oxford: Clarendon Press.

Becker, C., 1986. *Kastanas. Ausgrabungen in einem Siedlungshugel der Bronze und Eisenzeit Makedoniens, 1975-1979*. Prähistorische Archäologie in Sudosteuropa 5. Wissenschaftverlang Volger Spiess, Berlin.

Becker, K., 1997. Tierknochenfunde aus der Höhle Pilarou – Reste von Tieropfern oder profane Schlacht – und Speisebfäle. In W. Hoepfner, ed. *Das Dorische Thera V. Stadtgeschichte und Kultstätten am nördlichen Stadtrand*. Berlin: Schriften des Seminars für Klassische Archäologie der Freien Universität Berlin, 151-183.

Bekker-Nielsen, T., 2002. Fish in the ancient economy. In K. Ascani, V. Gabrielsen, K. Kvist and A. Holm Rasmussen, eds. *Ancient History Matters. Studies presented to Jens Erik Skydsgaard on his seventieth birthday*. Analecta Romana Instituti Danici, Suppl. 30, Roma: L'Erma di Bretschneider, 29-37.

Bekker-Nielsen, T., ed. 2005. *Ancient Fishing and Fish Processing in the Black Sea Region*. Black Sea Studies, Aarhus: Aarhus University Press.

Bernard, H.R., 1976. Kalymnos. In M., Dimen, and E., Friedl, eds. *Regional Variations in Modern Greece and Cyprus: toward a perspective on the ethnography of Greece*. New York: New York Academy of Science, 291-30.

Bevan. E., 1986. Representations of Animals in Sanctuaries of Artemis and Other Olympian Deities. BAR Internaitonal Serries315, Oxford: British Archaeological Reports.

Binford, L., 1972. *An Archaeological Perspective*. New York: Seminar Press.

Bintliff, J.L., 1976. The plain of Western Macedonia and the Neolithic site of Nea Nikomedeia. *Proceedings of the Prehistoric Society* 42, 241-262.

Bintliff, J.L., 1977. *Natural Environment and Human Settlement in Prehistoric Greece. Part 1 and 2*. BAR International Series, S28, Oxford: British Archaeological Reports.

Blount, G.B., 2005. Small-scale fisheries in the contemporary world: adoption and management. *Reviews in Anthropology*, 34, 1-29.

Blumer, H., 1920. Salsamenta. *Paulis Real Encyclopädie der Classischen Altertumwissenshaft*, band 1a, Stuttgart: J.B. Metzlersche Verlagsbuchhandlung, 2011-12.

Boardman, J., 1987. Very like a whale: Classical sea monsters. In A.E. Farkas, O.P. Harper and B.E. Harrison, eds. *Monsters and Demons in the Ancient and Medieval Worlds: papers presented in honour of Edith Porada*. Mainz on Rhine: P. von Zabern, 73-84.

Boeck, de F., 1994. "When hunger goes around the land": hunger and food among the aLuund of Zaire. *Man*, 29 (2), 257-282.

Boeckh, A., 1886. *Die Staatshausmaltung der Athener*. 3 Aufl., Berlin: Von Max Frankel.

Boessneck, J., 1994. Zooarchäologische Ergebnisse an den Tierknochen – und Molluskenfunden, In W. Hoepfner and E.-L. Schwandner, eds. *Haus und Stadt im Klassischen Griechenland*. München: Deutscher Kunstverlag, 175-179.

Bokonyi, S. 1986. Faunal Remains. In C. Renfrew, M. Gimbutas and E. Elster, eds. *Excavations at Sitagroi. A prehistoric village at NE Greece*, vol. 1, Los Angeles: California, 63-132,

Bookidis N. and Stroud, R.S., 1997. *The Sanctuary of Demeter and Kore: Topography and Architecture*. Corinth vol. 18, Princeton, N.J.: American School of Classical Studies at Athens.

Bookidis, N., Hansen, J., Snyder, L. and Goldberg, P., 1999. Dining in the sanctuary of Demeter and Kore at Corinth. *Hesperia*, 68 (1), 1-54.

Bourdieu, P., 1990. *The Logic of Practice*. Cambridge: Polity.

Bousoulengas, A., Katsiki A.V., Souvermetzoglou, E., Theocharis, A., Satsmatis, L., Mimikos, N., 1998. An intensive environmental study in the open waters of the Aegean and Ionian Seas – results of 1986. *Rapports. Commission Internationale pour la Mer Méditerranée*, 31, 145.

Bouffartigue, J. and Patillon, M., eds. 1977. *Porphyry. De l'abstinence*. Paris: Les Belles Letres.

Bourbou, C. and Richards, M.P. 2007. The Middle Byzantine menu: palaeodiatery information from isotopic analysis of humans and fauna from Kastela, Krete. *International Journal of Osteoarchaeology*, 17, 63-72.

Bowersock, G.W., ed. 1974. *Approaches to the Second Sophistic: papers presented at the 105th annual meeting of the American Philological Association*, University Park, Pa.: American Philological Association.

Bowie, E.L., 1970. Greeks and their past in the Second Sophistic. *Past and Present*, 46, 3-41.

Bradley, K., 1998. The Roman family at dinner. In I. Nielsen and H.S. Nielsen, eds., *Meals in a Social Context. Aspects of the communal meal in the*

Hellenistic and Roman world. Aarhus, Aarhus University Press, 36-55.

Braund, D., 1995. Fish from the Black Sea: Classical Byzantium and the Greekness of trade. In J. Wilkins, D. Harvey and M. Dobson, eds. *Food in Antiquity,* Exeter, University of Exeter Press, 162-170.

Braund, D. and Wilkins, J., eds, 2000. *Athenaeus and His World: reading Greek culture in the Roman Empire.* Exeter: University of Exeter Press.

Brewster, H., 1997. *The River Gods of Greece.* London, New York: I.B. Tauris Publishers.

Briggs, C., 2000. Literacy, reading and writing in the Medieval West. *Journal of Medieval History,* 26, 397-420.

Bristow, P. ed. 1992. *The Illustrated Encyclopedia of Fishes.* London: Chancellor Press.

Brown, D.E., 2002. The archaeology of the «Priest's House» and the houses of Cape Zoster and Aegina (abstract). The 103rd Annual Meeting. *American Journal of Archaeology,* 106 (2), 262.

Bruford, W.H., 1975. *The German Tradition of Self-cultivation: "Bildung" from Humboldt to Thomas Mann.* Cambridge: Cambridge University Press.

Bruneau, Ph. and Ducat, J., 2005. *Guide de Délos.* Paris: École Française d' Athène,

Brunet, M., Ph. Bruneau, Chankowski, V., Chevalier, L., Collin-Bouffier S., Delongevalle, R., Dandrau, A., Farnoux, A., Fouche, E., Hasenohr, Cl., Hellmann, M-Ch., Karvonis, P., Leguilloux, M., Moretti, J.Ch., Peulvast, J.-P., Siard, H., Siebert, G., Sinlès G., 2001. L' eau à Délos, un milieu naturel et son aménagement durant l'Antiquité. *Bulletin de Correspondance Hellénique,* 125, 620-627.

Brunet, M., Chevalier. L., Desruelles, St., Finker, M., Fouache, E., Hasenohr, Cl., Karvonis, P., Labarthe, J.-M., Moretti, J.-Ch., Peulvast, J.-P., Siard, H. and Sintès, G., 2002. L' eau à Délos, programme de recherches géoarchéologiques. *Bulletin de Correspondance Hellénique,* 126, 546.

Bruschi, T. and Wilkens, B., 1996. Conserves de poisson à partir de quatre amphores romaines. *Archaeofauna,* 5, 165-69.

Bucher, K., 1893. *Die Entstehung der Volkswirtschaft.* Tübingen: H. Lau

Bucholz, H.G., Johrens G. and Maull, I., 1973. *Jagd und Fischfang. Mit einem Anhang: Honiggewinnung.* Göttingen: Vandenhoeck und Ruprecht.

Bunsmann, L., 1911. *De piscatorum in Graecis atque Romanis litteris Usu.* Munster.

Buraselis, K., 2000. *Kos Between Hellenism and Rome. Studies on the political, institutional and social history of Kos from circa the middle second century B.C. until Late Antiquity.* Philadelphia: American Philosophical Society.

Burke, M.E., 1992. The economy of Athens in the Classical era: some adjustments to the primitivist model. *Transactions of the American Philological Association,* 122, 199-226.

Burkert, W., 1983. *Homo Necans. The anthropology of ancient Greek sacrificial ritual and myth.* Translated by P. Bing, Berkeley: University of California Press.

Burkert, W., 1985. *Greek Religion Archaic and Classical.* Translated by J. Raffan. Oxford: Blackwell.

Burr-Stebbins, E., 1929. *The Dolphin in the Literature and Art of Greece and Rome.* Menasha Wis.: George Banta Publishing.

Burt, J.O., 1954. *Minor Attic Orators.* Loeb Classical Library, London: William Heineman and Cambridge Mass.: Harvard University Press.

Buxton, R., 1994. *Imaginary Greece. The context of mythology.* Cambridge: Cambridge University Press.

Cahill, N.D., 2002. *Household and City Organization at Olynthus.* New Haven: Yale University Press. Also available from www.stoa.org.olyntus/ (accessed 1 October 2006).

Cambitoglou, A., Papadopoulos, J.K. and Jones, O. T, 2001. *Torone I: the Excavations of the 1975, 1976 and 1978.* Athens: H en Athinais Archaeologiki Etairia.

Campbell, A.C., 1982. *The Hamlyn Guide to Flora and Fauna of the Mediterranean Sea.* London: Hamlyn.

Cancian, F., 1980. Risk and uncertainty in agricultural decision-making. In P. Barlet, ed. *Agricultural decision makers.* London and New York: Academic Press, 161-176.

Caplan, P., 1994. *Feasts, fasts and famine. Food for thought.* Oxford, Providence USA: Berg.

Cartledge, P., 2002. The political economy of Greek slavery. In P. Cartledge, E. Cohen, and L. Foxhall, eds. 2002. *Money, labour and land: approaches to the economies of ancient Greece.* London: Routledge, 156-166.

Cartledge, P., Cohen, E. and Foxhall, L. eds. 2002. *Money, labour and land: approaches to the economies of ancient Greece.* London: Routledge.

Casteel, R., 1976. *Fish Remains in Archaeology and Paleoenvironmental Studies.* New York: Academic Press.

Chaniotes, A., 1996. *Die viertäzwischen Kretischen Poleis in der Hellenistischen Zeit.* HABES 24, Stuttgard: F. Steiner.

Chouliara-Raiou, E., 2003. *H Alieia stin Aigypto Ypo to fos ton Ellinikon Papyron.* Ioannina: University of Ioannina.

Clark, G.A., 1972. Forward. In E.S. Higgs, ed. *Papers in Economic Prehistory.* Cambridge: Cambridge University Press, i-xii.

Cole, G.S., 2004. *Landscape, Gender and Ritual Space: the ancient Greek experience.* Berkeley, Calif.: University of California Press.

Colley, S.M., 1987. Cooking fish on a fire: An experiment. Paper presented at the Fourth ICAZ meeting, York, UK, unpublished manuscript.

Cooney, G., 2003. Introduction: seeing land from the sea. *World Archaeology,* 35 (3), 323-328.

Corcoran, H.T., 1963. Roman fish sauces, *Classical Journal,* 58, 204-10.

Cough, H., 1935/36. Fishing in Homer. *Classical Journal,* 31, 303-314.

Cowan, J., 1991. Going out for coffee? Contesting the grounds of gendered pleasures in everyday sociability. In P. Loizos and E., Papataxiarchis, eds. *Contested Identities: gender and kinship in modern Greece.* Princeton: Princeton University Press, 180-202.

Craik, E., 2000. Hippocratic diaeta. In J. Wilkins, D. Harvey and M. Dobson, eds. *Food in Antiquity.* Exeter: University of Exeter Press, 343-350.

Curtis, I.R., 1979. The garum shop of Pompei. *Cronache Pompeiane*, 5, 5-23.

Curtis, I.R., 1983. In defense of Garum. *Classical Journal,* 78, 232-40.

Curtis, I.R., 1991. *Garum and Salsamenta. Production and commerce.* Studies in Medical History, Leiden: E.J. Brill.

Curtis, I.R., 2001. *Ancient Food Technology.* Leiden: Brill Academic Publications.

Curtis, I.R., 2005. Sources for production and trade of Greek and Roman processed fish. In Bekker-Nielsen, T., ed. *Ancient Fishing and Fish Processing in the Black Sea Region.* Black Sea Studies, Aarhus: Aarhus University Press, 31-46.

Curtius, C., 1870. Inschriften aus Ephesos. *Hermes*, 4, 174-228.

Cutting, C.L., 1955. *Fish Shaving. A history of fish preservation from ancient to modern times.* London: Leonard Hill.

Cutting, C.L., 1962. The influence of drying, slating and smoking on the nutritive value of fish. In E. Heen and R. Kreuzer, eds. *Fish in Nutrition.* London: FAO, Fishing News, 161-179.

Dalby, A., 1995b. Achaestratus where and when. In J. Wilkins, D. Harvey and M. Dobson, eds. *Food in Antiquity.* Exeter: University of Exeter Press, 400-412.

Dalby, A., 1996. *Siren Feast. A history of food and gastronomy in Greece.* London and New York: Routledge.

Dalby, A. and Grainger, S., 2000. *The Classical Cookbook.* London: British Museum Press.

Danforth, M.L., 1984. The ideological context of the search for continuities in Greek culture. *Journal of Modern Greek Studies*, 2, 53-85.

Danier, W.B., 1813. *Supplement to the Rural Sport,* London: T. Davidson

(http://books.google.com/books?hl=en&lr=&id=zD8AAAAAQAAJ&oi=fnd&pg=PA1&dq=fishing+as+genteel+sport&ots=rLYUnVjlss&sig=1wE-h8maKUz1XO-PC2pn-Uqf0rE) (Accessed 15-5-2007).

Dasen, V. and Piérart, M., eds. 2005. *Idia kai Demosia. Les cadres « privés » et «publics » de la religion grecque antique. Actes du IXem colloque du CIERGA, tenu à Fribourg du 8 au 10 septembre 2003.* Kernos, suppl. 15. Liège: Centre Internationale d Étude de la Religion Grecque Antique.

Davaras, K., 1974. Rock-cut fish tanks in Eastern Crete. *Annual of the British School at Athens*, 69, 87-93.

Davaras, K., 1975. Laxeuti ichthyodexameni sta Ferma Ierapetras. *Archaeologiko Deltio*, 30, 149-154.

Davidson, G.R., 1952. *The Minor Objects. Corinth XII.* Princeton: American School of Classical Studies at Athens.

Davidson, J., 1993. Fish, sex and revolution in Athens. *Classical Quarterly*, 43 (1), 53-66.

Davidson, J., 1995. *Opsophagia*: revolutionary eating in Athens. In J. Wilkins, D. Harvey and M. Dobson, eds. *Food in Antiquity.* Exeter: University of Exeter Press, 204-213.

Davidson, J., 1996. On the fish missing from Homer. In J. Wilkins, ed. *Food in European Literature.* Exeter: Intellect Books, 57-64.

Davidson, J., 1997. *Courtesans and Fish Cakes. The consuming passions of Classical Athens.* London: Fontana Press.

Degani, E., 1995. Problems in Greek gastronomic poetry: On Matro's *Attikon Deipnon* . In J. Wilkins, D. Harvey and M. Dobson, eds. *Food in Antiquity.* Exeter: University of Exeter Press, 413-28.

Demir, M. and A. Acara, 1955. Biological and hydrobiological factors controlling the migration of mackerel from the Black Sea to the sea of Marmora. *Proceedings of the General Fisheries Council of the Mediterranean*, 3, 127-134.

Denell, R.W., 1979. Prehistoric diet and nutrition: some food for thought. *World Archaeology*, 11 (2), 121-135.

Deonna, W., 1938. *Delos 18. Les mobilier Délien.* Paris: Editions de Boccard.

Desse-Berset, N., ed. 1982. *2nd Fish Osteoarchaeology meeting.* Paris: Editions du CNRS.

Desse, J. and Desse-Berset, N., 1989. Les Profils Rachidiens Globaux. Reconstitution de la Taile des Poissons et Appréciation du Nombre Minimal d'Individus a Partir des Pièces Rachidiennes. *Revue de Paléobiologie*, 8, 89-94.

Desse-Berset, N. and Desse, J., 2000. Salsamenta, garum et autres préparations de poissons. Ce qu'en disent les os. *Mélange de l École Française de Rome. Antiquité*, 112, 84-92.

Detienne, M., 1989a. Culinary practices and the spirit of sacrifice. In M. Detienne and J-P. Vernant, eds. *The Cuisine of Sacrifice Among the Greeks.* Translated by P. Wissing, Chicago, London: The University of Chicago Press, 1-20.

Detienne, M., 1989b. The violence of wellborn ladies: women in the *Thesmophoria.* In M. Detienne and J-P. Vernant, eds. *The Cuisine of Sacrifice Among the Greeks.* Translated by P. Wissing, Chicago, London: The University of Chicago Press, 129-147.

Diegues, A.C., 2002. Traditional fisheries knowledge and social appropriation of marine resources in Brazil, Paper presented at Mare Conference: People and the Sea, Amsterdam, August/September, 2001. Available from **http://www.usp.br/nupaub/amsten.doc** (accessed 10-10-2006).

Diels, H., 1974. *Die Fragmente der Vorsokratiker.* 1st edition 1903, Germany: Weidmann.

Dietler, M., 1996. Feasts and commensal politics in the political economy. Food, power and status in prehistoric Europe. In P. Wiessner and W., Schiefenhovel, eds. *Food and the Status Quest. An interdisciplinary perspective.* Oxford, Providence: Berghahm Books, 87-125.

Dietler, M. and Hayden, B., eds. 2001. *Feasts: archaeological and ethnographic perspectives on food, politics and power.* Washington, D.C.: Smithsonian Institution Press.

Diez, E., 1957. Delphin. *Reallexicon für Antike und Christentum*, band III. Stuttgart: Anton Hiersemann, 667-682.

Dilley, R., 1992. Contesting markets. A general introduction to market ideology, imagery and discourse. In R. Rilley, ed. *Contesting Markets: analyses of ideology, discourse and practice.* Edinburgh: Edinburgh University Press, 1-34.

Dölger, F.J., 1922. *ICHTHYS. Der Heilige Fisch in den Antiken Religionen und im Christentum.* Münster.

Dorandi, T., 1999. *Antigone de Caryste Fragments.* Paris: Les Belles Lettres.

Douglas, M., 1966. *Purity and Danger: an analysis of concepts of pollution and taboo.* London: Routledge and Kegan Paul.

Douglas, M., 1997. Deciphering a meal. In C. Counihan, and P. van Esterik, eds. *Food and Culture. A reader.* New York, London: Routledge, 36-54.

Douglas M. and Isherwood, B., 1996. *The World of Goods: towards an anthropology of consumption.* London: Routledge.

Douglas, N., 1974. *Birds and Beasts of the Greek anthology.* Chatham: Minerva Press.

Dumont, J.C., 1976/77. La pêche du thon a Byzance a l'époque hellénistique. *Revue des Etudes Anciennes,* 78/79, 96-119.

Dunant, C., 1956. Thassos. Petits objets. *Bulletin de Correspondance Hellénique*, 80, 425, fig. 33.

Durand, J.-L., 1989. Ritual as instrumentality. In M. Detienne and J-P. Vernant, eds. *The Cuisine of Sacrifice Among the Greeks.* Translated by P. Wissing, Chicago, London: The University of Chicago Press, 119-128.

Eberl, G., 1892. *Die Fischkonserven der Alten.* Stadfamhof.

Economidis, P.S., 1991. *Check list of the freshwater fishes of Greece (recent status of threats and protection*). Hellenic Society for the Protection of Nature, special publication.

Economidis, P.S., 1995a. Endangered freshwater fishes of Greece. *Biological Conservation,* 72, 201-211.

Economidis, P., 1995b. Ta psaria tou Aigaiou kai tis Symis. *Alieutika Nea*, 163, 85-88.

Economidis, P.S. and Sinis, A.I., 1988. A natural hybrid of *Leuciscus cephalus macedonicus x Chalcalburnus chalcoides macedonicus* (Pisces, Cyprinidae) from Lake Volvi (Macedonia, Greece). *Journal of Fish Biology*, 32, 593-605.

Economou, A.N., Zogaris, S., Giakoumi, S., Barbieri, R. and Petridis, D., 2002. *Developing a biotic river typology and defining reference conditions in the rivers of Greece: a spatially-based approach.* Available from http://fame.boku.ac.at/downloads/D9_13_SBM_Reports/ecoregion_6_SBA_Economou_etal.pdf (accessed 10 October 2006).

Edmondson, J.C., 1987. *Two Industries in Roman Lusitania. Mining and garum production.* BAR international Series 362, Oxford.

Ehrenberg, V., 1968. *Aristophanes und das Volk von Athen: eine Soziologie der altattischen Komödie.* Zürich, Stuttgart: Artemis Verlag.

Eliot, C.W.J., 1962. *Coastal Demes of Attika: A study in the policy of Kleisthenes.* Toronto: University of Toronto Press.

Ellen, R., 1982. *Environment, Subsistence and System.* Cambridge: Cambridge University Press.

Ellen, R., 1993. *The Cultural Relation of Classification.* Cambridge: Cambridge University Press.

Ellen, R., 1996. Introduction. In R. Ellen and K. Fukui, eds. *Redefining Nature. Ecology, culture and domestication.* Oxford, Washington DC: Berg, 1-36.

Evershed, R.P., Dudd, S.N., Lockhart, M.J. and Jim, S., 2001. Lipids in archaeology. In D.R. Brothwell and M.A., Pollard, eds. *Handbook of Archaeological Science.* New York: John Wiley, 331-350.

Faber, G., 1883. *The Fisheries of the Adriatic and the Fish Thereof.* London: Bernard Quaritch.

Falk, P., 1994. *The Consuming Body.* London: Sage.

Farb, P. and Armelagos, G., 1980. *Consuming Passions: the anthropology of eating*, Boston: Houghton Mifflin.

Farkas, E.A., Harper, P.O. and E.B. Harrison, eds., 1987. *Monsters and Daemons in the Ancient and Medieval Worlds.* Mainz on Rhine: Verlag Philipp von Zabern.

Farnell, L.R., 1977. The cult of the Greek states, vol. 4. New Rochelle, New York: Caratzas Brothers.

Feyel, M., 1936. Nouvelles Inscriptions d' Akraephia. *Bulletin de Correspondance Hellénique*, 60, 11-36.

Finley, I.M., 1973. *The Ancient Economy.* Berkeley and Los Angeles: University of California Press.

Finley, I.M., ed., 1979. *The Bucher-Meyer Controversy.* New York: Arno.

Finley, I.M., 1985. *Ancient History: evidence and models.* London: Chatto and Windus.

Firth, R., 1966. *Malay Fishermen. Their peasant economy.* London: Routledge and Kegan Paul.

Froese, R. and Pauly, D., eds. 2006. *FishBase.* World Wide Web electronic publication. http://www.fishbase.org, version (10/2006).

Fisher, N., 2000. Symposiasts, fish-eaters and flatterers: social mobility and moral concerns in Old Comedy. In D. Harvey and J. Wilkins, eds. *The Rivals of Aristophanes. Studies in Athenian Old Comedy.* London: Duckworth and Classical Press of Wales.

Foley, D., 1985. Optimality theory in anthropology. *Man,* 20, 222-242.

Forbes, M.H., 1976. Farming and foraging in Prehistoric Greece: a cultural ecological perspective. In M.

Dimen and E. Friedl, eds. *Regional Variations in Modern Greece and Cyprus: toward a perspective on the ethnography of Greece.* New York: New York Academy of Science, 127-142.

Forbes, H.A., 1989. Of grandfathers and grand theories: the hierarchised ordering of responses to hazard in a Greek rural community. In P. Halstead and J. O' Shea, eds. *Bad Year Economics: cultural responses to risk and uncertainty.* Cambridge: Cambridge University Press, 87-97.

Forbes, H.A., 1996. The uses of uncultivated landscape. In G., Shipley and J. Salmon, eds. *Human landscape in Classical Antiquity. Environment and Culture.* London and New York: Routledge, 68-97.

Forstenpointner, G., 2003. Promethean legacy: investigations into the ritual procedure of "Olympian" sacrifice. In E. Kotzabopoulou, Y. Hamilakis, P. Halstead, C. Gamble, and P. Elefanti, eds. *Zooarchaeology in Greece: recent advances.* BSA Studies 9, London: British School at Athens, 203-213.

Fossey, M.J., 1988. *Topography and Population of Ancient Boeotia.* Chicago: Ares Publishers.

Fotiadis, N., 1995. Modernity and the past-still-present: Politics of time in the birth of regional archaeological projects. *American Journal of Archaeology,* 99, 59-78.

Foxhall, L., 2000. The running sands of time: Archaeology and the short-term. *World Archaeology,* 31 (3), 484-498.

Foxhall, L., 2004. Field sports. Engaging Greek archaeology and history. In W.E. Sauer ed. *Archaeology and Ancient History: breaking down the boundaries.* London and New York: Routledge, 76-84.

Foxhall, L., forthcoming. House clearance: unpacking the "kitchen" in Classical Greece. In R. Westgate, N. Fisher and J. Whitley, eds. *Building Communities: house, settlement and society in the Aegean and beyond.* Conference volume, British School at Athens.

Foxhall, L. and Forbes, H., 1982. Sitometria: the role of grain as a staple food in Classical antiquity. *Chiron,* 12, 41-90.

Fraser, T.M., 1960. *Rusembilan: A Malay fishing village in South Thailand.* Ithaca, New York: Cornell University Press.

Frayn, J., 1995. The Roman meat trade. In J. Wilkins, D. Harvey and M. Dobson, eds. *Food in Antiquity.* Exeter: University of Exeter Press, 107-114.

Friedl, H., 1984. *Tierknochenfunde aus Kassope/Griechenland (4.-1 Jh.v.Chr.).* Inaugural – Dissertation zur Erlangung der tiermrdizinischen Doktorwürde der Tierärztlichen Fakultät der Ludvig-Maximilians- Universität, München.

Frisch, P., 1983. *Die Inschriften von Parion.* Inschriften griechischer Städte aus Kleinasien, 25. Bon.

Frost, F.J., ed., 1971. *Greek Society.* Lexington, Mass., Toronto, D.C.: Heath.

Gallant, T., 1985. *A Fisherman's Tale: An analysis of the potential productivity of fishing in the ancient Mediterranean.* Brussels: Gent.

Gallant, T., 1991. *Risk and Survival in Ancient Greece: reconstructing the rural domestic economy.* Stanford: Stanford University Press.

Gallo, L., 1997. Lo sfruttamento delle risorse. In S. Settis, ed. *I Greci: storia, arte, societa.* Turin: G. Einaudi, 423-52.

Gamble, S.C., 1979. Surplus and self-sufficiency in the Cycladic subsistence economy. In J.L. Davis and J.F. Cherry, eds. *Papers in Cycladic Prehistory.* Institute of Archaeology, Los Angeles: University of California, 122-133.

Gamble, S.C., 1982. Animal husbandry, population and urbanisation. In C. Renfrew and M. Wagstaff, eds. *An Island Policy: the archaeology of exploitation in Melos.* Cambridge: Cambridge University Press, 161-171.

Garioti, G.L., 1977. *Oi Palioi Parades tou Pineioou.* Larisa: Ekdoseis Periodikou "Sparmos".

Garnsey, P., 1988. *Famine and Food Supply in the Greco-Roman World: responses to risk and crisis.* Cambridge: Cambridge University Press.

Garnsey, P., 1999. *Food and Society in Classical Antiquity.* Cambridge: Cambridge University Press.

Garnsey, P. and Saller, S., 1987. *The Roman Empire, economy, society and culture.* London: Duckworth.

Gärtner, H., ed. 1970. *Rufus von Epheusus. Quaestiones medicinales.* Bibliotheca scriptorium graecorum et romanorum Teubneriana, Leipzig: B.G. Teubner Verlagsgesellschaft.

Gebhard, E.R, 1992. The early stadium at Isthmia and the founding of the Isthmian games. In H. Kyrieleis and W.D.E. Coulson, eds., *Symposium on the Olympic Games.* Athens: Deutsches Archäologisches Institut Athen, 73-79.

Gebhard, E.R., 1993. Isthmian Games and the sanctuary of Poseidon in the Early Empire. In T.E. Gregory, ed. *The Corinthia in the Roman Period. Journal of Roman archaeology,* suppl. 8, Ann Arbor, MI., pp. 78-94.

Gebhard, R.C., 1998. Small dedications in the Archaic Temple of Poseidon at Isthmia. In R. Hägg, ed. *Ancient Greek Cult Practice from the Archaeological Evidence. Proceedings of the fourth international seminar on ancient Greek cult, organized by the Swedish Institute at Athens 22-24 October 1993.* Stockholm: Swedish Institute at Athens, 91-115.

Georgacas, D., 1977. *Ichthyological Terms for the Sturgeon and Etymology of the International Terms Botargo, Caviar and Congeners: a linguistic, philological and culture-historical study.* Athens, Grafeio Dimosieumaton Akadimias Athinon.

Giannopoulou – Konsolaki, E., 1990. *Glyphada. Istoriko Parelthon kai Mnimeia.* Athina: Greek Ministry of Culture.

Gifford-Gonzalez, D., 1989. Ethnographic analogues for interpreting modified bones: some cases from East Africa. In R. Bonnichsen and M.H. Sorg, eds. *Bone*

Modification. Orono, Maine: Centre for the Study of the First Americans, 179-246.

Gilula, D., 1995. Comic food and food for comedy. In J. Wilkins, D. Harvey and M. Dobson, eds. *Food in Antiquity*. Exeter: University of Exeter Press, 386-99.

Godelier, M., 1972. *Rationality and Irrationality in Economics*. Translated by B.Pierce, London, New York: Monthly Review Press.

Goldberg, Y.M., 1999. Spatial and behavioural negotiation in Classical Athenian city houses. In M.P. Allison, ed., *The Archaeology of Household Activities*. London and New York: Routledge, 142-161.

Goodman, J., Lovejoy, P.E and Sherratt, A., eds. 1995. *Consuming Habits: drugs in history and anthropology*. London and New York: Routledge.

Goody, J., 1982. *Cooking, Cuisine and Class: A study in comparative sociology*. Cambridge: Cambridge University Press.

Gordon, E.A., 1993. Screen size and differential faunal recovery: a Hawaiian example. *Journal of Field Archaeology*, 20, 453-459.

Gosden, C. and Marshall, Y., eds., 1999. *The Cultural Biography of Objects*. *World Archaeology*, 31, London: Routledge.

Grant, M., 1995. Oribasios and Medical Dietetics or the three Ps. In J. Wilkins, D. Harvey and M. Dobson, eds. *Food in Antiquity*. Exeter: University of Exeter Press, 371-385.

Greco, E., Kalpaxis, Th., Schnapp, A., and Viviers, D., 1997. Travaux menés en collaboration avec l École française en 1996, *Bulletin de Correspondance Héllenique*, 121, 809-24.

Greco, E., Kalpaxis, Th., Schnapp, A., and Viviers, D., 1998. Travaux menés en collaboration avec l École française en 1997 – Itanos (Crete oriental), *Bulletin de Correspondance Héllenique*, 122, 585-602.

Greco, E., Kalpaxis, Th., Schnapp, A., Papadakis, N. and Viviers, D., 2001. Travaux menés en collaboration avec l École française en 2000 – Itanos (Crete oriental), *Bulletin de Correspondance Héllenique*, 125, 636-644.

Greco, E., Kalpaxis, Th., Schnapp, A., and Viviers, D., 2003. Travaux menés en collaboration avec l École française en 2002 – Itanos (Crete oriental), *Bulletin de Correspondance Héllenique*, 127, 546-553.

Green, M. and Depew, D., 2004. *The Philosophy of Biology. An episodic history*. Cambridge University Press, Cambridge.

Greenwood, P.H. and Howes, G.J., 1973-74. Fish remains. In Karageorgis, V., *Excavations in the Necropolis of Salamis 3*. Salamis 5, Nicosia, 259-67.

Greaves, A., 1999. *The History of Miletus to the End of the Archaic Period*. Ph.D. dissertation, Leeds: Leeds University.

Guest-Papamanoli, A., 1984. Une pêche au guet: Le "Taliani", origine, évolution et distribution géographique. In *Actes du Colloque: L'exploitation de la mer de l'antiquité a nos jours. La mer, lieu de production*, Antibes. Musée Archéologique d'Antibes. Centre de recherches archéologiques du C.N.R.S., 185-204.

Guest-Papamanoli, A., 1985. Archéologie, Ethnographie ou Ethnoarchéologie des ressources maritimes de sites côtiers. Le cas de la pêche aux muges dans les lagunes de la Grèce Occidentale. In *Actes du Colloque, L'exploitation de la mer de l'antiquité a nos jours. La mer, lieu de production*. Antibes. Musée Archéologique d'Antibes. Centre de recherches archéologiques du C.N.R.S., 281-304

Guthrie, W.K.C., 1993. *Orpheus and Greek Religion. A study of the orphic movement*. Princeton, N.J., Princeton University Press.

Hadjidaki, E., 1988. *The Classical and Hellenistic Harbour of Phalasarna in Western Crete*. PhD Thesis, University of California, Santa Barbara, Ann Arbor, Mich., University Microfilms International.

Hadjidaki, E. and Stephanakis, M., 2004. Mystika tis Phalasarnas. *Kritiko Panorama*, 2, 100-135.

Haedrich, R.L. and Barnes, S.M., 1997. Changing over time of the size structure in an exploited shelf fish community. *Fisheries Research,* 31(3), 229-239.

Hägg, R. and Alroth, B. eds. 2005. *Greek Sacrificial Ritual, Olympian and Chthonian: Proceedings of the sixth international seminar on ancient Greek cult, organized by the Department of Classical archaeology and ancient history, Göttenborg University, 25-27 April 1997*. Stockholm, Swedish Institute in Athens.

Hall, M., 2000. *Archaeology and the Modern World: colonial transcripts in South Africa and the Chesapaeke*. London and New York, Routledge.

Halstead, P., 1981a. From determinism to uncertainty: social storage and the rise of the Minoan Palace. In G. Bailey, ed. *Economic Archaeology*. BAR International Series 96, Oxford, 187-213.

Halstead, P., 1981b. Counting sheep in Neolithic and Bronze Age Greece. In I. Hodder, G. Isaac and N. Hammond, eds. *Pattern of the Past: studies in honour of David Clarke*, Cambridge: Cambridge University Press, 307-339.

Halstead, P., 1987. Traditional and ancient rural economy in Mediterranean Europe: plus ça change? *Journal of Hellenic Studies,* 107, 77-87.

Halstead, P. and Jones, G., 1989. Agrarian ecology in the Greek islands: time stress, scale and risk. *Journal of Hellenic Studies* 109, 41-55.

Halstead, P. and O'Shea, J., eds. 1989. *Bad Year Economics. Cultural responses to risk and uncertainty*. New Directions in Archaeology. Cambridge, N. York, N. Rochelle, Melbourne, Sydney, Cambridge University Press.

Halverson, J., 1976. Animal categories and terms of abuse. *Man (N.S.)*, 11 (4), 505-516.

Hamilakis, Y., 1995. *Strategies for Survival and Strategies for Domination: wine, oil and "social complexity" in Bronze Age Crete*. PhD dissertation, Sheffield: Department of Archaeology and Prehistory.

Hamilakis, Y., 1998. Eating the dead: mortuary feasting and the politics of memory in the Aegean Bronze Age societies. In K. Branigan, ed. *Cemetery and Society in the Aegean Bronze Age*, Sheffield: Sheffield Academic Press, 115-132.

Hamilakis, Y., 1999a. Food technologies / technologies of the body: the social context of wine and oil production and consumption in Bronze Age Crete. *World Archaeology* 31(1), 38-54.

Hamilakis, Y., 1999b. The anthropology of food and drink consumption and Aegean archaeology. In S.J. Vaughan and W.D.E. Coulson, eds. *Paleodiet in the Aegean*. Oxford, Oxbow, 55-63.

Hamilakis, Y., 2001a. Re-inventing environmental archaeology. A comment on economic prehistory or environmental archaeology? On gaining a sense of identity. In U. Albarella, ed. *Environmental Archaeology: meaning and purpose*. Dordrecht, Boston, London, Kluwer Academic Publishers, 29-38.

Hamilakis, Y., 2001b. The tyranny of vision (commentary). *Journal of the Royal Anthropological Institute*, 7(1), 153-154.

Hamilakis, Y., 2002. The past as oral history. In Y. Hamilakis, M. Pluciennik and S. Tarlow, eds. *Thinking Through the Body. Archaeologies of corporeality*. New York, Boston, Dordrecht, London, Moscow, Kluwer Academic/Plenum Publishers, 121-136.

Hamilakis, Y., Pluciennik, M. and Tarlow, S., eds. 2002a. *Thinking Through the Body. Archaeologies of corporeality*. New York, Boston, Dordrecht, London, Moscow, Kluwer Academic/Plenum Publishers.

Hamilakis, Y., Pluciennic, M. and Tarlow, S., 2002b. Introduction. Thinking through the body. In Y. Hamilakis, M. Pluciennik and S. Tarlow, eds. *Thinking Through the Body. Archaeologies of corporeality*. New York, Boston, Dordrecht, London, Moscow: Kluwer Academic/Plenum Publishers, 1-21.

Hanson, C.K., 1997. Galilean fishing economy. The Galilean fishing economy and the Jesus tradition. *Biblical Theology Bulletin*, 27, 99-111.

Harris, M., 1980. *Cultural Materialism. The struggle for a science of culture*. New York, Vintage Books.

Harris, M., 1986. Foodways: historical overview and theoretical prolegomenon. In M. Harris and E.B. Ross, eds. 1986. *Food and Evolution. Towards a theory of food habits*. Philadelphia: Temple University Press, 57-90.

Harris, M. and Ross, E.B., eds., 1986. *Food and Evolution. Towards a theory of food habits*. Philadelphia, Temple University Press.

Hartog, F., 1988. *The Mrror of Herodotus. The representation of the other in the writing of history*. Translated by J. Lloyd, Berkeley: University of California Press.

Hasluck F.W., 1904. Unpublished inscriptions from the Cyzicus neighbourhood. *Journal of Hellenic Studies*, 24, 20-40.

Hawkes, C., 1954. Archaeological theory and method: some suggestions from the Old World. *American Anthropologist*, 56, 155-68.

Hayden, B., 1996. Feasting in Prehistoric and traditional societies. In P. Wiessner and W. Schiefenhövel, eds. *Food and the Status Quest. An interdisciplinary perspective*. Province, Oxford: Berghahn Books, 127-148.

Hayden, B. and Cannon, A., 1983. Where the garbage goes: refuse disposal in the Maya Highlands. *Journal of Anthropological Archaeology*, 2, 117-163.

Head, B.V., 1911. *Historia numorum: A manual of Greek numismatics*. Oxford: Clarendon Press.

Heichelheim, F.M., 1930. *Wirtschaftliche schwankungen der Zeit von Alexander bis Augustus*. Jena, Verlag von Gustav Fischer.

Heliotis, D.F., 1988. An inventory and review of the wetland resources of Greece. *Wetlands*, 8, 15-31.

Helms, M., 1988. *Ulysses' Sail: an ethnographic odyssey of power, knowledge and geographical distance*. Princeton, N.J.: Princeton University Press.

Helskog, K., 1999. The shore connection: cognitive landscape and communication with rock carvings in Northernmost Europe. *Norwegian Archaeological Review*, 32(2), 73-94.

Henning, D., 1983. Die "heiligen Häuser" von Delos. *Chiron*, 13, 411-495.

Herzfeld, M., 1982. *Ours Once More: folklore, ideology and the making of modern Greece*. Austin, University of Texas Press.

Herzog, R., 1931. *Die Wunderheilungen von Epidauros: Ein Beitrag zur Geschichte der Medizin und der Religion*. Philologus Suppl., 22 (3), Leipzig: Dieterich.

Heuzey, L., 1860. *Le mont Olympe et l' Acarnanie : Exploration des ces Deux Régions, avec l' Etude de leurs Antiquités, de leurs Populations Anciennes et Modernes, de leur Géographie et de leur Histoire*. Paris, Librairie Firmin Didot Frères.

Hicks, E.L., 1890. *Greek Inscriptions in the British Museum* III. Oxford.

Higginbotham, J., 1997. *Piscinae: artificial fishponds in Roman Italy*. Chapel Hill, London: University of North Carolina Press.

Higgs, E.S., ed., 1972. *Papers in Economic Prehistory*. Cambridge: Cambridge University Press.

Higgs, E.S., and Jarman, M.R., 1975. Paleoeconomy. In S.E. Higgs, ed. *Paleoeconomy*. New York, London: Cambridge University Press.

Hill, S. and Wilkins, J., 1996. Mithaikos and other Greek cooks. In H. Walker, ed., *Proceedings of the Oxford symposium on food in antiquity – cooks and other peoples*. Totnes (Devon): Prospect Books, 144-8. Available from (http://latis.ex.ac.uk/classics/undergraduate/food2/mithaikos.htm) (accessed 10 October 2006).

Hinde, R.A., 1991. A biologist looks at Anthropology. *Man*, 26, 583-608.

Hjohlman, J., forthcoming. Donations, *hiketeia* and the people of Kalaureia. In B. Wells and A. Pentinenn, eds. *The Physical Environment and Daily life at the Temple on Kalaureia (Poros).* Swedish Institute at Athens.

Hjohlman, J., Penttinen A. and Wells, B., eds. 2005. *Pyrgouthi. A rural site in the Berbati valley from the Early Iron Age to Late Antiquity. Excavations by the Swedish Institute at Athens 1995-1997.* Stockholm, Acta Instituti Atheniensis Regni Sueciae, Series in 4°. LII, Swedish Institute in Athens.

Hodder, I., 1998. *Theory and Practise in Archaeology.* London and New York: Poutledge.

Hoffman, H. and Halverson, J., 1977. Animal categories and terms of abuse. *Man (N.S.),* 12 (3/4), 527-528.

Hoffman, H.D. and Jordan, D.S., 1892. A catalogue of the fishes of Greece, with notes on the names now in use and those employed by classical authors. *Proceedings of the Academy of National Sciences at Philadelphia*, 230-285.

Højte M.J., 2005. The archaeological evidence fro fish processing in the Black Sea region. In T. Bekker-Nielsen, ed. *Ancient fishing and fish processing in the Black Sea region.* Black Sea Studies 2, Aarhus, Aarhus University Press, 133- 60.

Homolle, Th. 1882. Comptes des Hiéropes du Temple D'Apollon Déliene. *Bulletin de Correspondance Hellénique,* 6, 1-167.

Homolle, Th. 1890. Des Temples Déliennes en l'année 279. *Bulletin de Correspondance Hellénique,* 14, 389-511.

Hood, S. and Leatham, J., 1958-9. Sub - marine exploration in Crete. *Annual of the British School at Athens,* 53-54, 264-275.

Hopper, J.R., 1979. *Trade and Industry in Classical Greece.* London: Thames and Hudson.

Horden, P. and Purcell, N., 2000. *The Corrupting Sea: a study of Mediterranean history.* Oxford: Blackwell.

Hornblower, S. and A. Spawforth, eds. 2003. *The Oxford Classical Dictionary.* Oxford, New York: Oxford University Press.

Horsley, G., 1989. A fishing cartel in first-century Ephesos. In *New Documents Illustrating Early Christianity*, 5, 95-114.

Hunger, H., 1978. *Die hocksprachiche profane Litteratur der Byzantiner.* Munich.

Ingold, T., 2000a. Evolving skills. In H. Rose and S. Rose, eds. *Alas, Poor Darwin: arguments against evolutionary psychology.* London: Jonathan Cape, 225-46.

Ingold, T., 2000b. *The Perception of the Environment. Essays on livelihood, dwelling and skill.* London and New York: Routledge.

Isaksson, S., forthcoming. Traces of a Feast. Organic Residues at the Sanctuary of Poseidon, Kalaureia, Greece. In B. Wells and A. Penttinen, eds. *The Physical Environment and Daily life at the Temple on Kalaureia (Poros).* Swedish Institute at Athens.

Isenberg, M., 1975. The sale of sacrificial meat. *Classical Philology,* 4, 271-273.

Jacob, C., 1997. *Geografia kai ethnografia stin archaia Ellada.* Translated by E. Toundasaki, Athens: Gutenberg.

Jacobsen, A.L.J., 2005. The reliability of fishing statistics as a source for catches and fish stocks in Antiquity. In T. Bekker-Nielsen, ed. *Ancient Fishing and Fish Processing in the Black Sea Region.* Black Sea Studies 2, Aarhus: Aarhus University Press, 97-104.

Jacobsen, T.W., 2000. Background of the Franchthi Project. In W.R. Farrand, ed. *Franchthi Cave: stratigraphy, sedimentology and chronology.* Bloomington and Indianapolis: Indiana University Press, 1-10.

Jameson, M.H, 1969. Excavations at Porto Cheli and vicinity, preliminary report, I: Halieis, 1962-1968. *Hesperia*, 38 (3), 311-342.

Jameson, M.H., 1988. Sacrifice and animal husbandry in Classical Greece. In C.R. Whittaker, ed. *Pastoral Economies in Classical Antiquity.* The Cambridge Philological Society, Suppl. 14, Cambridge: Cambridge University Press.

Jameson, M.H., 1990. Private space and the Greek city. In O. Murray and S. Price, eds. *The Greek City from Homer to Alexander.* Oxford: Oxford University Press.

Jameson, H.M, Runnels, C. N. and Munn M. H., 1994. *Countryside. The southern Argolid from prehistory to present day.* Stanford, California: Stanford University Press.

Jamot, P., 1889. Borne- Frontière des territoires de Kopai et d' Akraiphiai, *Bulletin de Correspondance Hellénique*, 13, 407-8.

Jardin, C., 1961. Garum e sauces de poissons de l' antiquite. *Rivista di Studi Liguri*, 27, 70-96.

Jarman, M.R., 1973. Preliminary report on the animal bones. In J.N. Coldstream, ed. *Knossos. The sanctuary of Demeter.* BSA suppl. 8, London: The British School of Archaeology, 177-179.

Jensen, B.L., 1963. Royal purple of Tyre. *Journal of Near Eastern Studies*, 22 (2), 104-118.

Jochim, A.M., 1979. Breaking down the system: recent ecological approaches to archaeology. *Advances in Archaeological Method and Theory*, 2, 77-117.

Jochim, A.M., 1998. *A Hunter-Gatherer Landscape. SW Germany in the Late Paleolithic and Mesolithic.* New York and London: Plenum Press.

Johnson, H. M., 1999. Rethinking historical archaeology. In A. P. P. Funari, M. Hall and S, Jones, eds. *Historical Archaeology. Back from the Edge.* London and New York: Poutledge, 23-36

Jones, A.K.G., 1984. Some effects of the mammalian digestive system on fish bones. In N. Desse-Bersett, ed. *2nd Fish Osteoarchaeology Meeting, CNRS.* Centre de recherches archéologiques. *Notes et Monographies Techniques*, 16, 61-65.

Jones, A.K.G., 1986. Fish bone survival in the digestive systems of the pig, dog and man: Some experiments. In D.C. Brinkhuizen and A.T. Clason, eds. *Fish and*

Archaeology. Studies in osteometry, taphonomy, seasonality and fishing methods. BAR International Series 294, Oxford.

ICAZ Fish Working Group Web page 2007. http://www.alexandriaarchive.org/workfish.htm (accessed 15 May 2007).

Kahn, H.C., 1974. Pythagorian philosophy before Plato. In A.P.D. Mourelatos, ed. *The Presocratics. A collection of critical essays.* New York: Anchor Books.

Kalitsounakis, J.E., 1926. *Opson* und *Opsarion.* In *Festschrift für Paul W. Kretschmer Beiträge zur Griechischen Sprachforschung.* Wien - Leipzig - New York, 96-106.

Kallergis, I., 1953. Trofae kai pota eis Proto-Byzantinous papurus. In *Epetiris Eterias Byzantinon Spoudon*, year 13th, 44-75.

Kallianiotis, A., Sophronidis, K., Vidoris, P. and Tselepides, A., 2000. Density, biomass and assemblages of demersal macrofauna off Iraklion Gulf, Cretan Sea. *Progress in Oceanography,* 46, 429-455.

Kallianiotis, A., Vidoris, P. and Sylaios, G., 2004. Fish species assemblages and geographical sub-areas in the North Aegean Sea, Greece. *Fisheries Research,* 68, 171-187.

Kalpaxis, Th., 1989/90. Anaskafes Eleuthernas 1988. *Kritiki Estia*, 3, 271-4.

Kalpaxis, Th., 2000/1. Anaskafes Eleuthernas 1995-7. *Kritiki Estia*, 8, 306-9.

Karlsson H., ed. 2001. *Its about Time: the concept of time in archaeology*. Göteborg: Bricoleur Press.

Kaspiris, F.P., 1973. *Symvoli eis tin Meletin ton Osteichthyon tou Korinthiakou kai Patraikou Kolpou kai tou Ioniou Pelagous*. PhD Thessis, University of Patras, Patra.

Katsadorakis G. and Paragamian, K., 2006. *Oi Ygrotopoi tou Aigaiou*. Athens: WWF Hellas.

Keene, S.A., 1983. Biology, behaviour and borrowing: a critical examination of optimal foraging theory in archaeology. In J.A. Moor and A.S. Keene, eds. *Archaeological Hammers and Theories*. New York, London: Academic Press, 137-155.

Keller, O., 1909. *Die antike Tierwelt*. Leipzig, W. Engelmann.

Kent, J.H., 1966. *The Inscriptions 1926-1950, Corinth 8(3)*. Princeton, New York: American School of Classical Studies in Athens.

Kent, J.H., 1948. The Temple estates of Delos, Rheneia and Myconos. *Hesperia*, 17 (4), 243-338.

Keramopoulos, A., 1919. Anaskafi en Aixoni Attikis (Pyrnari kai Alyki Glyfadas). *Praktika tis En Athinais Archaeologikis Etairias*, 32-43.

Keramopoulos, D.A., 1918. Mykenaika: Ichthyophagia. *Archaiologiko Deltio,* 88-94.

Kerschner, M., 2004. Phokäische Thalassokratie oder Phantom-Phokäer? Die frühgriechischen Keramikfunde im Süden Der Iberischen Halbinsel aus der ägäischen Perspektive. In K. Lomas, ed. *Greek Identity in the Western Mediterranean. Papers in Honour of Brian Shefton.* Leiden: 115-148.

Keydell, R., 1939. Oppian. In *Paulis* Real *Encyclopädie der Classischen, Altertumwissenshaft*, band IA, Stuttgart: J.B. Metzlersche Verlagsbuchhandlung, 703.

Kirk, G.S., Raven, J.E. and Schoefield, M., 1983. *The Presocratic Philosophers: a critical history with a selection of texts*. Cambridge, New York: Cambridge University Press.

Knapp, A.B., ed. 1992. *Archaeology, Annales and Ethnohistoy*. Cambridge, New York: Cambridge University Press.

Knapp, A.B. and Ashmore, W., 2000. Archaeological landscapes: constructed, conceptualized, ideational. In W. Ashmore and A. B. Knapp, eds. *Archaeologies of Landscape. Contemporary perspectives.* Malden Mass, Oxford: Blackwell, 1-30.

Koinski A. and Olsen, B., 2001. The Burkert interpretation of animal sacrifice. Available from: http://inside.bard.edu/academic/specialproj/ritual/Rituals/Burkert/00.html (accessed 1 September 2006).

Konsolaki–Giannopoulou, E., 1990. *Glyfada. Istoriko parelthon kai mnimeia*. Athens.

Konstan, D. 1995. *Greek Comedy and Ideology.* N. York and Oxford: Oxford University press.

Kopytoff, I., 1986. The cultural biography of things: commoditization as process. In A. Appaduarai, ed. *The social Life of Things. Commodities in cultural perspective*. Cambridge: Cambridge University Press, 64-91.

Koromila, M., 2001. *Oi Ellines sti Mavri Thalassa apo tin Epoxi tou Chalkou os tis Arches tou 20ou Aiona.* Athina: Politistiki Etaireia « Panorama ».

Kotsakis, K., 1991. The powerful past: theoretical trends in Greek archaeology. In Hodder, I., ed. *Archaeological Theory in Europe. The last three decades*. London, New York: Routledge, 65-90.

Koukoules, F., 1948. Ek tou alieutikou viou ton Byzantinon. *Epetiris Etairias Byzantinon Spoudon*, 18, 28-41.

Kourouniotis, K., 1927-8. To Ieron tou Apollonos tou Zostiros. *Archaeologiko Deltion*, 9-52.

Koutrakis, M., 1999. Anaskopisi ton Istorikon anaforon kai tis systimatikis ton Kefalon (Mugilidae) tis Messogeiou. *Geotechnika Epistimonika Themata,* 10.4i.4, 365-374.

Koutrakis, T.E., Tsikliras, C.A., and Sinis, A.I., 2005. Temporal variability of the ichthyofauna in a Northern Aegean lagoon (Greece). Influence of environmental factors. *Hydrobiologia,* 543, 245-257.

Kruglov, M., 1994. The Black Sea Ichthyofauna. Composition and contemporary condition. *Bulletin of Marine Science*, 55 (2-3), 1342.

Kus, S., 1992. Towards an archaeology of body and soul. In J.-C. Gardin, and C. Peebles, eds. *Representations in Archaeology*. Bloomington and Indianaposis: Indiana University Press, 168-177.

Kuznetsov, V.D., 2001. Archaeological ivestigations in the Tamman Peninsula. In G.R. Tsetskhladze, ed.

North Pontic Archaeology. Recent discoveries and studies. Leiden, Boston, Köln: Brill, 319-344.

Kuznetsov, V.D., 2003. KEPOI - Phanagoria –Taganrog. In D.V. Grammenos and E.K.Petropoulos, eds. *Ancient Greek Colonies in the Black Sea.* Publications of the Archaeological Institute of Northern Greece no 4, Thessalonike: Archaeological Institute of Northern Greece, 896-955.

Kyriakidou – Nestoros, A., 1978. *H Theoria tis Ellinikis Laografias*. Athens: Etaireia Spoudon Neoellinikou Politismou kai Genikis Paideias.

Lacroix, L., 1937a. *La Faune Marine dans la Décoration des Plats a Poissons: étude sur la céramique Grecque d' Italie méridionale*. Verviers: Lacroix.

Lacroix, L., 1937b. Noms de Poissons et noms d'oiseaux en Grec Ancien. *Antiquité Classique*, 6, 265-302.

Lampadaridis, S.N., 1973. H psariki sta Marmaronisia. *Elliniki Laiki Techni,* 11, 110-120.

Larson, J., 2001. *Greek Nymphs. Myth, cult, lore.* Oxford: Oxford University Press.

Laubscher, H.P., 1982. *Fischer und Landleute: Studien zur Hellenistischen Genreplastik.* Mainz am Rhein: Philipp von Zabern.

Lauffer, S., 1976. Inschriften aus Boiotien. *Chiron* 6, 11-51.

Lazarides, D., 1965. Archaeotites kai Mnimeia Anatolikis Makedonias. *Archaeologiko Deltio*, 20 (chronica), 443-70.

Laurence, R., 1994. *Roman Pompeii. Space and Society.* London and New York: Routledge.

Lazarides, D., 1971. *Avdera and Dikaia.* Athens: Athinaikos Technologikos Omilos.

Leach, E., 1967. Twins, birds and vegetables. *Man (N.S.),* 2 (1), 129-130.

Leach, E., 1970. *Levi-Strauss*. London: Fontana.

Lee, H.D.P., 1948. Place-names and the date of Aristotle's biological works. *Classical Quarterly*, 42 (3/4), 61-67.

Lee, H.D.P., 1985. The fishes of Lesbos again. In A. Gottlef, ed. *Aristotle on Nature and Living Things. Philosophical and historical studies presented to David H. Balme on his seventieth birthday*. Bristol, Pittsburg: Mathesis Publications and Pensylvania and Bristol Classical Press, 3-8.

Leguilloux, M., 2003. The Delian chora in Classical and Hellenistic times: an island landscape planned for pastoralism. In E. Kotzabopoulou, Y. Hamilakis, P. Halstead, C. Gamble, and P. Elefanti, eds. *Zooarchaeology in Greece: recent advances.* BSA Studies 9, London: British School at Athens, 250-256.

Leonard A. Jr., 1972. Kouphonisi revisited. *Archaeologia Classica,* 24, 353-363.

Leone, M. and P. Potter, 1988. Introduction: issues in historical archaeology. In M. Leone and P. Potter, eds. *The Recovery of Meaning*. Washington DC: Smithsonian Institution Press, 1-22.

Levi, P. 1971. *Pausanias. A Guide to Greece.* Baltimore: Harmondsworth.

Levi-Strauss, C., 1963. *Totemism*. Boston: Beacon Press.

Levi-Strauss, C., 1969. *The Raw and the Cooked.* New York: Harper and Row.

Levi-Strauss, C., 1997. The culinary triangle. In C. Counihan and P. van Esterik, eds. *Food and Culture. A reader*. New York, London: Routledge, 28-35.

Lewis, S., 2003. Representation and perception. Athenian pottery in its Italian context. In J. Wilkins and E. Herring, eds. *Inhabiting Symbols: symbol and image in the ancient Mediterranean.* London: Accordia Research Institute, University of London, 175-192.

Lightfoot, J.L., 2003. *Lucian on "The Syrian Goddess".* Oxford: Oxford University Press.

Lindenlauf, A., 2003. The sea as a place of no return in ancient Greece. *World Archaeology*, 35(3), 416-433.

Linder, E., 1996. Human apprehension of the sea. In E.E. Rice, ed. *The sea and history*. Phoenix Mill, Thrupp, Stroud, Gloucestershire: Sutton Publishing, 15-22.

Linder, E., and Laenhardt, O., 1964. Recherches d'archéologie sous marine sur le côte Méditerranéenne d' Israël. *Revue Archéologique (N.S.)*, 7 (1), 47-51.

Linders, T., 1994. The Sacred Menus on Delos. In R. Hägg, ed. *Ancient Greek Cult Practice from the Epigraphical Evidence: proceedings of the second international seminar on ancient Greek cult, organized by Swedish Institute at Athens, 22-24 November 1991.* Stockholm: Swedish Institute at Athens, 71-79.

Lloyd, J., 1986. Why should historians take archaeology seriously? In J. Bintliff and C. Gaffney, eds. *Archaeology at the Interface: studies in Archaeology's relationships with history, geography, biology and physical science.* Oxford: British Archaeological Reports, 40-51.

Lockwood, W.B., 1955. Word taboo in the language of the Faroese fishermen. *Transactions of the Philological Society*, 53, 1-24.

Longo, O., 1989. *Le Forme della Predazione. Cacciatori e pescatori della Grecia antica.* Napoli: Liguori.

Loukopoulou, L., 2004. Thrace from Nestos to Hebros. In M.H. Hansen and T.H. Nielsen, eds. *An Inventory of Archaic and Classical Poleis.* Oxford: Oxford University Press, 870-884.

Lund, J. and Gabrielsen, V., 2005. A fishy business. Transport amphorae of the Black Sea Region as a source for the trade in fish and fish products in the Classical and Hellenistic periods. In T. Bekker-Nielsen, ed. *Ancient Fishing and Fish Processing in the Black Sea Region.* Black Sea Studies 2, Aarhus: Aarhus University Press, 161-170.

Lupton, D., 1996. *Food, the Body and the Self.* London, Thousand Oaks, New Delhi: Sage Publications.

Lupu, E., 2005. *Greek Sacred Law. A collection of new documents (NGSL).* Leiden, Boston: Brill.

Lykousis, V. and Chronis, G., 1989. Mechanisms of sediment transport and deposition: sediment sequences and accumulation during the Holocene on the Thermaikos Plateau, the continental slope and

Basin (Sporades basin), NW Aegean Sea, Greece. *Marine Geology*, 87, 15-26.

Lykousis, V., Chronis, G., Tselepides, A., Price, N.B., Theocharis, A., Siokou-Frangou, I., Van Wambeke, F., Danovaro, R., Stavrakakis, S., Duineveld, G., Georgopoulos, D., Ignatiades, L., Souvremezoglou, A. and Voutsinou-Taliadouri, E., 2002. Major outputs of the recent multidisciplinary biochemical researches undertaken in the Aegean Sea. *Journal of Marine Systems*, 33-34, 313-334.

Mair, A.W., 1963. Introduction. In A.W. Mair, *Oppian, Colluthus, Thryphiodorus*. Loeb Classical Collection. Cambridge Mass., London: Harvard University Press and William Heinemann.

Maniatis, Y., Jones, R.E., Whitbread, I. K., Kostikas, A., Simopoulos, A., Karakalos, Ch., and Williams II, C. K., 1984. Punic amphoras found in Corinth, Greece: an investigation of their origin and technology. *Journal of Field Archaeology*, 11, 207-222.

Marangou, Ch., 2002. Rocks and itineraries: sea and land perspectives on an Aegean island. In H.W. Waldren and J.A. Ensenyat, eds. *World Islands in Prehistory. International insular investigations.* BAR International Series 1095, Oxford: BAR Publishing, 7-18.

Marchiori, A., 2000. Between *Ichthyophagists* and Syrians. In D. Braund and J. Wilkins, eds. *Athenaeus and his World: reading Greek culture in the Roman Empire*. Exeter: University of Exeter Press, 327-338.

Margaritis, E., 2006. *Archaeobotanical Investigations During the Hellenistic Period in Macedonia, Greece.* PhD dissertation, University of Cambridge.

Marinatos, N., 1993. What were Greek sanctuaries? A synthesis. In N. Marinatos and R. Hägg, eds. *Greek Sanctuaries: new approaches.* London: Routledge, 228-233.

Marshall, E., 2000. Synecius, Homer and the sea. The sea and the representation of Lybians in Synecius's Letters. In G.J.R. Oliver, R. Brock, T.J. Cornell and S. Hodkinson, eds. *The Sea in Antiquity.* BAR international Series 899, Oxford: BAR Publishing, 13-18.

Martin, L., 1987. *Hellenistic Religions: an introduction.* New York: Oxford University Press.

Mau, A., 1982. *Pompei, Its life and art.* Translated by F.W. Kelsey, first edition 1902, New Rochelle, N.York: Caratzas Brothers.

McDonald, A.W., 1972. The problems and the program. In W.A. McDonald and G.R. Rapp, eds. 1972. *The Minnesota Messenia Expedition. Reconstructing a Bronze Age regional environment.* Minneapolis: The University of Minnesota Press, 3-17.

McDonald, W.A. and Rapp, G.R., eds. 1972. *The Minnesota Messenia Expedition. Reconstructing a Bronze Age regional environment.* Minneapolis: The University of Minnesota Press.

McDowell, D.M., 1971. *Aristophane's Wasps. Text and commentary.* Oxford: Clarendon Press.

McNiven, J.I., 2003. Saltwater People: spiritscapes, maritime rituals and the archaeology of Australian indigenous seascapes. *World Archaeology*, 35 (3), 329-339.

McPhee, I. and Trendall, A.D., 1987. *Greek Red-Figured Fish Plates.* Basel: Vereinigung der Freunde Antiker Kunst and Archäologisches Seminar der Universität.

Megas, G.A., 2004. *Ellinikes Giortes kai Ethima tis Laikis Latreias.* 1st edition 1956, Athens: Estia.

Meillassoux, C., 1974. *Anthropologie Economique des Gouro de Cote d'Ivoire: de l'economie de subsistance a l'agriculture commerciale.* Paris: Mouton.

Mennell, S., 1985. *All Manners of Food: eating and taste in England and France from the Middle Ages to the present.* Oxford: Blackwell.

Mennell, S., 1997. On a civilizing of appetite. In C. Counihan and P. van Esterik, eds. *Food and Culture. A reader.* New York, London: Routledge, 315-337.

Mentzou, K., 1975. *Symvolae eis tin Meletin tou Oikonnomikou kai Koinonikou Viou tis Proimou Byzantinis Periodou.* PhD Dissertation, University of Athens.

Messer, E., 1984. Anthropological perspectives on diet. *Annual Review of Anthropology*, 13, 205-249.

Mette, H.J., 1977. *Urkunden Dramatischen Aufführungen in Griechenland.* Berlin, New York: W. de Gruyter.

Michel, H., 1957. *Economics of Ancient Greece.* 2nd edition, Cambridge: W. Heffer.

Miller, G.S., 1976. Excavations at Nemea, 1975. *Hesperia*, 45, 178-205.

Miller, G.S., 1980. Excavations at Nemea, 1979. *Hesperia*, 49, 174-202.

Millett, M., 1993. Samian from the sea: Cala Culip shipwreck IV, *Journal of Roman Archaeology*, 6, 415-419.

Millett, P., 1990. Sale, credit and exchange in Athenian low and Society. In P. Cartledge, P. Millett and S. Todd, eds. *Nomos: essays in Athenian low, politics and society.* Cambridge, New York: Cambridge University Press, 167-194.

Mintz, W.S., 1985. *Sweetness and Power: the place of sugar in modern history.* New York: E. Sifton-Peguin.

Mintz, W.S., 1993. The changing roles of food in the study of consumption. In J. Brewer and R. Porter, eds. *Consumption and the World of Goods.* London and New York: Routledge, 261-273.

Moatsos, P.G., 1932. *Porfyra me Deka ex Eikones.* Alexandria.

Montserrat, D. ed. 1998. *Changing Bodies, Changing Meanings: Studies on the human body in antiquity.* London: Routledge.

Morales, A. and Roselund, K., 1979. *Fish Bone Measurements: an attempt to standardize the measuring of fish bones from archaeological sites.* Copenhagen: Steenstrupia.

Moreland, J., 2001. *Archaeology and Text.* London: Duckworth.

Morin, P.J., 1960. *The Cult of Dea Syria in the Greek World.* PhD dissertation, Ohio State University: Ann Arbor University Microfilms International.

Morrill, W.T., 1967. The ethno-ichthyology of Cha-Cha. *Ethnology*, 6, 405–417.

Morris, I., ed. 1994a. *Classical Greece: ancient histories and modern archaeologies*. Cambridge: Cambridge University Press.

Morris, I., 1994b. Archaeologies of Greece. In I. Morris, ed. *Classical Greece: ancient histories and modern archaeologies*. Cambridge: Cambridge University Press, 8-48.

Morris, I., 1994c. The community against the market in Classical Athens. In C. Duncan and D.W. Tandy, eds. *From Political Economy to Anthropology. Situating economic life in past societies*. Montréal, New York: Black Rose Books, 52-79.

Morris, I., 1999a. The social and economic archaeology of Greece. In R.F. Dockers and E.M. Moormann, eds. *Proceedings of the 15th International Congress of Classical Archaeology. Classical archaeology towards the third millennium: reflections and perspectives*. Allard Pierson Series v. 12, Amsterdam: Allard Pierson Museum, 27-33.

Morris, I., 1999b. Forward. In M. Finley, *The Ancient Economy*. Berkeley: University of California Press, i-xxxvi.

Morris, I., 2001a. Hard surfaces. In P. Cartledge, E. Cohen and L. Foxhall, eds. *Money, Labour, and Land in Ancient Greece*. London: Routledge, 8-43.

Morton, J., 2001. *The Role of Physical Environment in Ancient Greek Seafaring*. Mnemosyne 213, Boston, Köln, Leiden: Brill.

Morton Braund, S., 2001. *Latin Literature*. New York: Routledge.

Mossé, C., 1969. *The Ancient World at Work*. Translated by J. Lloyd, London: Chatto and Windus.

Müller, K., 1965. *Geographi Graeci Minores*. Hildsheim: Georg Olms Verlagsbuchhandlung.

Müller, K. 1979. *The Fragments of the Lost Historians of Alexander the Great*. Chicago: Ares Publishers.

Murcott, A., 1988. Sociological and social anthropological approaches to food and eating. *World Review of Nutrition and Diet*, 55, 1-40.

Murray, T., ed. 1994. *Time and Archaeology*. Routledge One World Archaeology 37, London, New York: Routledge.

Mylona, D., 1997. Etudes zooarchaeologique, in E. Greco, Th. Kalpaxis, A. Schnapp and D. Vivieers, Travaux menes en collaboration avec l' Ecole française en 1996. *Bulletin de Correspondance Hellénique*, 121 (2), 809-824.

Mylona, D., 2000. Representations of fish and fishermen on the Thera wall paintings in light of the fish bone evidence. In S. Sherratt, ed. *International Symposium "The wall Paintings of Thera", Thera 30th of August - 4th of September 1997*. London: Thera Foundation, 561-7.

Mylona, D., 2003a. Archaeological fish remains in Greece: general trends of the research and a gazetteer of sites. In E. Kotzabopoulou, Y. Hamilakis, P. Halstead, C. Gamble, and P. Elefanti, eds. *Zooarchaeology in Greece: recent advances*. BSA Studies 9, London: British School at Athens, 193-200.

Mylona, D., 2003b. Fishing in Late Antiquity. In E. Kotzabopoulou, Y. Hamilakis, P. Halstead, C. Gamble, and P. Elefanti, eds. *Zooarchaeology in Greece: recent advances*. BSA Studies 9, London: British School at Athens, 103-110.

Mylona, D., 2003c. The exploitation of fish resources in the Mesolithic Sporades: fish remains from the Cave of Cyclope, Youra. In N. Galanidou and C. Perles eds. *The Greek Mesolithic: problems and perspectives*. London: The British School at Athens, 181-188.

Mylona, D., 2004. Fish remains. In J. Soles and C. Davaras, eds. *Mochlos IC, Period III, Neopalatial Remains on the Coast: The Artisans' Quarters and the Chalinomouri Farmhouse*. Philadelphia: INSTAP Academic Press, 121-5.

Mylona, D., 2005. The animal bones from the Pyrgouthi in Berbati Valley in the Argolid. In J. Hjohlman, A. Penttinen and B. Wells, eds. 2005. *Pyrgouthi. A rural site in the Berbati valley from the Early Iron Age to Late Antiquity. Excavations by the Swedish Institute at Athens 1995-1997*. Acta Instituti Atheniensis Regni Sueciae, Series in 4°. LII, Stockholm: Swedish Institute in Athens, 301-308.

Mylona, D., forthcoming. Mammal and fish remains from the Sanctuary of Poseidon on Kalaureia, Poros. In B. Wells and A. Penttinen, eds. *The Physical Environment and Daily life at the Temple on Kalaureia (Poros)*. Swedish Institute at Athens.

Mylonopoulos, J., 2003. *Peloponnisos Oikitirion Poseidonos: Heiligtumer und Kulte des Poseidon auf der Peloponnes*. Liège: Centre International d' Etude de la Religion Grecque Antique.

Nettle, D. and Romaine, S., 2000. *Vanishing Voices. The extinction of the world languages*. Oxford: Oxford University Press.

Nevett, L.C., 1999. *House and Ssociety in the Ancient Greek World*. Cambridge: Cambridge University Press.

Nicholson, A.R., 1995. Out of the frying pan into the fire: what value are burned bones to archaeology? *Archaeofauna*, 4, 47-64.

Nikitsky, A., 1964. ΑΝΕΠΙΒΑΣΙΑ. *Hermes*, 38, 406-413.

Nobis, G., 1993. Zur antiken Wild- und Haustierfauna Kretas - nach Studien an Tierresten aus den archäologischen Grabungen Poros bei Iraklion und Eleutherna bei Arkadi. *Tier und Museum*, 3 (4), 109-120.

Nobis, G., 1994. Die Tierreste aus dem antiken Messene - Grabung 1990-91. In M. Kokabi, ed. *Beitrage zur Archaozoologie und prähistorischen Anthropologie*. Arbeitstreffen der Osteologen, Kostanz 1993. Stuttgart: Andenken an Joachim Boessneck, 297-313.

Nobis, G., 1997. Tieropfer aus einem Heroen- und Demeterheiligtum des antiken Messene (SW-Peloponnes, Griechenland) – Grabungen 1992 bis 1996. *Tier und Museum*, 5/4, 97-111.

Noble, R. and Cowx, I., 2002. *Development of a River-Type Classification System (D1).* Available from: http://www.boku.ac.at/fame/downloads/D1_2_typology_and%20species_classification.pdf (accessed 15 October 2006).

Nutton, V., 2000. Travellers' fair. In J. Wilkins, D. Harvey and M. Dobson, eds. *Food in Antiquity.* Exeter: University of Exeter Press, 359-370.

O'Connell T.C., Levine M.A., and Hedges R.E.M., 2003. The importance of fish in the diet of central Eurasian peoples from the Mesolithic to the Early Iron Age. In: M.A. Levine, A.C. Renfrew, and K. Boyle, eds. *Prehistoric Steppe Adaptation and the Horse,* McDonald Institute Monograph. Cambridge: McDonald Institute for Archaeological Research, 253-268

O'Connor, J.J. and Robertson, E.F., 2003. *D' Arcy Thompson.* Available from: http://www-history.mcs.st-andrews.ac.uk/Mathematicians/Thompson_D'Arcy.html. (accessed 15 February 2005).

Ogilvie, A.G., 1943. Population density in Greece. *The Geographical Journal,* 5/6, 251-260.

Orlove, B.S. and Rutz, H.J., 1989. Thinking about consumption: a social economy approach. In J.H. Rutz and B.S. Orlove, eds. *The Social Economy of Consumption.* Lanham: University Press of the America, 1-57.

Osborne, C., 1990. Boundaries in nature: Eating with animals in the 5th c. B.C. *Bulletin of the Institute of Classical Studies,* 37, 15-29.

Osborne, C., 1995. Ancient vegetarianism. In J. Wilkins, D. Harvey and M. Dobson, eds. *Food in Antiquity.* Exeter: University of Exeter Press, 214-224.

Osborne, R., 1987. *Classical Landscape with Fgures: The ancient Greek city and its countryside.* Dobbs Ferry, N.Y. : Sheridan House.

Palombi, A. and Santorelli, M., 1960. *Gli Animali Commestibili dei Mari d' Italia.* 2nd edition, Milan: Hoepli.

Pálsson, G., 1993. Household words: attention, agency and the ethnography of fishing. In G. Pálsson, ed. *Beyond Boundaries. Understanding, translation and anthropological discourse.* Oxford, Providence: Berg.

Pálsson, G., 1994a. Enskilment at sea. *Man (N.S.),* 29, 901-927.

Pálsson, G., 1994b. The idea of fish: land and sea in the Icelandic world view. In R. Willis, ed. *Signifying Animals. Human meaning in the natural world.* London: Routledge, 119-133.

Papadakis, N., 1983. Kouphonisi Siteias. *Archaeologiko Deltio* (chronika), B-2, 379-381.

Papadakis, N., 1989. *Siteia. I patrida tou Mysona kai tou Kornarou. Odigos gia tin istoria, archaeologia kai ton politismo tis.* Siteia: Dimos Siteias.

Papadiamantis, A., 1892. *Sto Christo sto Kastro.* (Available at www.phys.uoa.gr/~nektar/arts/tributes/alexandros_papadiamantis/sto_xristo_sto_kastro.htm - 97k (accessed 12 November 2006)

Papadopoulos, J. and Paspalas, S.A., 1999. Mendean as Chalkidian wine. *Hesperia,* 68 (1), 161-188.

Papadopoulos, J.K. and Ruscillo, D., 2002. A *ketos* in early Athens: an archaeology of whales and sea monsters in the Greek world. *American Journal of* Archaeology, 106 (2), 187-227.

Papakonstantinou, C., 1988. *Check-List of Marine Species of Greece.* Fauna Greciae IV, Athens: National Centre for Marine Research – Hellenic Zoological Society.

Papathanasopoulos, F.G.A. 1970. Makrisia. *Archaiologiko Deltio* Chron. 2, 191-3.

Papendic, A., 1926. *Die Fischnamen in griechisch-latinischen Glossaren.* Königsberg.

Paraskeuopoulou, A., date unknown. H psariki stin Sozopoli. *Arxeion tou Thrakikou Laografikou kai Glossikou Tysaurou,* 2, 65-95.

Paris, J., 1916. Contributions a l'étude des ports antiques du monde Grec. *Bulletin de Correspondance Hellénique,* 40, 5-73.

Parker, R., 1983. *Miasma: pollution and purification in early Greek religion.* Oxford: Clarendon Press.

Paton A.R. and E. L. Hicks, 1990. *The Inscriptions of Cos.* Hildesheim, Zürich, New York: Georg Olms Verlag.

Payne, S., 1973. Animal bones. In Jacobsen W.T., Excavation in the Franchthi cave 1996-1971, part 1. *Hesperia,* 42 (1), 59-66.

Peek, W., 1969. *Inschriften aus dem Asklepieion von Epidauros.* Abhandlungen der Sächsischen Akademie der Wissenschaften zu Leipzig, Philologisch-historische Klasse, 60 (2), Berlin.

Pellegrin, P., 1986. *Aristotle's Classification of Animals: biology and the conceptual unity of the Aristotelian corpus.* Translated by A. Preus, Berkeley: University of California Press.

Pennywise, S., 2004. The old man and the sea: a field study of change in the Greek fishing industry based on interviews with fishermen on the island of Santorini. *International Journal of Entrepreneurship and Small Business,* 1 (1-2), 206-210.

Perlman, P., 1996. Πόλις υπήκοος. The dependent Polis and Crete. In M.H. Hansen, ed. *Introduction to an Inventory of Poleis. Symposium August 23-26, 1995.* Acts of the Copenhagen Polis Centre, vol. 3, Copenhagen: Royal Danish Academy of Science and Letters, 233-285.

Petrakos, V., 1987. *H En Athinais Archaeologiki Etereia. I istoria ton 150 chronon tis, 1837-1987.* Athens: H En Athinais Arxaiologiki Etairia.

Pfeiffer, R., 1976. *History of Classcial Scholarship from 1300 to 1850.* Oxford: Clarendon Press.

Pfister, F., 1951. *Die Reisebilder des Herakleides.* Wien: R.M. Rohrer.

Pickard-Cambridge, A., 1968. *The Dramatic Festivals of Athens.* Oxford: Clarendon Press.

Pirazzoli, P.A., 1987. Submerged remains of Ancient Megisti in Castelorizo island (Greece): preliminary

survey. *International Journal of Nautical Archaeology*, 16 (1), 57-66.
Places de, E, ed. 1982. *Porphyrie. Vie de Pythagore, Lettre à Mercella*, Paris : Les Belles Lettre.
Plassart, A., 1916. Fouilles de Délos (1912-13). *Bulletin de Correspondance Hellénique*, 40, 145-256.
Poggie, J.J. Jr. and Gersuny, C., 1972. Risk and ritual. An interpretation of fishermen's folklore in a New England community. *Journal of American folklore*, 85 (335), 66-72.
Polanyi, K., 1944. *The Great Transformation.* Boston: Beakon Press.
Pollitt, J.J., 1994. *I Techni stin Ellinistiki Epochi.* Ttranslated by A. Gazis, Athens: Papadimas.
Poplin, F., 1991. La Fauna, in Amandry, R., Temple d' Apollon: les fondations de la cella. *Bulletin de Correspondance Hellénique*, 115, 686-700.
Porter, J. 2001. Ideals and Ruins. Pausanias, Longinus and the Second Sophistic. In S. Alcock, J.F. Cherry and J. Elsner, eds. *Pausanias. Travel and memory in Roman Greece.* Oxford: Oxford University Press, 63-92.
Poulaki–Pantermali, E., 2001. 1997-2001: Erga ethnika kai alla stin periochi tou Makedonikou Olympou. *To Archaeologiko Ergo stin Makedonia kai Thraki,* 15, 331-345.
Powell, J., 1996. *Fishing in the Prehistoric Aegean.* Jonsered: Paul Åström.
Powell, J., 2003. The fish bone assemblage from the cave of Cyclope, Youra: evidence for continuity and change. In N. Galanidou and C. Perles, eds. *The Greek Mesolithic: problems and perspectives.* London: The British School at Athens, 173-180.
Powell, O., 2003. *Galen and the Properties of Foodstuffs.* Cambridge: Cambridge University Press, i-xxvi.
Price, S. and Kearns, E., eds. 2003. *The Oxford Dictionary of Classical Myth and Religion.* Oxford: Oxford University Press.
Prummel, W., 2003a. Animal remains form the Hellenistic town of New Halos in the Almyros plain, Thessaly. In E. Kotzabopoulou, Y. Hamilakis, P. Halstead, C. Gamble, and P. Elefanti, eds. *Zooarchaeology in Greece: recent advances.* BSA Studies 9, London: British School at Athens, 153-160.
Prummel, W., 2003b. Animal husbandry and mollusc gathering. In H.R. Reinders and W. Prummel, eds. *Housing in New Halos: a Hellenistic town in Thessaly, Greece.* Exton Pa.: Swets and Leitlinger Publishers, 172-223.
Psaroudakis, K., 2004. "Ta Iera Kopadia tou Iliou". Oikonomiko-koinonikes parametroi ton ieron zoon stin arxaiotita. *Arxaiologiko Deltio* (meletes), 55, 23-36.
Purcell, N., 1995. Eating fish. The paradoxes of seafood. In J. Wilkins, D. Harvey and M. Dobson, eds. *Food in Antiquity.* Exeter: University of Exeter Press, 132-149.
Pyke, M. 1964. *Food Science and Technology.* London: Murray.
Rabinovitch, M., 1947. *Der Delphin in Sage und Mythos der Griechen.* Dornach, Germany/Basel, Switzerland: Hybernia-Verlag.
Radcliffe, W., 1926. *Fishing from the Earliest Times.* New York: Burt Franklin.
Raubitchek, I.K., 1998. *The Metal Objects, 1952-1989. Isthmia v. 7.* Princeton, N.J.: Princeton University Press.
Rauth, K.N., 1993. *The Sacred Bonds of Commerce: religion, economy and trade society at Hellenistic and Roman Delos, 166-87 B.C.* Amsterdam: J.C. Gieben.
Reese, D., 1984. Shark and ray remains in Aegean and Cypriot Archaeology, *Opuscula Atheniensia*, 15, 188-192.
Reese, D., 1989. Faunal remains from the Altar of Aphrodite Ourania, Athens. *Hesperia*, 58 (1), 63-70.
Reese, D., 2000a. The bird remains. In J.W. Shaw and M.C. Shaw, eds. *Kommos IV. The Greek sanctuary.* Princeton: Princeton University Press, 560-570.
Reese, D., 2000b. The marine invertebrate. In J.W. Shaw and M.C. Shaw, eds. *Kommos IV. The Greek sanctuary.* Princeton: Princeton University Press, 571-642.
Reese, D., and Ruscillo, D., 2000. The mammal remains. In J.W. Shaw and M.C. Shaw, eds. *Kommos IV. The Greek sanctuary.* Princeton: Princeton University Press, 415-646.
Reese, D., Rielley, K. and Rose, M.J. 1987. A bone assemblage at Corinth of the second century after Christ, *Hesperia*, 56 (3), 255-274.
Reinders, H.R., 1988. *New Halos: A Hellenistic town in Thessaly, Greece.* Utrecht: HES Publishers.
Reinders, H.R. and Prummel, W., eds. 2003. *Housing in New Halos: a Hellenistic town in Thessaly, Greece.* Exton Pa.: Swets and Leitlinger Publishers.
Reinhold, M., 1970. *History of purple as a status symbol in Antiquity.* Latomus 116, Brussels.
Renfrew, C., 1980. The Great Tradition versus the Great Divide: Archaeology as Anthropology. *American Journal of Archaeology*, 84, 287-298.
Renfrew, J.M., Whitehead, P., and Greenwood, P.H., 1968. Appendix VIII: the fish bones. In J.D. Evans and C. Renfrew, eds. *Excavations at Saliagos near Antiparos.* BSA supplementary Volume 5, London: Thames and Hudson, 118-121.
Rhode, P., 1892. *Thynnorum Captura. Quanti Fuerit apud Veteres Momenti.* Leipzig: Druck und Verlag von Teubner.
Richardson, B.E., 1969. *Old Age Among the Ancient Greeks: the Greek portrayal of old age in literature, art and inscriptions, with a study of the duration of life among the ancient Greeks on the basis of inscriptional evidence.* New York: Greenwood Press.
Rizakis, A.D. 1997. Roman colonies in the province of Achaia: territories, land and population. In S.A. Alcock, ed. *The Early Roman Empire in the East.* Oxbow Monograph 95, Oxford: Oxbow Books.
Robertson, N., 2005. Sacrifice to the sea: a custom prior to the "Olympian" and "Chthonian" categories? In R. Hägg and B. Alroth, eds. *Greek Sacrificial Ritual,*

Olympian and Chthonian: Proceedings of the sixth international seminar on ancient Greek cult, organized by the department of Classical archaeology and ancient history. Göteborg University, 25-27 April 1997. Stockholm: Swedish Institute in Athens, 85-98.

Robinson, M.D., 1936. The residential district and the cemeteries at Olynthos. *American Journal of Archaeology*, 36 (2), 118-138.

Robinson, D.M., 1941. *Excavations at Olynthos. Part X. Metal and minor miscellaneous finds*. Baltimore, Maryland: Johns Hopkins University Press.

Romero Recio, M. 2000. *Cultos Maritimos y Religiosidad de Navegantes en el Mundo Griego Antiguo*. BAR Int. Series 897. Oxford.

Rose, M., 1994. *With Line and Glittering Bronze Hook: fishing in the Aegean Bronze age*. PhD dissertation, Program in Classical Archaeology, Indiana University.

Rose, M., 1995a. Fish remains: Building AD Centre. In P. Betanourt and C. Davaras, eds. *Pseira I. The Minoan buildings on the west side of area A. P. P. B.* Philadelphia: The University Museum, 130-140.

Rose, M., 1995b. The fish remains. In J.W. Shaw and M.C. Shaw, eds. *Kommos. An excavation on the south coast of Crete. The Kommos region and houses of the Minoan Town. The Kommos region, ecology and Minoan industries*. Princeton: New Jersey, Princeton University Press, 204-239.

Rose, M., 1995c. Fishing in a changing environment: Franchthi cave, Greece, 11.000-5.000 BP. *Old World Archaeology Newsletter*, 18 (3), 21-26.

Rose, M., 1998. The fish bones. In P. Betancourt and C. Davaras, eds. *Pseira I. The Minoan Buildings on the west side A*. Philadelphia, Pa.: University of Pennsylvania Museum, 145-148.

Rose, M., 2000. The fish remains. In J.W. Shaw and M.C. Shaw, eds. *Kommos IV. The Greek sanctuary*. Princeton: Princeton University Press, 495-560.

Rostovtseff, M.I., 1926. *The Social and Economic History of the Roman Empire*. Oxford, Clarendon.

Rostovtzeff, M.I., 1941. *The Social and Economic History of the Hellenistic World*. Oxford: Oxford University Press.

Rotroff, S.I and Oakley, J.H., 1992. *Debris from a Public Dining Place in the Athenian Agora*. Hesperia Supplements 25, Athens: The American School of Classical Studies at Athens.

Rouse, W.H.D., 1976. *Greek Votive Offerings*. 1st ed. 1902, Hildesheim, New York: Georg Olms Verlag.

Ruscillo, D., 1993. Faunal remains from the Akropolis site, Mytilene. *Echoes du Monde Classique*, 37, 201-210.

Ruscillo, D., 2001. When gluttony ruled! Bones recovered form a Roman villa attest an age of culinary hedonism. *Archaeology*, 54 (6), 20-25.

Said, S. and M. Trédé, 1999. *A Short History of Greek History*. London and New York: Routledge.

Sahlins, M., 1976. *Culture and Practical Reason*. Translated by C. Selous and others, Chicago: Chicago University Press.

Sakellarakis, J.A., 1974. Le thème du pécheur dans l'art préhistorique de l' Egeé. *Archaeologika Analekta Athenon*, 7, 370-390.

Sallares, R., 1991. *The Ecology of the Ancient Greek World*. London: Duckworth.

Salmon, H.M., 1989. Efficient explanations and efficient behaviour. In V. Pinsky and A. Wylie, eds. *Critical Traditions in Contemporary Archaeology*. Cambridge: Cambridge University Press, 10-13.

Salviat, Fr. and Vatin, C., 1971. *Inscriptions de Grèce Centrale*. Paris: Editions de Boccard.

Sarpaki, A., 1999. The study of palaeodiet in the Aegean: food for thought. In S.J. Vaughan and W.D.E. Coulson, eds. *Paleodiet in the Aegean*. Oxford: Oxbow, 115-121.

Sarpaki, A., forthcoming. The Kalaureia Sanctuary site: archaeobotanical whispers. In B. Wells and A. Penttinen, eds. *The Physical Environment and Daily life at the Temple on Kalaureia (Poros)*. Swedish Institute at Athens.

Sauer, W.E., 2004. The disunited subject. Human history's split into "history" and "archaeology". In W.E. Sauer, ed. *Archaeology and Ancient Hstory: breaking down the boundaries*. London and New York: Routledge, 17-45.

Schama, S., 1995. *Landscape and Memory*. New York: Knopf.

Schaps, D.M., 1985-8. Comic inflation in the market place. *Scripta Classica Israelica*, 8-9, 66-73.

Schaps, D.M., 1987. Small change in Boeotia. In W. Eck, H. Engelmann, D. Hagedorn, R. Kassel, L. Koenen, W.D. Lebek. and R. Merkelach, eds. *Zeitschrift für Papyrologie und Epigraphik*. Bonn: Rudolf Habelt GMBH, 293-296.

Scheidel, W., and von Reden, S., eds. 2002. *The Ancient Economy*. Edinburgh Readings in the Ancient World. Edinburgh: Edinburgh University Press.

Schlager H., Blackman D.J. and Schafer J., 1968. Der Hafen von Anthedon. *Archäologischer Anzeiger*, 1, 21-98.

Schmidt, M., 1965. *Hesychius Alexandrinus Lexicon*. Amsterdam: Adolf M. Hakkert Publishers.

Schneider, K., 1892. *Der Fischer in der Antiken Literatur*. Aachen.

Schwab, A.K., 2000. Bronze, lead and bone implements. In In J.W. Shaw and M.C. Shaw, eds. *Kommos IV. The Greek Sanctuary*. Princeton: Princeton niversity Press, 391-395.

Schwandener, E.-L., 1999. Kassope: eine spätklassische Streifenstadt in Nordwestgriechenland. In Hoepfner, W., ed. *Geschichte des Wohnens*. Band 1, Stuttgart: Deutsche Verlags-Anstalt, 368-383.

Scott, C.J., 1976. *The Moral Economy of the Peasant: rebellion and subsistence in Southeast Asia*. New Heaven, Conn.: Yale University Press.

Scott, J.A., 1916/17. Homeric heroes and fish. *Classical Journal*, 12, 328.

Scranton, R., 1978. The harbor-side commercial facilities. In R. Scranton, J.W. Shaw and L. Ibrahim,

Kenchreai. Eastern port of Corinth. Topography and architecture. Leiden: E.J.Brill, 39-52.

Scranton, R., Shaw, J.W. and Ibrahim, L., 1978. *Kenchreai. Eastern port of Corinth. Topography and architecture*. Leiden: E.J.Brill.

Seremetakis, N., 1994. *The Senses Still: perceptions and memory as material culture in modernity*. Boulder: Westview Press.

Serjeantson, D., 2000. Good to eat and good to think with: classifying animals from complex sites. In P. Rowley-Conwy, ed. *Animal Bones, Human Societies*. Oxford: Oxbow, 179-189.

Shaffer, B.S. and Sanchez, J.L.J., 1994. Comparison of 1/8 inch and ¼ inch mesh recovery of controlled samples of small-to-medium-sized mammals. *American Antiquity*, 59, 525-530.

Shapiro, H.A., 1989. Poseidon and the tuna. *L' Antiquite Classique*, 58, 32-43.

Shaps, D., 1987. Small change in Boeotia, *Zeitschrift für Papyriologie und Epigraphik*, 69, 293-296.

Shaw, B.D., 1982-83. "Eaters of flesh, drinkers of milk": the ancient Mediterranean ideology of the pastoral nomad. *Ancient Society*, 13/14, 5-31.

Shaw, J.W., 1978. The harborage. In R. Scranton, J.W. Shaw, and L. Ibrahim, *Kenchreai. Eastern port of Corinth. Topography and architecture*. Leiden: E.J.Brill, 13-38.

Shaw, J.W. and Shaw, M.C., eds. 2000. *Kommos IV. The Greek sanctuary*. Princeton: American School of Classical Studies in Athens.

Shay, T. and Shay J.M., 2000. The charcoal and seeds from Iron Age Kommos. In J.W. Shaw and M.C. Shaw, eds. *Kommos IV. The Greek Sanctuary*. Princeton: Princeton University Press, 647-668.

Shear, T.L. Jr., 1975. The Athenian Agora. Excavations 1973-74. *Hesperia*, 44 (4), 331- 374.

Shear, T.L. Jr., 1984. The Athenian Agora Excavations of 1980-1982. *Hesperia*, 53 (1), 1-57.

Shepard, K., 1940. *The Fish-Tailed Monster in Greek and Etruscan Art*. PhD dissertation, New York: Bryn Mawr College.

Sherratt, A. and Sherratt, S., 1991. From luxuries to commodities: the nature of Mediterranean Bronze Age Trading Systems. In N.H. Galle, ed. *Bronze Age Trade in the Mediterranean. Papers presented at the conference held at Rewley House, Oxford, in December 1989*. Jonsered: Paul Astroms Forlag, 351-386.

Sherwin-White, M.S., 1978. *Ancient Cos. An historical study from the Dorian settlement to the Imperial period*. Hypomnemata 51, Göttingen: Vandenhoeck and Ruprecht.

Shewan, A., 1927. Fishing with rod in Homer. *Classical Philology*, 22, 170-183.

Simoons, F., 1994. *Eat not this Flesh: food avoidances from prehistory to the present*. Madison: University of Winsconsin Press.

Skoulikidis, T.N., Bertahas, I., and Koussouris, T., 1998. The environmental state of freshwater resources in Greece (rivers and lakes). *Environmental Geology*, 36 (1-2), 1-17.

Slastenenko, E.P., 1955. Review of the Black Sea fish fauna and general marine life conditions. *Copeia*, 3, 230-235.

Small, B.D., ed. 1995a. *Methods in the Mediterranean: historians' and archaeologists' views on the use of texts and archaeology*. Leiden.: E.J. Brill.

Small, B.D., 1995b. Monument, laws and analysis: Combining archaeology and text in ancient Athens. In. B. D. Small, ed. *Methods in the Mediterranean: historians' and archaeologists' views on the use of texts and archaeology*. Leiden: E.J. Brill, 143- 174

Small, B.D., 1999. The tyranny of the text: lost social strategies in current historical period archaeology in the classical Mediterranean. In A.P.P. Funari, M. Hall and S. Jones, eds. *Historical Archaeology: back from the edge*. London and New York: Routledge, 122-136.

Smidth, J.K., 1875. Historical observations on the conditions of fisheries among the Ancient Greeks and Romans, and on their mode of salting and pickling fish. *Miscellaneus Documents*, 108, 3-20.

Smith, G.K. and Dartwall, W.R.T., 2005. *The Status and distribution of freshwater fish endemic to the Mediterranean basin*, IUCN Freshwater Biodiversity Assessment Programme. Available from: http://www.uicn.org/places/medoffice/documentos/med_fw_fish_report.pdf (accessed 22 October 2006).

Snodgrass, A., 2002. A paradigm shift in classical archaeology? *Cambridge Archaeological Journal*, 12 (2), 179-194.

Sokolowski, F., 1962. *Lois sacrées des cités grecques*, Supplément. Paris: E. de Boccard.

Solmsen, F., 1978. The fishes of Lesbos and their alleged significance for the development of Aristotle. *Hermes*, 106 (3), 467-480.

Souvermezoglou E., 1989. Distribution of Nutrients and Oxygen in Eastern Mediterranean Sea. In the *Proceedings of the UNESCO/OC second scientific workshop, Trieste, Italy*. POEM Scientific Reports 3, Cambridge, Harvard: University Publication, 85-102.

Souvermezoglou, E., Krasakopoulou, E. and Pavlidou, A., 1999. Temporal variability in oxygen and nutrient concentrations in the southern Aegean Sea and the Straits of the Cretan Arc. *Progress in Oceanography*, 44, 573-600.

Sparkes, B.A. 1962. The Greek kitchen. *Journal of Hellenic Studies*, 82, 121- 137.

Sparkes, B.A., 1995. A pretty kettle of fish. In J. Wilkins, D. Harvey and M. Dobson, eds. *Food in Antiquity*. Exeter: University of Exeter Press, 150-161.

Sparkes, A.B. and Talcott, L., 1970. *Black and Plain Pottery of the 6th, 5th and 4th c. B.C. The Athenian Agora*, XII. Princeton: American School of Classical Studies in Athens.

Spyridakis, S., 1979. *Ptolemaic Itanos and Hellenistic Crete*. Berkeley: University of California Press.

Sraffa, P., 1972. *Production of Commodities by Means of Commodities; prelude to a critique of economic theory*. Cambridge: Cambridge University Press.

Stabolidis, Ch. N., 1993. *Eleutherna*. Rethymno: Ekdoseis Panepistimiou Kritis.

Stanley, J., Perissoratis, C., 1977. Aegean ridge barrier and basin sedimentation patterns. *Marine Geology*, 24, 97-107.

Stavropoulos, D.F., 1938. Ieratiki Oikia en Zostiri tis Attikis. *Archaeologiki Ephimeris*, 1-31.

Steinhauer, G., 1994. Inscription agoranomique du Pirée. *Bulletin de Correspondance Hellénique*, 118, 51-68.

Stillwell, R., 1979. *The Princeton Encyclopaedia of Classical Sites*. Princeton, NJ: Princeton University Press.

Stromberg, R., 1943. *Studien zur Etymologie und Bildung der griechischen Fischnamen*. Göteborgs Hogskolas Arsskrift 49.

Struck, A., 1908. *Makedonische Fahrten. Zur Kunde der Balkanhalbinsel, Reisen und Beobachtungen*. Serajewo: Herc. Inst. F. Balkanoforschungun in Serajewo.

Struder, J. 1994. Roman fish sause in Petra, Jordan. In W. van Neer, ed. *Fish Exploitation in the Past. Proceedings of the 7th Meeting of the ICAZ fish remains working group.* Annales du Musée Royal de l Afrique Centrale, Sciences Zoologiques no 274, Tervuren, 191-196.

Sutton, E.D., 2001. *Remembrance of Repasts: an anthropology of food and memory*. Oxford: New York, Berg.

Svetovidov, A.N., 1964. *Ribi Tehornogo Moria*. Akademia Nauk SSSR, Morcow: Izd. "Nauka"

Swain, S., 1996. Hellenism *and Empire. Language, classicism and power in the Greek world, AD*. 50-250. Oxford: Clarendon Press.

Syrides, G. forthcoming. Sea shells form the Sanctuary of Poseidon at Kalaureia. In B. Wells and A. Pentinenn, eds. *The Physical Environment and Daily life at the Temple on Kalaureia (Poros).* Swedish Institute at Athens.

TalentCruiser, 2006. www.cruiser.gr/en_main.html (accessed 15-November 2006).

Tarlow, S., 1999. Strangely familiar. In S. Tarlow and S. West, eds. *The Familiar Past: Archaeologies of Later Historical Britain*. London, New York: Routledge, 263-72.

Tarlow, S., 2000. Emotion in archaeology. *Current Anthropology*, 41 (5), 713-46.

Teodorson, S. T. 1989. A Commentary on Putarch's "Table Talks". Göteborg, Sweden: Acta Universitatis Gotheburgensis.

Terray, E., 1969. *Le Marxisme devant les Societes "primitives"*. Paris : Maspero.

Themelis, P., 1971. *Brauron, Guide to the Site and Museum*. Athens: Apollo Editions.

Themelis, P., 1999. *Archaia Messene*. Athens: TAPA editions.

Themelis, P. 2002. *Eleutherna*. Athens: TAPA editions.

Theodoropoulou, T., in prep. *L'Exploitation des Faunes Aquatiques en Grec Septentrionale au Néolithique et a l'Age du Bronze*. PhD dissertation, Université de Paris I Panthéon- Sorbonne

Thiercy, P., 1993. Les odeurs de la Polis ou le 'nez' d' Aristophane. In A.H. Sommerstein, S. Halliwell, J. Henderson and B. Zimmermann, eds. *Tragedy, Comedy and the Polis*. Papers from the Greek Drama Conference, Nottingham, 18-20 July 1990. Bari: Levante editori, 505-26.

Thompson, D.A.W., 1947. *A Glossary of Greek Fishes*. Oxford: Oxford University Press.

Thomas, J., 1993. The politics of vision and archaeologies of landscape. In B. Bender, ed. *Landscape: politics and perspectives*. Oxford: Berg, 73-97.

Tilley, C., 1995. *A Phenomenology of Landscape. Places, paths and monuments*. Oxford, Providence: Berg.

Trakadas, A., 2005. The archaeological evidence for fish processing in the Western Mediterranean. In T. Bekker-Nielsen, ed. *Ancient Fishing and Fish Processing in the Black Sea Region,* Black Sea Studies 2, Aarhus: Aarhus University Press, 47-82.

Trantalidou, C., 1990. Animals and human diet in Prehistoric Aegean. In D. Hardy, ed. *Thera and the Aegean World III: proceedings of the third international congress, Santorini, Greece, 3-9 September 1989*. London: Thera Foundation, 392-405.

Trendall, A.D., 1967. *The Red Fgured Vases of Lucania, Campania and Sicily*. Oxford: Clarendon Press.

Trendall, A.D., 1989. *Red Figure Vases of South Italy and Sicily: a handbook*. World of Art, New York: Thames and Hudson.

Tsamis, V., forthocming. *Body, Senses and Space in Late Bronze Age - Early Iron Age Macedonia, Greece*. PhD dissertation, Archaeology, University of Southampton.

Tserpes, G., Peristeraki, P., Potamias, G., Tsimenides, N., 1999. Species distribution in the Southern Aegean Sea based on Bottom-trawl surveys. *Aquatic Living Resources*, 12 (3), 167-175.

Tsimbidou-Avloniti, M., 2005. *Makedonikoi Tafoi ston Foinika kai ston Agio Athanasio Thessalonikis*. Dimosieymata tou Arxaiologikoy Deltioy no 91, Athens: Ministry of Culture.

Tsimenides, N., Tserpes, G., Machias, A., Kallianiotis, A., 1991. Distribution of fishes on the Cretan shelf. *Journal of Fish Biology*, 39, 661-672.

Tsountas, C., 1893. *Mykenai kai Mykenaïkos Politismos*. Athens: Estia.

Tsountas, C., 1899. Kykladica. *Archaiologiki Ephimeris*, 137-212.

Unluata, U., Oguz, T., Latif, M.A., Ozsoy, E., 1990. On the physical oceanography of the Turkish straits. In L.J. Pratt ed. *The Physical Oceanography of the Sea Straits.* Dordrecht: Kluwer Academic Publishers, 25-60.

Vafeiadou, V., 1974. Ithi kai ethima Sozopoleos. *Laografia,* 29, 114-226.

Vale, D. and Gargett, H.R., 2002. Size matters: 3-mm sieves do not increase richness in a fishbone assemblage from Arrawarra I, an aboriginal Australian Shell Midden on the Mid-north Coast of New South Wales, Australia. *Journal of Archaeological Science*, 29, 57-63.

Valmin, N., 1928-29. Inscriptions de la Messénie. *Bulletin de la Société Royale des Lettres de Lund,* 22 (2), 19-24.

van den Veen, M., 2003. When is food a luxury? *World Archaeology*, 34 (3), 405-427.

van Effenterre, H., Kalpaxis, Th., Petropoulou A., Stavrianopoulou E., Furtwangler, A. and Schnapp, A., 1991. *Eleutherna, Tomeas II. Epigrafes apo to Pyrgi kai to Nisi.* Rethymno: Ekdoseis Panepistimiou Kritis.

van Nijf, M.O., 1997. *The Civic World of Professional Associations in the Roman East.* Amsterdam: J.C. Gieben.

Vatin, C., 1971. Le tarif des poissons d' Akraiphia. In F. Salviat and C. Vatin, eds. *Inscriptions de Grèce Centrale.* Paris: Editions de Boccard, 95-109.

Vayda P.A. and McCay, B.J., 1975. New directions in ecology and ecological anthropology. *Annual Review of Anthropology,* 4, 293-306.

Vermeule, E.T. 1979. *Aspects of Death in the Early Greek Art and Pottery.* Berkeley: University of California Press.

Vermeule, E.T., 1996. Archaeology and philology: the dirt and the word. *Transactions of the American Philological Association*, 126, 1-10.

Vidali, S., 1997. *Archaische Delphindarstellungen.* Wurzburg: Konigshausen and Neumann.

Vidalis, K., and Tsimenidis, N., 1997. Distribution of the picarel (*S. smaris*, L.) on the Cretan continental shelf. *Proceedings of the 5th Symposium on Oceanography and Fisheries, Kavala, Greece, April 17-18, 1997, vol. II.* Athens: National Center of Marine Research, 59-62.

Viviers, D., 1999. Territorial dynamics in Crete from the Archaic to Hellenistic period. In A. Chaniotes, ed. *From Minoan Farmers to Roman Ttraders: sidelights on the economy of ancient Crete.* Stuttgart: F. Steiner, 221-233.

von Rudloff, R., 1999. *Hekate in Ancient Greek Religion.* Victoria – Canada: Owl Publishing.

Walters, I., 1988. Fire and bones: patterns of discard. In B. Meehan and R. Jones, eds. *Archaeology with Ethnography. An Australian perspective.* Canberra: Australian National University, 215-221.

Warren, P., 1979. Review of the "Natural environment and human settlement in Prehistoric Greece based on original fieldwork" by J. Bintliff. *Proceedings of the Prehistoric Society*, 45, 344-346.

Wells, B. and Penttinen, A., forthcoming. *The Physical Environment and Daily life at the Temple on Kalaureia (Poros).* Swedish Institute at Athens.

Wells, B., Penttinen A. and Hjohlman, J., forthcoming. The Kalaureia excavation project: the 2004 and 2005 seasons. *Opuscula Atheniensia*, 31 (2006).

Wells, B., Penttinen and M.-Fr. Billot, 2003. Investigaitons in the sanctuary of Poseidon on Kalaureia, 1997-2001. *Opuscula Atheniensia*, 28, 29-87.

Wells, B., Penttinen, A., Hjohlman, J. and Savini, E., 2005. The Kalaureia excavation project: the 2003 season. *Opuscula Atheniensia*, 30.

Wessels, A. and R. Krumeich, 1999. "Dychtyoulkoi". In R. Kumeich, N. Rechstein and B. Seidensticker, eds. *Das Griechisch Satyrspiel.* Darmstadt: Wissenschaftliche Buchgesellschaft, 107-124.

West, A.B., 1931. *Corinth VIII, ii. Latin inscriptions, 1896-1926.* Cambridge Massachusetts: Harvard University Press.

Westerdahl, C., 2005. Seal on Land, Elk at Sea: Notes on the Applications of the ritual landscape at the seaboard. *The International Journal of Nautical Archaeology*, 34 (1), 2-23.

Wheeler, A. and Jones, G.A., 1989. *Fishes*. Cambridge: Cambridge University Press.

Wheeler, A., and Locker, A., 1985. The estimation of size in sardines (*Sardina pilchardis*) from the amphorae in a wreck at Randello, Sicily. *Journal of Archaeological Science*, 12, 97-100.

Wiessner, P. and Schiefenhovel, W., eds. 1996. *Food and the Status Quest. An interdisciplinary perspective.* Providence, Oxford: Berghahm Books.

Wigglesworth, N., 1996. *The Evolution of English Sport,* London: Routledge.

Wilkins J., Harvey, D. and Dobson, M., eds. 2000. *Food in Antiquity.* Exeter: University of Exeter Press.

Wilkins, J., 1993. Social status and fish, in Mars, V., and Mars G., eds. *Food, culture and history.* London: London Food Seminar, 191-203.

Wilkins, J., 1994. Archaestratos: The life of Luxury. Introduction. In J. Wilkins and S. Hill, eds. *Archaestratos: The life of luxury.* Totnes (Devon): Prospect, 9-33.

Wilkins, J., 1998. Greek Gastronomy? *The Classical Review (N. S.)*, 48 (2), 387-388.

Wilkins, J., 2000. *The Boastful Chef. The discourse of food in ancient Greek comedy.* New York: Oxford University Press.

Wilkins, J., 2003. Forward. In O. Powell, *Galen and the Properties of Foodstuffs.* Cambridge: Cambridge University Press, i-xxvi.

Wilkins, J. 2005. Fish as a source of food in Antiquity. In T. Bekker-Nielsen, ed. *Ancient Fishing and Fish Processing in the Black* Sea *region.* Black Sea Studies 2, Aarhus: Aarhus University Press, 21-30.

Williams, C. K., II, 1978. Corinth 1977: Forum Southwest. *Hesperia,* 47 (1), 1-39.

Williams, C. K., II, 1979. Corinth, 1978: Forum Southwest. *Hesperia,* 48 (2), 108-118.

Williams, C. K., II, 1980. Corinth excavations, 1979. *Hesperia,* 49 (1), 107-134.

Williams, C.K. II, 1993. Roman Corinth as a commercial centre. In T.E. Gregory, ed. *The Corinthia in the Roman Period.* Journal of Roman Archaeology, Suppl. Ser. 8, Ann Arbor, Mi.: *Journal* of Roman Archaeology, 31-46.

Williams, C. K., II, and Russell, P., 1981. Corinth: Excavations 1980. *Hesperia*, 50 (1), 1-44.

Williams, C. K., II, and Zervos, O.H., 1986. Corinth 1985: East of Theater. *Hesperia* 55 (2), 129-205.

Wittenburg, A., 1990. *Il Testamento di Epikteta*, Trieste: Universita degli Studi di Trieste.

Wood, A.F., 1927. Greek fish names, Part I. *American Journal of Philology*, 48 (4), 297-325.

Wood, A.F., 1928a. Greek fish names, Part II. *American Journal of Philology*, 49 (1), 36-56.

Wood, A.F., 1928b. Greek fish names, Part III. *American Journal of Philology*, 49 (2), 167-187.

Wright, G.R.H., 1990. Of fishes and men. Fish symbols in Ancient religion. *Journal of Prehistoric Religion*, 3-4, 30-44.

Wylie, A., 1985. The reaction against analogy. In M., Schiffer, ed. *Advances on Archaeological Method and Theory* (8), New York: Academic Press, 63-111.

Xanthopoulou, M., 2005. Palaiochristianiki kerameiki tis archaeas Itanou, In A. Ausilio, ed. *Creta Romana e Protobizantina. Atti del Congresso Internationale (Iraklion, 23-30 Settembre 2000).* Padova: Bottega d' Erasmo, 1013-1027.

Yonge, C.D. Unknown date. The *lives and opinions of eminent philosophers. Pythagoras* (Book VIII). Available from: http://classicpersuasion.org/pw/diogenes/ (accessed 27 October 2006)

Yorke, R.A., and Dallas, M.F., 1968. Underwater survey of North Africa, Jougoslavia and Italy. *Underwater Association Report 1968*, 21-34.

Yuce, H., 1995. Northern Aegean water masses. *Estuarine Coastal and Shelf Science*, 41, 325-343.

Zachariou-Mamaliga, E., 1986. *Oi Psarades tis Symis. Oikonomiki, koinoniki, politiki opsi.* Rhodes: Unknown publisher.

Zapheropoulou, F., 1991. *Ta Mnimeia tis Dilou. Prosfata erga syntirisis kai stereosis.* Number 111, Athens: Vivliothiki tis en Athinais Archaeologikis Etairias.

Zappas, T., 1968. Alieutiko chroniko tou Euboikou. *Archeio Euvoikon Meleton*, 14, 1-86.

Zgusta, L., 2002. Demetrius John ("Mimis") Georgacas (1908-1990). Available from http://www.wtsn.binghamton.edu/onoma/. (accessed 15-November 2006).

Zimmerman Munn, M.-L., 1983. *Corinthian Trade with the West in the Classical Period.* PhD thesis, Bryn Mawr College.

Zimmerman Munn, M.-L., 2003. Corinthian trade with the Punic West in the Classical period. In N. Bookidis, ed. *CORINTH. Results of Excavations Conducted by the American School of Classical Studies at Athens.* Princeton: Princeton University Press, XX, 195-217.

Zindel, C., 1998. *Meeresleben und Jenseitsfahrt : die Fischteller der Sammlung Florence Gottet*, Kilcberg, Zürich: Akanthus.

www.ingramcontent.com/pod-product-compliance
Lightning Source LLC
LaVergne TN
LVHW071212210926
848482LV00015B/784
9781407301938